Atmospheric Composition Analysis and Meteorology: Instrumentation, Principles and Applications

Atmospheric Composition Analysis and Meteorology: Instrumentation, Principles and Applications

Edited by **Smith Paul**

SYRAWOOD
PUBLISHING HOUSE

New York

Published by Syrawood Publishing House,
750 Third Avenue, 9th Floor,
New York, NY 10017, USA
www.syrawoodpublishinghouse.com

Atmospheric Composition Analysis and Meteorology: Instrumentation, Principles and Applications
Edited by Smith Paul

International Standard Book Number: 978-1-68286-021-2 (Hardback)

Printed in the United States of America.

Contents

Preface

This book aims to highlight the current researches and provides a platform to further the scope of innovations in this area. This book is a product of the combined efforts of many researchers and scientists from different parts of the world. The objective of this book is to provide the readers with the latest information in the field.

The variations in the different components of the atmosphere have an effect on the air quality, regional climate and weather. This book is a valuable compilation of topics such as atmospheric layers and stratification, application of advanced equipment in meteorology, weather forecasting, assessment of various atmospheric components and particles, that range from the basic to the most complex advancements in this field. For all readers who are interested in this field, the researches included in this book will serve as an excellent guide to develop a comprehensive understanding.

I would like to express my sincere thanks to the authors for their dedicated efforts in the completion of this book. I acknowledge the efforts of the publisher for providing constant support. Lastly, I would like to thank my family for their support in all academic endeavors.

Editor

Geostatistical merging of ground-based and satellite-derived data of surface solar radiation

M. Journée and C. Bertrand

Royal Meteorological Institute of Belgium, Brussels, Belgium

Abstract. In this paper, we demonstrate the benefit of using observations from Meteosat Second Generation (MSG) satellites in addition to in-situ measurements to improve the spatial resolution of solar radiation data over Belgium. This objective has been reached thanks to geostatistical methods able to merge heterogeneous data types. Two geostatistical merging methods are evaluated against the interpolation of ground-data only and the single use of satellite-derived information. It results from our analysis that merging both data sources provides the most accurate mapping of surface solar radiation over Belgium.

1 Introduction

Knowledge of the local solar radiation is essential for many applications, including design, planning and operation of solar energy systems, architectural design, crop growth models and evapotranspiration estimates. Traditionally, solar radiation is observed by means of networks of meteorological stations. Costs for installation and maintenance of such networks are very high and national networks comprise only few stations. Consequently the availability of observed solar radiation measurements has proven to be spatially inadequate for many applications. Mapping solar radiation by interpolation/extrapolation of measurements is possible but usually leads to large errors, except for dense networks (Zelenka et al., 1992; Hay, 1981, 1984; Hay and Hanson, 1985; WMO, 1981; Perez et al., 1997).

Because several authors have shown the potentialities of the images of the Earth taken by polar-orbiting and geostationary satellites for mapping the global irradiation impinging on a horizontal surface at the ground level (e.g., Zelenka et al., 1992, 1999; Perez et al., 1997, 2002; Pinker et al., 1995), we evaluate in the present paper the benefit of using space-based observations as an additional information source when interpolating the ground measurements. More specifically, we consider surface incoming global short-wave radiation products derived from Meteosat Second Generation (MSG, Schmetz et al., 2002) in order to improve the spatial resolution of daily surface solar radiation data over Belgium.

To reach that objective, we implemented two geostatistical methods able to merge heterogeneous data types (i.e., kriging with external drift and regression kriging) and evaluate these methods against mappings derived from a single source of data (i.e., either in-situ or satellite data).

2 Solar radiation data

2.1 Ground-based solar radiation measurements

The Royal Meteorological Institute of Belgium (RMIB) is currently performing measurements of global solar irradiance (in Wm^{-2}) by means of CNR1 and CM11 pyranometers of Kipp & Zonen at 13 sites well-distributed over Belgium (see Fig. 3). Measurements are made with a 5 s time step and time-integrated on a 10 min basis in the RMIB data warehouse. The 10-min solar irradiation data (in Whm^{-2}) are then subject to a set of semi-automatic quality assessment tests and gaps in the time series are filled by model estimations (Journée and Bertrand, 2011).

2.2 MSG-derived surface solar irradiance

Within the Satellite Application Facility (SAF) network, the down-welling short-wave irradiance at the Earth's surface is operationally retrieved from MSG imageries by three decentralized SAFs: the Ocean and Sea Ice SAF (OSI-SAF, www.osi-saf.org), the Land Surface Analysis SAF (LSA-SAF, land-saf.meteo.pt) and the SAF on Climate Monitoring (CM-SAF, www.cmsaf.eu). To retrieve the same parameter, the different SAFs use their own algorithms and different ancillary input data.

Sea surface being out of the scope of this study, we focused our investigation on the LSA-SAF and CM-SAF products. The LSA-SAF surface solar radiation product (Geiger et al., 2008) is generated every 30 min and distributed to the users in near real-time at the pixel spatial resolution of the MSG spectral imager (i.e., about 6 km in NS direction and 3.3 km in EW direction over Belgium). The operational CM-SAF surface solar radiation product (Mueller et al., 2009) is an off-line product provided on a 15×15 km sinusoidal grid in daily and monthly average. The monthly mean diurnal cycle is also provided. Because of the relatively coarse spatial and temporal resolution of the CM-SAF operational product, intermediate CM-SAF values (R. Mueller, personal communication, 2010) were considered in the present study, namely instantaneous hourly CM-SAF solar surface irradiance remapped onto 3×3 km, 9×9 km and 15×15 km grids.

3   Geostatistical mapping methods

In this study, we considered three geostatistical mapping methods: ordinary kriging (OK), kriging with external drift (KED) and regression kriging (RK). The aim of these methods is to interpolate a random field, G, (e.g., the spatial distribution of surface solar radiation) from observations, $G(x_i)$, at selected locations $x_i|_{i=1,...,N}$ (e.g., the ground measurements). Based on the knowledge of the spatial dependence of the random field, the kriging estimation $\hat{G}(x_0)$ at an unobserved location x_0 is computed as a linear combination of the observations, $\hat{G}(x_0) = \sum_{i=1}^{N} w_i G(x_i)$, where the weights, w_i, are chosen in such a way that the estimator is unbiased and the error variance is minimized. The spatial correlation of the random field between two locations x_i and x_j is described by means of the variogram $\gamma(x_i, x_j) = \mathbb{E}[(G(x_i) - G(x_j))^2]$, which is often chosen as isotropic, meaning that it is function only of the distance d between x_i and x_j, i.e., $\gamma(x_i, x_j) = \gamma(d)$. When the number of observation locations is sufficient, a model can be fitted to the empirical variogram derived from the observed data. The variogram has otherwise to be assumed.

While in OK the random field G is assumed to have a constant, albeit unknown, mean at all locations, the KED and RK techniques rely on the knowledge of a densely sampled auxiliary variable g (e.g., the SAFs' products) to model G as a non-stationary random field of the form $\mathbb{E}[G(x)] = a_0 + a_1 g(x)$. Although the KED and RK methods are aimed toward a similar objective, they differ in the way to compute the parameters a_i and the weights w_i (Hengl et al., 2003). In KED, a_i and w_i are derived together by forcing an exact interpolation of the auxiliary variable, $g(x_0) = \sum_{i=1}^{N} w_i g(x_i)$. In RK, the parameters a_i are first computed by linear regression from data at the observed locations. The regression residuals are then interpolated by OK. From a computational point of view, all three kriging methods require to solve linear systems of equations. We refer to Wackernagel (1995), Hengl

et al. (2003) and references therein for more details on these techniques.

4   Cross validation analysis

This study is focused on the mapping of surface solar radiation data over Belgium on a daily basis. Daily totals of the 10-min RMIB ground data are obtained by simple summation, while the instantaneous satellite data are integrated by trapezoidal integration. The considered interpolation and merging methods are evaluated by leave-one-out cross-validation (CV) on the basis of two years of quality-controlled data (2008 and 2009). In total, we used a set of 491 days for which the ground data and both SAFs' satellite data were available at all stations and over the entire diurnal cycle. The performance of the different methods is assessed by the average on these 491 instances of three indices derived from the bias between the cross-validation prediction $\hat{G}$ and the actual measurement G at the N locations $x_i|_{i=1,...,N}$:

- the cross-validation mean bias error
  $\mathrm{MBE_{cv}} = \frac{1}{G_{avg} N} \sum_{i=1}^{N} (\hat{G}(x_i) - G(x_i))$,

- the cross-validation mean absolute error
  $\mathrm{MAE_{cv}} = \frac{1}{G_{avg} N} \sum_{i=1}^{N} |\hat{G}(x_i) - G(x_i)|$ and,

- the cross-validation root mean square error
  $\mathrm{RMSE_{cv}} = \frac{1}{G_{avg}} \sqrt{\frac{1}{N} \sum_{i=1}^{N} (\hat{G}(x_i) - G(x_i))^2}$.

where $G_{avg} = \frac{1}{N} \sum_{i=1}^{N} G(x_i)$ is the average solar radiation over all stations.

Since surface solar radiation is measured at only 13 stations, variograms can hardly be estimated from the ground data. Hence, we chose a fixed variogram model, e.g., an exponential variogram $\gamma(d) = 1 - \exp(-d/\bar{d})$. The evolution of $\mathrm{RMSE_{cv}}$ as a function of the range parameter $\bar{d}$ indicates that the best performance is reached when $\bar{d} = 500$ km for OK and $\bar{d} = 50$ km for KED and RK (see Fig. 1). This difference in optimal values of $\bar{d}$ results from the high spatial resolution of the satellite data used by KED and RK. Even if the variogram is expected to vary with the sky conditions, we used these values of $\bar{d}$ for all the 491 days.

Because of the possible non-uniformity of the surface global radiation within the MSG pixel, the SAFs' satellite products have been used at various spatial resolutions as auxiliary information for the KED and RK methods (i.e., from 1 pixel to 3×3 pixels aggregates for the LSA-SAF product and from 3×3 km to 15×15 km areas for the CM-SAF product). The geostatistical interpolation of ground data and the geostatistical merging of ground data with satellite information are compared against the SAFs' estimations in Table 1. Since the $\mathrm{MBE_{cv}}$ error is in overall very small for all methods, performance comparison relies essentially on the $\mathrm{MAE_{cv}}$ and $\mathrm{RMSE_{cv}}$ indices. First, the largest $\mathrm{RMSE_{cv}}$ and $\mathrm{MAE_{cv}}$ are found for OK of the ground measurements and

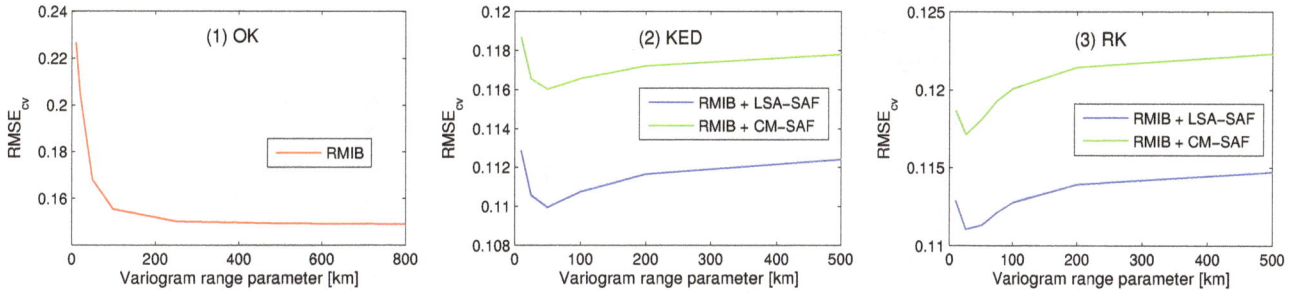

Figure 1. Distribution of the cross-validation root mean square error ($RMSE_{cv}$) as a function of the variogram range parameter $\bar{d}$ for the three kriging methods (left panel: OK; center panel: KED; right panel: RK). The SAFs' products are used at the finest spatial resolution as auxiliary information for KED and RK (i.e., MSG pixel resolution for the LSA-SAF and 3×3 km resolution for the CM-SAF).

Table 1. Cross validation mean bias error (MBE_{cv}), mean absolute error (MAE_{cv}) and root mean square error ($RMSE_{cv}$) for the geostatistical interpolation by OK of the RMIB ground data, the SAFs' products used at various spatial resolutions, and the geostatistical merging by KED and RK of the RMIB ground data with the SAFs' products.

Method & Data		MBE_{cv}	MAE_{cv}	$RMSE_{cv}$
OK	RMIB	−0.0032	0.116	0.149
–	LSA-SAF (1×1 px)	−0.0004	0.120	0.144
–	LSA-SAF (2×2 px)	0.0008	0.120	0.144
–	LSA-SAF (3×3 px)	0.0006	0.120	0.140
–	CM-SAF (3×3 km)	−0.0078	0.112	0.136
–	CM-SAF (9×9 km)	−0.0310	0.112	0.135
–	CM-SAF (15×15 km)	−0.0321	0.110	0.133
KED	RMIB+LSA-SAF (1×1 px)	0.0006	0.087	0.110
KED	RMIB+LSA-SAF (2×2 px)	0.0001	0.087	0.111
KED	RMIB+LSA-SAF (3×3 px)	0.0006	0.087	0.105
KED	RMIB+CM-SAF (3×3 km)	−0.0005	0.092	0.116
KED	RMIB+CM-SAF (9×9 km)	−0.0001	0.089	0.112
KED	RMIB+CM-SAF (15×15 km)	0.0003	0.088	0.111
RK	RMIB+LSA-SAF (1×1 px)	0.0007	0.087	0.111
RK	RMIB+LSA-SAF (2×2 px)	0.0002	0.087	0.112
RK	RMIB+LSA-SAF (3×3 px)	0.0007	0.087	0.106
RK	RMIB+CM-SAF (3×3 km)	−0.0003	0.094	0.118
RK	RMIB+CM-SAF (9×9 km)	0.0002	0.091	0.114
RK	RMIB+CM-SAF (15×15 km)	0.0003	0.088	0.111

the unmerged LSA-SAF mappings, respectively, while the best performance is observed once ground and satellite data are merged together. The KED and RK merging methods exhibit virtually identical results. Using both ground-based and satellite-derived information provides a significant improvement with respect to mappings based solely on one of these data sources (e.g., in case of the LSA-SAF product, the $RMSE_{cv}$ is reduced by about 25% and 30% with respect to the satellite and the OK mappings, respectively). Regarding the satellite information used as auxiliary information for KED and RK, the both SAF products provide comparable results although the best scores are obtained by KED with the

LSA-SAF data. Finally, the spatial resolution of the satellite products appears to have little impact on the resulting performance, albeit that slight improvements are obtained with larger resolutions.

In Fig. 2, we investigate the impact of sky conditions on the mapping performance. Sky-type classification is made upon both the mean and standard deviation over all stations of the daily clearness index (i.e., the ratio of the daily totals of surface and top-of-the-atmosphere incoming solar radiation). First, as far as the average sky condition over Belgium is concerned (see Fig. 2, left panel), the OK interpolation of ground data outperforms the single use of SAFs' data for overcast and very clear skies. In overcast conditions, it is well-known that the SAF's data overestimate the surface incoming solar radiation, while the exact mechanism that causes this overestimation is still unclear (Ineichen et al., 2009; Journée and Bertrand, 2010). The geostatistical merging of ground and satellite data exhibits the best performance for all types of sky. The improvement is however less pronounced for very clear skies. Second, regarding the influence of the spatial variability in sky conditions, the benefit of using the SAF's products is the largest for sky conditions that are highly inhomogeneous over the country (see Fig. 2, right panel).

5 Maps of surface solar radiation

Figure 3 compares the spatial distribution over Belgium of the average daily clearness index as computed by the OK interpolation of ground data, by the single use of LSA-SAF data, and by the KED merging of ground and LSA-SAF data. The daily clearness index at a specific location is inferred by means of the geostatistical interpolation and merging methods or directly from the satellite data for each of the 491 days selected in this study. Averages on these 491 instances are then computed.

All maps clearly highlight the global south-east to north-west positive gradient in clearness index. Satellite-based information is however needed to capture more regional features. The values derived from the LSA-SAF data only

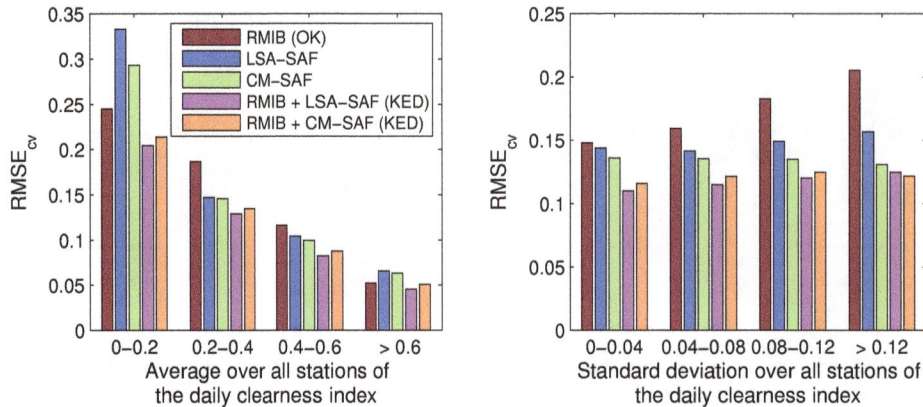

Figure 2. Distribution of the cross-validation root mean square error $RMSE_{cv}$ as a function of sky conditions. Sky-type classification is made upon the average (left-hand plot) as well as the standard deviation (right-hand plot) over all stations of the daily clearness index. The SAFs' products are used at the finest spatial resolution (i.e., MSG pixel resolution for the LSA-SAF and 3×3 km resolution for the CM-SAF). The results obtained with RK are not represented since they are virtually identical to those of KED.

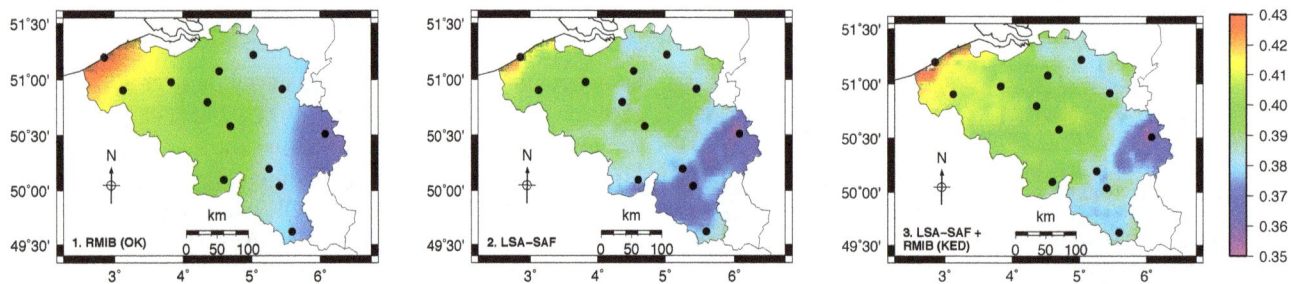

Figure 3. Spatial distribution over Belgium of the average daily clearness index as computed (**1**) by the OK interpolation of ground data, (**2**) by the single use of satellite-based estimations, and (**3**) by the KED merging of ground and satellite data. The displayed values of the daily clearness index are averages on the 491 days selected in this study. Dots indicate the ground stations' locations.

(Fig. 3, middle panel) are at most stations slightly below those derived from ground measurements only (Fig. 3, left panel). Hence, merging the two data sources (Fig. 3, right panel) enables to take advantage of both the accuracy of ground measurements and the global spatial coverage of satellite observations.

6 Conclusions

In this paper, we demonstrated the benefit of using MSG satellites imageries in addition to in-situ measurements to improve the spatial resolution of solar radiation data over Belgium. Regarding the MSG-derived information, we considered two products delivered by the LSA-SAF and the CM-SAF, respectively. The variograms used in the implemented geostatistical interpolation/merging methods were estimated to provide the best cross-validation scores over 13 sites in Belgium. Differences by a factor of 10 were observed for the optimal variogram range parameter, depending on whether

the high spatial resolution satellite data are involved or not in the interpolation process.

The best performance has been observed once ground observations are merged with the LSA-SAF product. No significant difference has been noted between the two implemented geostatistical merging methods (kriging with external drift and regression kriging). This reflects the fact that both methods are conceptually very close. Concerning the impact of sky conditions, mappings inferred by merging ground and SAFs' data were systematically more accurate than when using each data source separately, whatever the sky-type, while the benefit of the method is less apparent in case of very clear skies. Finally, the benefit of using satellite information appeared to be the largest in case of highly variable sky conditions over the studied area. In overall, merging ground and satellite data enables to take advantage of both the high accuracy of ground data and the global spatial coverage of satellite information.

Acknowledgements. The authors are grateful to Richard Mueller (German Meteorological Service, Germany) for providing us with 2 years of hourly CM-SAF solar surface irradiance remapped onto regular latitude-longitude grids at various spatial resolutions (3×3 km, 9×9 km and 15×15 km). This study was supported by the Belgian Science Policy under the research project: Contribution de l'IRM au développement de l'énergie renouvelable en Belgique.

Edited by: M. Dolinar
Reviewed by: J. Walawender and another anonymous referee

References

Geiger, B., Meurey, C., Lajas, D., Franchistéguy, L., Carrer, D., and Roujean, J.-L.: Near real-time provision of downwelling short-wave radiation estimates derived from satellite observations, Meteorol. Appl., 15, 411–420, 2008.

Hay, J. E.: The mesoscale distribution of solar radiation at the Earth's surface and the ability of satellites to resolve it. In Proceedings of the First Workshop on Terrestrial Solar resource Forecasting and on Use of Satellites for Terrestrial Solar Resource Assessment, Washington D.C., 2–5 February, 1981.

Hay, J. E.: An assessment of the mesoscale variability of solar radiation at the Earth's surface, Sol. Energy, 32, 425–434, 1984.

Hay, J. E. and Hanson, K. J.: Evaluating the solar resource: a review of problems resulting from temporal, spatial and angular variations, Sol. Energy, 34, 151–161, 1985.

Hengl, T., Geuvelink, G. B. M., and Stein, A.: Comparison of kriging with external drift and regression-kriging, Technical note, ITC, Enschede, Netherlands, 2003.

Ineichen, P., Barroso, C. S., Geiger, B., Hollmann, R., Marsouine, A., and Mueller, R.: Satellite Application Facilities irradiance products: hourly time step comparison and validation over Europe, Int. J. Remote Sens., 30, 5549–5571, 2009.

Journée, M. and Bertrand, C.: Improving the spatio-temporal distribution of surface solar radiation data by merging ground and satellite measurements, Remote Sens. Environ., 114, 2692–2704, 2010.

Journée, M. and Bertrand, C.: Quality control of solar radiation data within the RMIB solar measurements network, Sol. Energy, 85, 72–86, 2011.

Mueller, R. W., Matsoukas, C., Gratzki, A., Behr, H. D., and Hollmann, R.: The CM-SAF operational scheme for the satellite based retrieval of solar surface irradiance – A LUT based eigenvector hybrid approach, Remote Sens. Environ., 113, 1012–1024, 2009.

Perez, R., Seals, R., and Zelenka, A.: Comparing satellite remote sensing and ground network measurements for the production of site/time specific irradiance data, Sol. Energy, 60, 89–96, 1997.

Perez, R., Ineichen, P., Moore, K., Kmiecik, M., Chain, C., George, R., and Vignola, F.: A new operational model for satellite-derived irradiances: Description and Validation, Sol. Energy, 73(5), 307–317, 2002.

Pinker, R. T., Frouin, R., and Li, Z.: A review of satellite methods to derive surface shortwave irradiance, Remote Sens. Environ., 51, 105–124, 1995.

Schmetz, J., Pili, P., Tjemkes, S., Just, D., Kerkmann, J., Rota, S., and Ratier, A.: An introduction to Meteosat Second Generation (MSG), B. Am. Meteorol. Soc., 83, 977–992, 2002.

Wackernagel, H.: Multivariate geostatistics: an introduction with applications, Springer-Verlag, Berlin, 1995.

World Meteorological Organization (WMO): Meteorological aspects of the utilization of solar radiation as an energy source, Annex: World maps of relative global radiation, Technical Note No. 172, WMO-No. 557, Geneva, Switzerland, 298 pp., 1981.

Zelenka, A., Czeplak, G., d'Agostino, V., Josefson, W., Maxwell, E., and Perez, R.: Techniques for supplementing solar radiation network data. Technical Report, International Energy Agency, IEA-SHCP-9D-1, Swiss Meteorological Institute, Krahbuhlstrasse, 58, CH-8044 Zürich, Switzerland, 1992.

Zelenka, A., Perez, R., Seals, R., and Renné, D.: Effective accuracy of satellite-derived hourly irradiances. Theor. Appl. Climatol., 62, 199-207, 1999.

Large-eddy simulation of plume dispersion under various thermally stratified boundary layers

H. Nakayama[1]**, T. Takemi**[2]**, and H. Nagai**[1]

[1]Japan Atomic Energy Agency, Ibaraki, Japan
[2]Disaster Prevention Research Institute, Kyoto University, Kyoto, Japan

Correspondence to: H. Nakayama (nakayama.hiromasa@jaea.go.jp)

Abstract. Contaminant gas dispersion in atmospheric boundary layer is of great concern to public health. For the accurate prediction of the dispersion problem, the present study numerically investigates the behavior of plume dispersion by taking into account the atmospheric stability which is classified into three types; neutral, stable, and convective boundary layers. We first proposed an efficient method to generate spatially-developing, thermally-stratified boundary layers and examined the usefulness of our approach by comparing to wind tunnel experimental data for various thermal boundary layers. The spreads of plume in the spanwise direction are quantitatively underestimated especially at large downwind distances from the point source, owing to the underestimation of turbulence intensities for the spanwise component; however, the dependence of the spanwise spreads to atmospheric stability is well represented in a qualitative sense. It was shown that the large-eddy simulation (LES) model provides physically reasonable results.

1   Introduction

Contaminant gas dispersion resulting from accidental release from industrial areas or intentional release of CBRN (chemical, biological, radiological, or nuclear) agent is of great concern to public health and social security. Plume dispersion within the atmospheric boundary layer is influenced by roughness elements, terrain, and thermal stability. In terms of thermal stability, atmospheric boundary layers are in general classified into three types; neutral boundary layer (NBL), stable boundary layer (SBL), and convective boundary layer (CBL). In an NBL turbulence is generated and maintained by wind shear, while in an SBL turbulence is not only maintained by wind shear but also constrained by negative buoyancy. In a CBL, turbulence is mainly produced by shear and/or buoyancy.

For simulating plume dispersion under various thermal conditions, there are typically two approaches: one is a wind tunnel experimental technique, and the other is a computational fluid dynamics (CFD) approach. It is well known that wind tunnel experiments are a reliable tool. There have been many studies of plume dispersion in various thermally-stratified boundary layers over the past 30 years

(Fackrell and Robins, 1982; Deardorff and Willis, 1984; Fedorovich and Thäter, 2002). Ohya et al. (1997) and Ohya and Uchida (2004) also investigated turbulence structures in thermally-stratified boundary layers. On the other hand, with the rapid development of computational technology, the CFD technique also has come to be regarded as a useful tool. In particular, there are two different approaches; the Reynolds-Averaged Navier–Stokes (RANS) and LES models. Hanna et al. (2004), Demael and Carissimo (2007), and Vervecken et al. (2013) carried out RANS simulations of atmospheric dispersion and compared to observed data of mean concentrations. Sykes and Henn (1992) and Henn and Sykes (1992) performed LESs of plume dispersion in neutral or convective boundary layers and investigated the instantaneous structure of a plume and the characteristics of concentration fluctuations.

As mentioned-above, there are a relatively small number of LES studies on plume dispersion in boundary layers with a wide range of thermal stability. Applicability of LES technique to plume dispersion under different thermal stability has not been fully demonstrated. In this study, we first propose an efficient method to generate various thermally-stratified boundary layers such as neutral, stable, and unstable boundary layers, and then examine the usefulness of our approach in comparison to wind tunnel experimental data.

2 Numerical model

The model used here is LOcal-scale High-resolution atmospheric DIspersion Model using LES (LOHDIM-LES) developed by Japan Atomic Energy Agency (Nakayama et al., 2011, 2012, 2013). The basic equations are the filtered continuity equation and the Navier–Stokes equation as follows;

$$\frac{\partial \overline{u}_i}{\partial x_i} = 0 \tag{1}$$

$$\frac{\partial \overline{u}_i}{\partial t} + \overline{u}_j \frac{\partial \overline{u}_i}{\partial x_j} = -\frac{1}{\rho} \frac{\partial \overline{p}}{\partial x_i} + \frac{\partial}{\partial x_j} \nu \left(\frac{\partial \overline{u}_i}{\partial x_j} + \frac{\partial \overline{u}_j}{\partial x_i} \right)$$
$$- \frac{\partial}{\partial x_j} \tau_{ij} - g\beta \left(\overline{\theta} - \theta_0 \right) \delta_{i3} \tag{2}$$

$$\tau_{ij} - \frac{1}{3} \delta_{ij} \tau_{kk} = -\nu_{\text{SGS}} \overline{S}_{ij} \quad \nu_{\text{SGS}} = \left(C_s \overline{\Delta} \right)^2 \left(2\overline{S}_{ij} \overline{S}_{ij} \right)^{\frac{1}{2}} \tag{3}$$

$$\overline{S}_{ij} = \left(\partial \overline{u}_i / \partial x_j + \partial \overline{u}_j / \partial x_i \right) / 2 \tag{4}$$

where u_i, t, p, ρ, ν, τ_{ij}, g, β, θ, θ_0, δ_{ij}, and $\overline{\Delta}$ are the wind velocity, time, pressure, density, kinematic viscosity, subgrid-scale (SGS) Reynolds stress, gravitational acceleration, thermal expansion coefficient, temperature, reference temperature, Kronecker delta, and grid-filter width, respectively. The subscript i stands for coordinates. ν_{SGS} and C_s are the eddy viscosity coefficient and the model constant of the flow field, respectively. Upper bars () denote application of the spatial filter. The subgrid-scale turbulent effect is represented by the standard Smagorinsky (1963) model with the constant value of 0.1. It should be noted that the application of the static Smagorinsky model requires some caution. For example, Moin and Kim (1982) mentioned that it is difficult to capture near-wall turbulent behaviors from LESs of turbulent channel flows using the conventional Smagorinsky model. Basu et al. (2008) conducted LESs of a diurnally varying atmospheric boundary layer and mentioned that the static type of SGS model doesn't reproduce a clear low-level jet in the nighttime although no significant differences between the static and dynamic type of models are found in the daytime. On the other hand, we have performed LESs of turbulent flows over a two-dimensional hill (Nakayama and Nagai, 2010), in building arrays with different obstacle densities (Nakayama et al., 2013), and in an actual urban area (Nakayama et al., 2014), and showed that the standard Smagorinsky model can produce reasonable results in comparison to wind tunnel experimental data. Therefore, the standard Smagorinsky model is used in our LES model because of its wide applicability to various environmental flows.

The filtered temperature and concentration transport equations are as follows;

$$\frac{\partial \overline{\theta}}{\partial t} + \overline{u}_j \frac{\partial \overline{\theta}}{\partial x_j} = \frac{\partial}{\partial x_j} \left(\alpha \frac{\partial \overline{\theta}}{\partial x_j} \right) - \frac{\partial}{\partial x_j} h_j \tag{5}$$

$$h_j = -\frac{\nu_{\text{SGS}}}{Pr_{\text{SGS}}} \frac{\partial \overline{\theta}}{\partial x_j} \tag{6}$$

$$\frac{\partial \overline{c}}{\partial t} + \overline{u}_j \frac{\partial \overline{c}}{\partial x_j} = \frac{\partial}{\partial x_j} \left(\kappa \frac{\partial \overline{c}}{\partial x_j} \right) - \frac{\partial}{\partial x_j} s_j \tag{7}$$

$$s_j = -\frac{\nu_{\text{SGS}}}{Sc_{\text{SGS}}} \frac{\partial \overline{c}}{\partial x_j} \tag{8}$$

where c is concentration. Pr_{SGS}, Sc_{SGS}, α, and κ are the turbulent Prandtl and Schmidt numbers, and molecular diffusivity for temperature and concentration fields, respectively. The subgrid-scale scalar flux is also parameterized by an eddy viscosity model. Although the turbulent Prandtl and Schmidt numbers also should be dynamically determined depending on the flow type, both numbers are set to a constant value 0.71.

The coupling algorithm of the velocity and pressure fields is based on the marker-and-cell method with the second-order Adams–Bashforth scheme for time integration. The Poisson equation is solved by the successive over-relaxation method. For the spatial discretization in the basic equations, a second-order accurate central difference scheme is used. However, for the advection term of the concentration transport equation, cubic interpolated pseudo-particle (Takewaki et al., 1985) is used.

3 Test simulations

3.1 Wind-tunnel experimental data

In this study, we compare the simulated results with the experimental data of Fackrell and Robins (1982) for a NBL case, Ohya et al. (1997) for a SBL case, and Ohya and Uchida (2004) for a CBL with a capping inversion case. The Reynolds numbers Re defined as $U_\infty \delta / \nu$ are 320 000 for the NBL case, 110 000 for the SBL case and 19 400 for the CBL case. U_∞ and δ are a free-stream velocity and boundary layer thickness, respectively. The bulk Richardson number Ri_δ defined as $g\delta \Delta\Theta / \Theta_0 U_\infty^2$ is 0.12 and -0.45 for the SBL and CBL cases, respectively. Here, Θ_0, and $\Delta\Theta$ are average absolute temperature in boundary layer, and temperature difference between temperature of ambient air Θ_∞ and surface temperature Θ_s. The bulk Richardson number based on the boundary layer thickness is generally defined using the potential temperature. However, within a scale of wind tunnel experiments which have a height of a few meters, the atmospheric pressure can be considered to be constant in the vertical direction, which indicates that the absolute temperature equals the potential temperature. Therefore, the Richardson number is defined using the absolute temperature here.

In the experiment of Fackrell and Robins (1982), a neutral boundary layer flow was generated by spires and a plume was released from the point source located at the downwind position of 6.7δ and elevated at the height of 0.19δ. In the experiment of the SBL case by Ohya et al. (1997), first, air flow with a temperature profile was imposed and a turbulent boundary layer was generated by an obstacle. Then, a stable boundary layer was produced by cooling the floor at a large downwind distance from the entrance of the test section. In their experiment, a downwind distance of 36.2δ was required to obtain a fully developed stable boundary layer. In the CBL with a capping inversion experiment by Ohya and Uchida (2004), first air flow with a temperature profile was imposed at the entrance of the test section. The convective boundary layer was produced by the obstacle and the entirely heated floor. In their experiment, a downwind distance of 24.1δ was required to obtain a fully developed convective boundary layer flow.

3.2 Computational settings

A driver region is set to generate a basic turbulent boundary layer flow by a recycling technique of Kataoka and Mizuno (2002) at the upstream part of model domain for each case. In this turbulent inflow technique, only fluctuating components are extracted at a recycling station and recycled back to the inlet boundary.

At the inlet boundary, mean wind velocity profile of the power law 0.14 for the NBL and CBL cases, and 0.25 for the SBL case is imposed. At the recycle station, a nearly target temperature profile is imposed for the SBL and CBL cases, and a thermally-stratified boundary layer is spatially developed. For the velocity field, the Sommerfeld radiation condition is applied at the exit. At the top, a free-slip condition is imposed for streamwise and spanwise velocity components, and vertical velocity component is set to be zero. At the side, a periodic condition is imposed. At the ground surface, a no-slip condition is imposed for each velocity component. For temperature and concentration fields, zero-gradient is imposed at all boundaries. However, for a temperature field, $\theta_s = 0$ and $\theta_s = \Theta_\infty$ is set at the ground surface for the SBL and CBL cases, respectively. θ_s is instantaneous surface temperature. The time step interval $\Delta t\, U_\infty/\delta$ is about 0.003 (Δt: time step). The maximum Courant–Friedrich–Levy number is about 0.1. The length of the simulation run to calculate the time averaged values $T\, U_\infty/\delta$ (T: averaging time) is 200. The length of the simulation run before releasing a plume is 300.

The sizes of the computational models for the NBL and SBL cases are $26\delta \times 2\delta \times 2\delta$ in the streamwise, spanwise and vertical directions. The number of grid points is $650 \times 100 \times 90$ for both cases. The model size and the number of grid points for the CBL case is $26\delta \times 5\delta \times 3\delta$ and $650 \times 250 \times 96$ in the streamwise, spanwise and vertical directions, respectively. Because large-scale convective

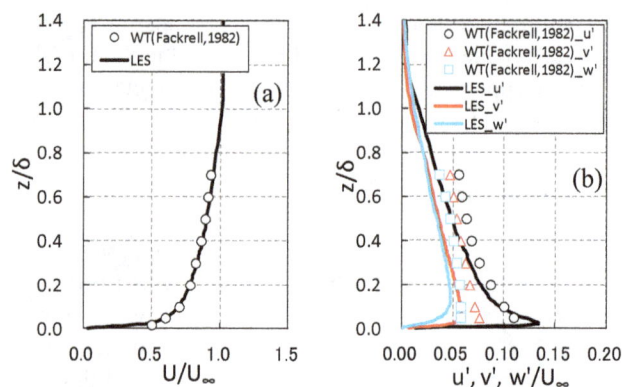

Figure 1. Vertical profiles of (a) mean wind velocity and (b) turbulence intensities for each wind component of a NBL flow.

motions are formed in CBL, a model domain size should be large enough to capture them. Therefore, more number of grid points is set than that of the NBL and SBL cases. The streamwise and spanwise grid spacing is uniform and the vertical grid spacing is stretched from 0.002δ to 0.09δ for each case. Re is set to 12 000 for each case and Ri_δ is set to the same value as that of the experiments for the SBL and CBL cases. The release point of a tracer gas is set at a distance of 11.0δ downstream from the inlet boundary and elevated with a height of $z/\delta = 0.19$ for each case.

4 Results

4.1 Turbulence statistics under various thermal conditions

Figure 1 compares the LES results of mean wind velocity (U) and turbulence intensities (u', v', w') with the wind tunnel experimental data of Fackrell and Robins (1982) for the NBL case. The mean wind velocity profile of the LES is consistent with the experimental data. Although the turbulence intensities of the LES are a little underestimated, the shape of each vertical profile is similar to those of the experimental data.

Figure 2 compares the LES results with the experimental data of Ohya et al. (1997) for the SBL case. They conducted wind tunnel experiments of SBL flows with a wide range of Ri_δ and showed that SBLs show two different distribution patterns of vertical profiles of turbulence statistics based on Ri_δ; a weak stability flow for $Ri_\delta < 0.25$ and a strong stability flow for $Ri_\delta > 0.25$. Based on this classification, the present LES case for the SBL flow corresponds to a weak type. The mean wind velocity profile of the LES is consistent with the experimental data. The turbulence intensity especially for the vertical component of the LES is much smaller than the experimental data of $Ri_\delta = 0.12$. It should be noted that the ratio of the vertical component turbulence intensity to the free-stream velocity of the stable boundary layer flow by Ohya et al. (1997) is comparable to that of the neutral

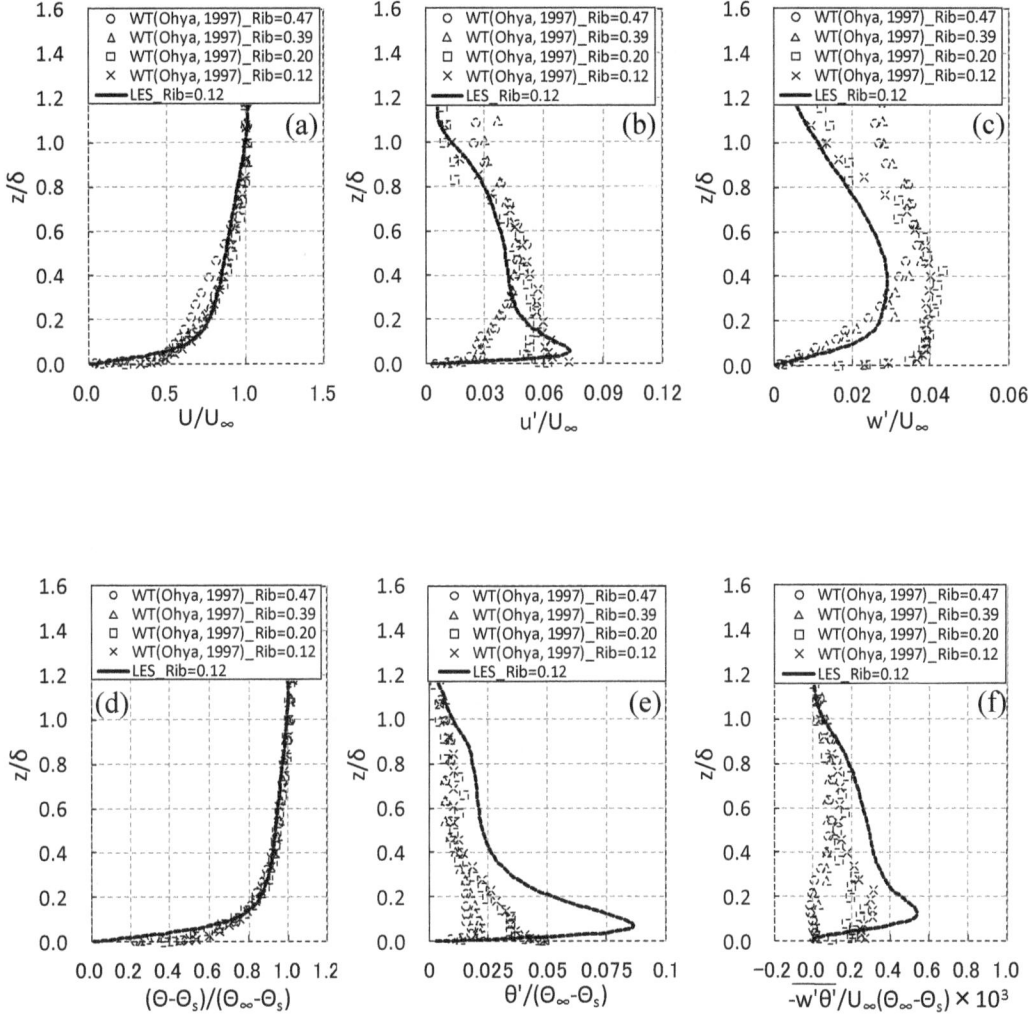

Figure 2. Vertical profiles of (**a**) mean wind velocity, (**b**) turbulence intensity for streamwise component, (**c**) turbulence intensity for vertical component, (**d**) mean temperature, (**e**) r.m.s. temperature, and (**f**) vertical heat flux of a SBL flow.

boundary layer flow by Fackrell and Robins (1982). However, the shape of the vertical profile corresponds to that of the experiments of a weak stability type $Ri_\delta < 0.25$. The vertical profile of the mean temperature is in good agreement with experimental data of Ohya et al. (1997). The r.m.s. (root mean square) temperature and vertical heat flux are entirely overestimated and underestimated in comparison to the experimental data of $Ri_\delta = 0.12$. As described in Sect. 2, it is pointed out that the conventional Smagorinsky model cannot accurately capture near-wall turbulent behaviors especially for stable boundary layers (Basu et al., 2008). These differences are due to the use of the static Smagorinsky model. However, the shape of the vertical profile corresponds well with that of the experiments for a weak type. The values of the ratio u_*/U_∞ and $\theta_*/\Delta\Theta$ in this LES are 0.021 and 0.026, while those are 0.026 and 0.025 in the experiments of Ohya et al. (1997). Here, u_* and θ_* are friction velocity and friction temperature, respectively. Those values are found to be

comparable to those of the experiments. However, in order to examine a turbulent Reynolds number effect in detail, turbulence statistics profiles in inner scaling have to be investigated in comparison to wind tunnel experimental data.

The LES results for the CBL case are shown in Fig. 3. U_m is mean wind velocity in the range where the values are constant in the middle part of the CBL. w_* is convective velocity scale defined as $(g\,Q_s\,\delta/\Theta_0)^{1/3}$. Here, Q_s is the maximum value of vertical heat flux $\overline{w'\theta'}$ near a ground surface. The mean wind velocity and turbulence intensities are in good agreement with the experimental data of Ohya and Uchida (2004). The characteristics such as a constant profile of mean temperature in the main portion of the CBL and the patterns of vertical distributions of the r.m.s. temperature and vertical heat flux are similar to those of the experiments. Deardorff (1972) showed that turbulent eddies have tendencies to become elongated by the mean wind shear and form rolls for a weakly unstable boundary layers $-z_i/L < 4.5$

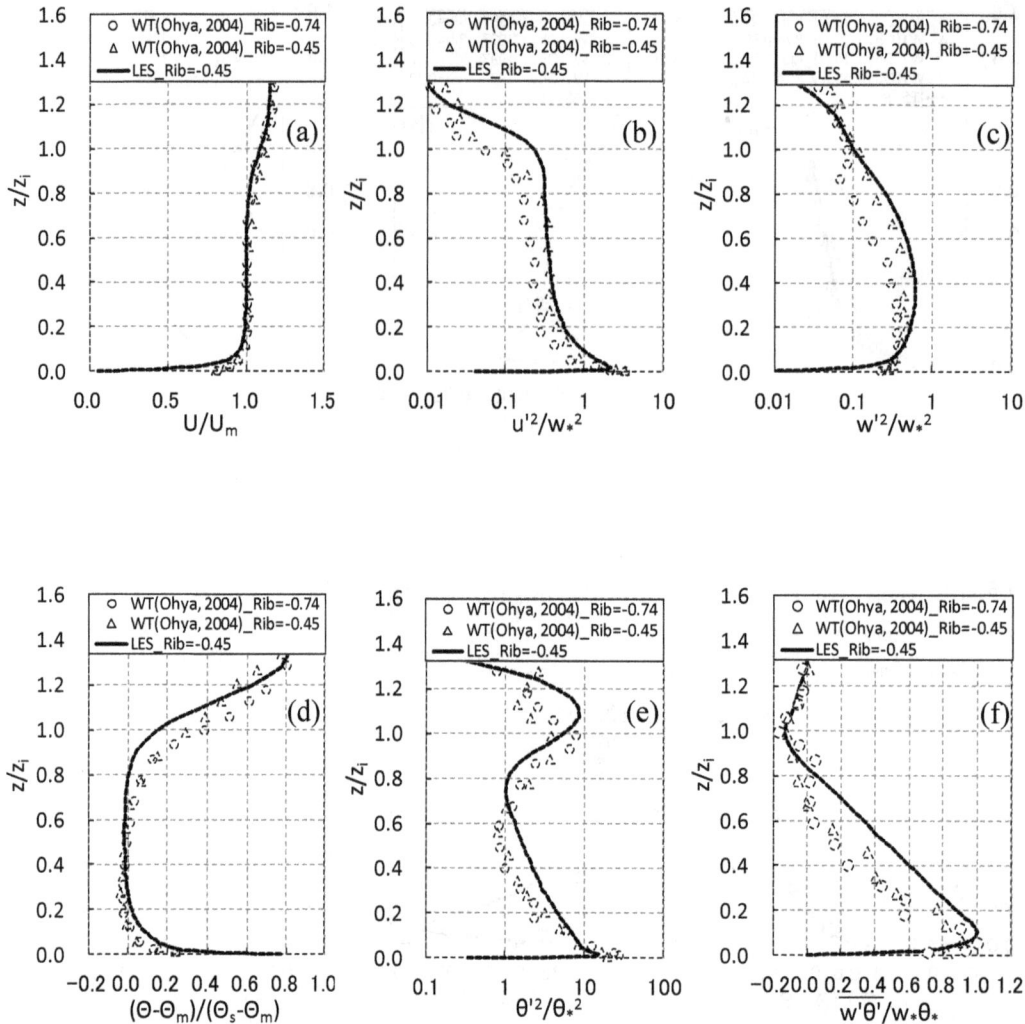

Figure 3. Vertical profiles of (**a**) mean wind velocity, (**b**) turbulence intensity for streamwise component, (**c**) turbulence intensity for vertical component, (**d**) mean temperature, (**e**) r.m.s. temperature, and (**f**) vertical heat flux of a CBL with a capping inversion flow.

(corresponding to $u_*/w_* > 0.45$) while the ordering effect of the wind shear gives way to the random orientation of convective structures for strongly unstable boundary layers $-z_i/L > 10$ (corresponding to $u_*/w_* < 0.34$). z_i and L are the inversion height and the Obukhov length, respectively. Moeng and Sullivan (1994) performed LESs of planetary boundary layers including extremes of shear and buoyancy forcing and intermediate cases, and implied that CBL flows have shear- and buoyancy-driven flows with u_*/w_* around 0.65. In this study, u_*/w_* is 0.46 which slightly exceeds the limit shown by Deardorff. From their results, the CBL flow in our LES model is considered to be driven by not only buoyancy flows but also shear flows.

From these results, for the NBL case, the boundary layer is found to be reasonably simulated. For the SBL case, the vertical component turbulence intensity and vertical heat flux are underestimated, and the r.m.s. temperature is overestimated. However, the shape of those profiles is similar

to the experimental data for a weak type. For the CBL case, the turbulence characteristics are generally similar to the experimental data. These indicate that the LES model provides physically reasonable results depending on atmospheric stability.

4.2 Plume spreads under various thermal conditions

The most common stability classification scheme is the Pasquill–Gifford (P–G) (Turner, 1970), which defines six stability classes namely A (highly unstable), B (moderately unstable), C (slightly unstable), D (neutral), E (moderately stable), and F (extremely stable). A comparison of the plume spreads with Pasquill–Gifford curves and those of the wind tunnel experimental data (Fackrell and Robins, 1982) is shown in Fig. 4. The spanwise and vertical plume spreads are estimated by the following equations;

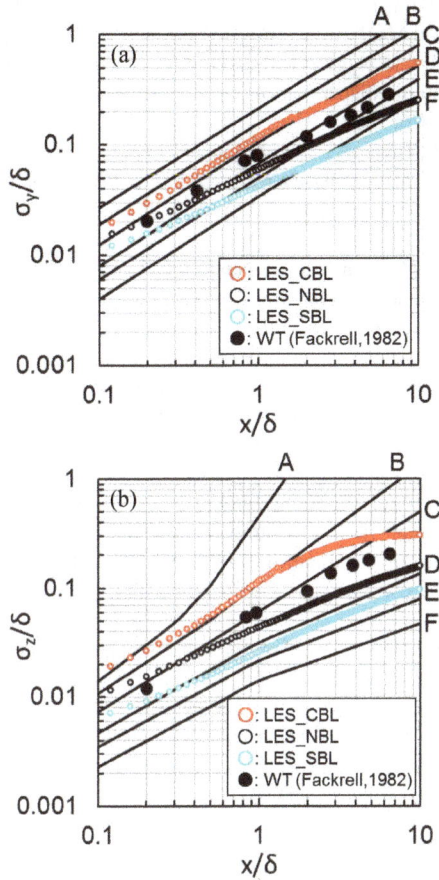

Figure 4. Streamwise variation of (**a**) spanwise and (**b**) vertical plume spreads under NBL, SBL, and CBL with a capping inversion flows. The characters A–F indicate the thermal stability classification scheme of the Pasquill–Gifford.

$$\sigma_y^2 = \int_{-\infty}^{\infty} c(y)(y-y_c)^2 \, dy / \int_{-\infty}^{\infty} c(y)\, dy \qquad (9)$$

$$\sigma_z^2 = \int_{0}^{\infty} c(z)(z-z_c)^2 \, dz / \int_{0}^{\infty} c(z)\, dz \qquad (10)$$

where σ_y and σ_z are spanwise and vertical plume spreads, respectively. y_c and z_c are the peak locations of the spanwise and vertical distributions of mean concentration at each downstream position, respectively. According to a classification for the P–G category by Snyder (1981), the bulk Richardson number values evaluated based on differences between 2 m and 10 m height of -0.02, -0.01, 0, 0.004 and 0.05 correspond to the categories B, C, D, E, and F. In this study, the bulk Richardson number values are estimated −0.018 and 0.025 for the CBL and SBL cases, which correspond to the categories B and between E and F, respectively. The σ_y and σ_z of plumes in the experimental data of Fackrell and Robins (1982) are found to be distributed

around the category D. On the other hand, those of the LES are overestimated especially near the point source for each case. Michioka et al. (2003) examined the sensitivity of the plume spreads to grid resolution for a point source which has 1.0 and 10 times the real diameters of the point source in comparison to wind tunnel experimental data. They showed that the plume spreads especially near the point source are largely overestimated in the coarse grid resolution, while those in the fine grid resolution are consistent with the experimental data. The point source is a tube with a diameter of $d_s/\delta = 0.007$ in the experiments of Fackrell and Robins (1982), while the mean size is $d_s/\delta = 0.016$ in the LES (d_s: a tube diameter of a point source). Therefore, it is considered that the overestimation especially near the point source is due to numerical diffusion by coarse grid resolution for the point source. The differences at large downwind distances from the point source are due to the underestimation of the turbulence intensities. These tendencies are observed for the SBL and CBL cases. The increases of the vertical spreads in the CBL case are constrained at a large downwind distance because of the capping effect of inversion. However, the distribution patterns of the spanwise and vertical plume spreads in response to the difference in atmospheric stability are well reproduced. This indicates that our LES model provides physically reasonable results.

5 Conclusions

We performed LESs of plume dispersion under various thermally-stratified boundary layers and examined the usefulness of our approach by comparing the simulated results with wind tunnel experimental data. For the NBL case, the turbulent boundary layer is reasonably produced. For the SBL case, the turbulence statistics such as the vertical component turbulence intensity, r.m.s. temperature, and vertical heat flux especially in the lower part of the boundary layer are quantitatively different from the experimental data. It is pointed out that the Smagorinsky model cannot reproduce near-wall turbulent behaviors by Moin and Kim (1982) and Basu et al. (2008). These differences are due to the use of the static Smagorinsky model. However, the shape of the distribution patterns corresponding to that of experimental data for a weak stability flow is obtained. This indicates that a weak-type SBL flow is successfully produced. For the CBL case, the turbulence characteristics are generally reproduced well.

Focusing on dispersion fields, the spanwise and vertical spreads of plumes are overestimated near the point source for each case. This is due to numerical diffusion by coarse grid resolution for the point source. At a large downwind distance from the point source, the spanwise plume spreads are underestimated for each case. This is due to the underestimation of the turbulence intensity for the spanwise component. However, the plume spreads of the LES are distributed nearly along the P–G curves depending on atmospheric stability.

From these results, the plume dispersion patterns are successfully simulated depending on atmospheric stability. It can be concluded that our LES model provides physically reasonable results.

Acknowledgements. This study is partly supported by General Collaborative Research #25G-05 provided by Disaster Prevention Research Institute, Kyoto University.

Edited by: M. Piringer
Reviewed by: two anonymous referees

References

Basu, S., Vinuesa, J. F., and Swift, A.: Dynamic LES modeling of a diurnal cycle, J. Appl. Meteorol. Clim., 47, 1156–1174, 2008.

Deardorff, J. W.: Numerical investigation of neural and unstable planetary boundary layers, J. Atmos. Sci., 29, 91–115, 1972.

Deardorff, J. W. and Willis, G. E.: Ground level concentration fluctuations from a buoyant and a non-buoyant source within a laboratory convective mixed layer, Atmos. Environ., 18, 1297–1984, 1984.

Demael, E. and Carissimo, B.: Comparative evaluation of an Eulerian CFD and Gaussian plume models based on prairie grass dispersion experiment, J. Appl. Meteorol. Clim., 47, 888–900, 2007.

Fackrell, J. E. and Robins, A. G.: Concentration fluctuations and fluxes in plumes from point sources in a turbulent boundary layer, J. Fluid Mech., 117, 1–26, 1982.

Fedorovich, E. and Thäter, J.: A wind tunnel study of gaseous tracer dispersion in the convective boundary layer capped by a temperature inversion, Atmos. Environ., 36, 2245–2255, 2002.

Hanna, S. R., Hansen, O. R. and Dharmavaram, S.: FLACS CFD air quality model performance evaluation with Kit Fox, MUST, Prairie Grass, and EMU observations, Atmos. Environ., 38, 4675–4687, 2004.

Henn, D. S. and Sykes, R. I.: Large-eddy simulation of dispersion in the convective boundary layer, Atmos. Environ., 26A, 3145–3159, 1992.

Kataoka, H. and Mizuno, M.: Numerical flow computation around aeroelastic 3D square cylinder using inflow turbulence, Wind Struct., 5, 379–392, 2002.

Michioka, T., Sato, A., and Sada, K.: Large-eddy simulation for the tracer gas concentration fluctuation in atmospheric boundary layer, Japan Soc. Mech. Eng., 69, 868–875, 2003.

Moeng, C.-H. and Sullivan, P. P.: A comparison of shear and buoyancy driven planetary-boundary-layer flows, J. Atmos. Sci., 51, 999–1022, 1994.

Moin, P. and Kim, J.: Numerical investigation of turbulent channel flow, J. Fluid Mech., 118, 341–377, 1982.

Nakayama, H. and Nagai, H.: Large-eddy simulation on turbulent flow and plume dispersion over a 2-dimensional hill, Adv. Sci. Res., 4, 71–76, 2010.

Nakayama, H., Takemi, T., and Nagai, H.: LES analysis of the aerodynamic surface properties for turbulent flows over building arrays with various geometries, J. Appl. Meteorol. Clim., 6, 79–86, 2011.

Nakayama, H., Takemi, T., and Nagai, H.: Large-eddy simulation of urban boundary-layer flows by generating turbulent inflows from mesoscale meteorological simulations, Atmos. Sci. Lett., 13, 180–186, 2012.

Nakayama, H., Jurcakova, K., and Nagai, H.: Development of local-scale high-resolution atmospheric dispersion model using large-eddy simulation Part 3: turbulent flow and plume dispersion in building arrays, J. Nucl. Soc. Technol., 50, 503–519, 2013.

Nakayama, H., Leitl, B., Harms, and F., Nagai, H.: Development of local-scale high-resolution atmospheric dispersion model using large-eddy simulation Part 4: turbulent flows and plume dispersion in an actual urban area, J. Nucl. Soc. Technol., 51, 628–638, 2014.

Ohya, Y. and Uchida, T.: Laboratory and numerical studies of the convective boundary layer capped by a strong inversion, Bound.-Lay. Meteorol., 112, 223–240, 2004.

Ohya, Y., Neff, D. E., and Meroney, R. N.: Turbulence structure in a stratified boundary layer under stable conditions, Bound.-Lay. Meteorol., 83, 139–161, 1997.

Smagorinsky, J.: General circulation experiments with the primitive equations, Mon. Weather Rev., 91, 99–164, 1963.

Snyder, W. H.: Guideline for fluid modeling of atmospheric diffusion, US EPA Report EPA-600/8-81-009, US EPA, Research Triangle Park, 1981.

Sykes, R. I. and Henn, D. S.: Large-eddy simulation of concentration fluctuations in a dispersing plume, Atmos. Environ., 26A, 3127–3144, 1992.

Takewaki, H., Nishiguchi, A., and Yabe, T.: Cubic Interpolated Pseudo-particle method (CIP) for solving hyperbolic-type equations, J. Comput. Phys., 61, 261–268, 1985.

Turner, D.B.: Workbook of Atmospheric Dispersion Estimates, No. AP-26, revised edition, US Environmental Protection Agency, Office of air programs publication, Research Triangle Park, 1970.

Vervecken, L., Camps, J., and Meyers., J.: Accounting for wind-direction fluctuations in Reynolds-averaged simulation of near-range atmospheric dispersion, Atmos. Environ., 72, 142–150, 2013.

3

Integration by identification of indicators

V. Giannini[1,2] **and C. Giupponi**[1,2]

[1]Ca' Foscari University of Venice, Italy
[2]Fondazione Eni Enrico Mattei, Venice, Italy

Abstract. The objective of the BRAHMATWINN research component described in this chapter is to develop integrated indicators with relevance to Integrated Water Resources Management (IWRM) and climate change for the Upper Danube and the Upper Brahmaputra River Basins (UDRB and UBRB), and to foster the integration process amongst the different research activities of the project. Such integrated indicators aim at providing stakeholders, NGOs and GOs with an overview of the present state and trends of the river basins water resources, and at quantifying the impacts of possible scenarios and responses to driving forces, as well as pressures from likely climate change. In the process the relevant indicators have been identified by research partners to model and monitor issues relevant for IWRM in the case study areas. The selected indicators have been validated with the information gathered through the NetSyMoD approach (Giupponi et al., 2008) in workshops with local actors. In this way a strong link between the main issues affecting the basins as perceived by local actors and the BRAHMATWINN activities has been created, thus fostering integration between research outcomes and local needs.

1 Introduction

This chapter describes the development of a set of integrated indicators to support IWRM, and to cover the environmental, social, economic, and governance spheres relevant for the project study areas. Indicators are used to simplify, quantify, communicate, and create order with complex data. They convey information that is synthesised, and that can therefore help to reveal complex phenomena. As indicators allow us to measure phenomena or monitor changes and progresses, they thus enable the establishment of a common ground, to compare different areas and situations, and draw conclusions.

A set of indicators that are able to describe in a concise but accurate manner the key environmental, social, economic, and governance aspects related to climate change impacts and IWRM in both Europe and Asia is needed to support the processes of IWRM strategies development, by providing countries with clear priorities. The literature on the selection of indicators is rich, and several international institutions active in various fields have proposed their own core sets of indicators that can be a useful starting point. These include, for instance:

– the EEA core set of indicators (EEA, 2005);

– the Indicators of Sustainable Development (DESA/DSD, 2001);

– OECD core set of indicators (OECD, 2007);

– The core sets of indicators for Eastern European, Caucasus and Central Asia (EECCA) by UNECE (www.unece.org);

– A core set of European Health Environmental indicators (www.euro.who.int).

2 Role within the integrated project

The creation of the Integrated Indicator Table (IIT), which will be described in this chapter and of its framework, should simplify exploring the data provided by the different models used in the BRAHMATWINN project. The IIT will also help researchers to compare and communicate the project's results in a concise but meaningful manner. The hierarchical structure of the IIT allows for flexibility, having different measures according to the different case study areas, and disciplines involved. The IIT is useful to integrate the results of research coming from the different disciplines represented, however, as any classification has some limitations and rigidity.

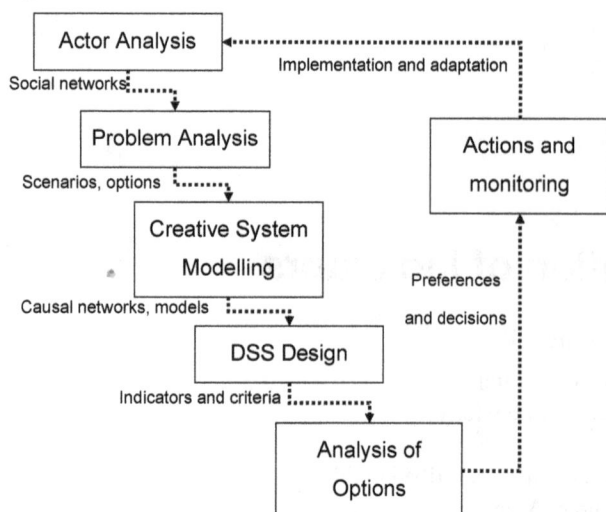

Figure 1. The NetSyMoD approach for participatory modelling and decision making (source: Giupponi et al., 2008).

3   Scientific methods applied

3.1   NetSyMoD

The NetSyMoD approach relies on the DPSIR (Driving Forces – Pressures – State – Impacts – Responses) framework (EEA, 1999) for problem conceptualisation and indicator selection. There are six steps envisaged within this methodology (Giupponi et al., 2008): two phases in particular have been used to build the Integrated Indicator Table (IIT), and are discussed here: (1) Problem Analysis, and (2) Creative System Modelling (see Fig. 1).

In the *Problem Analysis* phase the opinions of Local Actors (LA) are elicited to describe the problem, taking into account all possible aspects. *Creative System Modelling* (CSM) techniques facilitate the process of participatory modelling and elicitation of knowledge and preferences from actors, thus build a common understanding of the problem framed in the DPSIR framework.

The key actors identified in the first NetSyMoD phase – *Actor Analysis* – are now involved in the development of a shared vision of the human-environmental system. They may be involved in various ways, typically through a participatory workshop, during which creative thinking and cognitive mapping techniques are used to develop a shared model of the problem. The CSM workshop can have two main aims, depending on the case at hand:

1. building a shared model of the problem, based on cause-effect chains and using the DPSIR conceptual model, and

2. developing shared scenarios, investigating the potential evolutions of the system over time, or under different

policies (this is the objective of the CSM workshops organised for the activities described in Chapter 8).

3.2   Workshops

The list of concepts collected during the CSM workshops carried out in Assam (April 2007), Bhutan (October 2007), Austria (October 2008), Nepal (November 2008) and Austria-Germany (February 2009) have been included in the list of indicators and responses. Four fields in the Integrated Indicator Table (IIT) list the issues and responses elicited from local actors in each case study area, i.e. we allocated within the below described framework issues and responses arisen during the CSM workshops, establishing a two way relationship at the Sub-domain level. The framework of the IIT is shown in Table 1, while an extract of it is shown in Table 2.

The CSM workshops allowed to elicit from local actors involved ideas and concepts related to the main issues affecting the project case study areas, and existing or needed response strategies to cope with them. The issues/responses identified can be expressed as indicators and complete the list of indicators identified by researchers. The scope is, in fact, to compare the quantitative information provided by research partners with the qualitative information provided by local actors during the workshops, and carry out a validation process. This process makes sure that relevant indicators describing and characterising the local context are included in the set (see Table 2).

The identification of quantifiable indicators to monitor the evolution of the different components of the causal loop diagrams is also necessary to develop future plausible scenarios for the case study areas, which will in turn enable the assessment of existing and potential responses in the years to come.

3.3   The structure of the Integrated Indicator Table

The structure of the Integrated Indicator Table (IIT) was defined in agreement with the BRAHMATWINN research partners. The set of integrated indicators is designed as a multilevel list, a tool for integrated assessment. The structure in which the indicators are organised is composed of four categories: *Themes – Domains – Sub-domains – Indicators* (see Table 1).

The Themes aim at characterising a sustainability framework, and are:

– *Environmental* describes the state of the Natural Environment.

– *Social* guarantees that the Human Dimension is described.

– *Economic* describes the human economic activities.

Table 1. Integrated Indicator Table (Themes, Domains, Sub-domains) and allocation to DPSIR scheme.

Theme	Domain	Sub-Domain	ED	D	P	S	I	R
Environmental	Basin description	Basin morphology				1		
Environmental	Ecosystem /Biodiversity	Ecosystem functions				1		
Environmental	Ecosystem /Biodiversity	Biodiversity				1		
Environmental	Land use / Land use change	Land use				1		
Environmental	Land use / Land use change	Glaciology				1		
Environmental	Land use / Land use change	Permafrost				1		
Environmental	Forests	Forest management				1		
Environmental	Water	Water quality				1		
Environmental	Water	Water resources pressure			1			
Environmental	Water	Water resources state				1		
Environmental	Water	Water resources impact					1	
Environmental	Water	Water flow				1		
Environmental	Climate	Precipitation	1					
Environmental	Climate	Aridity					1	
Environmental	Climate	Evapotranspiration				1		
Environmental	Climate	Temperature	1					
Environmental	Environmental hazards	Vulnerability					1	
Social	Livelihoods/ Assets	Poverty			1			
Social	Livelihoods/ Assets	Water availability			1			
Social	Livelihoods/ Assets	Education /Information			1			
Social	Population	Population dynamics			1			
Social	Gender	Gender issues			1			
Social	Community structure	Age distribution			1			
Social	Health/ Sanitation	Morbidity and mortality					1	
Social	Health/ Sanitation	Sanitation system			1			
Social	Health/ Sanitation	Healthcare delivery			1			
Social	Settlements	Housing settlements			1			
Social	Settlements	Urban settlements			1			
Social	Infrastructure	Access to infrastructure			1			
Social	Infrastructure	Road infrastructure		1				
Social	Infrastructure	Water infrastructure		1				
Social	Infrastructure	Infrastructure pressures			1			
Economic	Wastes	Waste management			1			
Economic	Energy	Energy consumption			1			
Economic	Energy	Energy production			1			
Economic	Economic development	Agricultural production			1			
Economic	Economic development	Service sector		1				
Economic	Economic development	Construction sector			1			
Economic	Economic development	Industrial production		1				
Economic	Economic development	GDP/GNP		1				
Economic	Economic development	Employment		1				
Governance	Education	Capacity building						1
Governance	Education	Increase knowledge						1
Governance	Institutional and legislative frameworks	Decision making						1
Governance	Institutional and legislative frameworks	Public Participation						1
Governance	Institutional and legislative frameworks	Disaster preparedness						1
Governance	Institutional and legislative frameworks	IWRM /NRM						1
Governance	Institutional and legislative frameworks	General institutional and legislative frameworks						1
Governance	International relations	Transboundary issues						1

– *Governance* describes the legislative and institutional frameworks, including the degree of public participation, education and awareness of a population.

Indicators provide information about complex, typical or critical processes in social-ecological systems, and simplify communication about the issues addressed. Research activities within the BRAHMATWINN project are many and varied, they fall within different disciplines, and they make use of various models and assessment frameworks. As a consequence, a significant number of indicators and data sets are required to populate the different models and approaches. To facilitate the identification of integrated IWRM indicators and of intra-disciplinary linkages, within this project we have adopted the terminology of *domain* to identify a particular Environmental, Social, Economic or Governance issue (e.g. Land-use/Land-cover change, Environmental Hazards, Livelihoods/Assets, Health/Sanitation, Energy, Economic development, Education, Institutional and legislative frameworks...). *Sub-domains* have also been defined, for the identification of more specific categories of issues addressed by groups of – site specific – detailed *indicators*. For instance the domain *Climate* could be subdivided into four sub-domains: *Precipitation*, *Temperature*, *Aridity* and *Evapotranspiration*, each of them quantified by one or more indicator.

Table 2. IIT section: Example of matching local actors' opinions with BRAHMATWINN researchers' indicators.

Theme	Domain	RESEARCHERS' INDICATOR	Sub-Domain	LOCAL ACTORS' ISSUE
Environmental	Water	Number of water extraction & discharge	Water quality	Effluents treatment
		Water quality		Pollution
		Contamination of ground water		
		Water supply	Water resources pressure	
		Renewable rate		
		Total water extraction		Ground water level
		Amount of water resources in typical, wet and dry years	Water resources state	Water depth
		Water reservoirs		
		Wetland (beel)		Natural flushing of wetlands
		Lake area [m2] per 1000m grid cell		
		Retention area [m2] per 1000m grid		
		Percentage of extracted water to total water resources in typical, wet and dry years	Water resources impact	Extraction of water
		Relative water stress index (RWSI)		Impact on aquatic resources
		Discharge	Water flow	
		Dominant type of runoff generation		
		Drainage density		
		Form factor (Horton)		Physical characteristics of the river
		Water level exceedance		
		Monthly discharge (12 mean monthly discharge values per year and catchment outlet)		River flow
		Annual runoff pattern		Runoff

4   Results achieved and deliverables provided

The choice of the set of indicators is carried out keeping in mind that it should meet the needs and priorities of users (e.g. policy and decision makers, experts, civil society groups) in monitoring processes towards the implementation of IWRM principles in the Upper Danube and Upper Brahmaputra River Basins. The collection of all indicators used by, or relevant to, partners is the first phase planned for the development of a set of integrated indicators within the research phase described here. Therefore, indicator profile forms have been prepared and distributed among the BRAHMATWINN partners to be filled with the information of each indicator they have selected. The template used to define each indicator's profile is divided in three different sections:

1. *general information about the indicator:* requires providing the main information about the indicator (e.g. name, definition, domain of applicability);

2. *rationale for indicator selection:* collects synthetic information on the choice of the particular indicator in relation to its usability;

3. *data needs:* collects information on data needs and data availability for the indicator.

The task of partners was to suggest a way of measuring the list of domains provided in the forms, through indicators. All indicators have been selected because of their policy relevance, with respect to climate change and water resources management, availability of historical time series, data availability over a large part of the UDRB and UBRB and transparency (i.e. they can be easily understood by the policy-makers and the general interested public). The information collected defined a list of indicators, organised according to the domains and sub-domains of reference in the common framework described above, and for further evaluation within the consortium.

In this way the results from the precedent research components (described in Chapter 2, 3 and 4) have been integrated as follows:

1. In Chapter 2 indicators have been developed by the downscaling of Global Climate Models projections.

2. Within Chapter 3 indicators quantifying the assessment and classification of the components of the Natural Environment (NE), such as topography, hydrology and groundwater, snow and glacier cover, permafrost and slope stability, land use and land cover, water quality, eco-hydrology and biodiversity were evaluated.

3. Social, Economic, and Governance indicators have been identified and applied at the local scale in Chapter 4.

As an example, the Sub-domain *Precipitation* (Theme: Environment, Domain: Climate) is described by the following *indicators:*

- Average annual temperature, Extreme temperature indices, Annual mean temperature, Seasonal mean temperature

- Growing season length, Growing season onset

- Hot-day threshold, Cold-day threshold

- Frost days frequency, Longest heat wave

The diversity of each case study area is taken into account at the level of sub-domains through different sets of indicators, which are relevant for each area. Thus, the list of sub-domains constitutes the interface between the quantitative and qualitative data sets. It is at this level that the integration process between quantitative and qualitative information, the latter provided by local actors, takes place. It is difficult to present the whole IIT in a publication such as this one, please refer to Table 2 for an example of how local actors' opinions and BRAHMATWINN researchers' indicators have been matched.

4.1 Validation

Three Delphi Rounds were carried out for the development and validation of the Integrated Indicator Table (IIT). The first round was carried out by distributing a template (described above) for the collection of the indicators from each partner. The following two rounds, here described, resulted in the validation of the IIT by the project partners.

Delphi Round 2 consisted in a gap analysis. Confronting the indicators selected by the partners with the concepts elicited from the local actors, a gap analysis has been performed to verify whether the partners have provided indicators suitable to address, quantify, and describe the issues identified by the local actors. When no indicator within the list provided by partners corresponds to an issue or response strategy as expressed by local actors, a gap was identified, which was then filled by the research partners.

This gives information about the appropriateness of the set of indicators proposed to describe problems at the local scale. The consolidation of a list of concepts vs. indicators couples will enable and validate the partners' research outcomes (analysis and modelling) with the opinions provided by the local actors, describing with more detail the needs and issues they have to cope with at the local level. This process allows for the integration of the analysis (both within the human dimension and the natural environment) carried out in the previous phases of the project's implementation in a common framework. The latter will serve to support the decision making process, as a base for the evaluation of different alternative options, carried out later through the application of mDSS.

With Delphi Round 3 the IIT was validated by the BRAHMATWINN partners. It must be said, however, that the IIT must not be thought of a rigid and definitive table, but more of a flexible structure within which indicators can be added or modified according to research needs and new findings.

5 Contribution to sustainable IWRM

Indicators have been identified by researchers according to their model outcomes. The indicators selected by researchers have been organised and listed in the IIT. This preliminary list has been integrated, compared and validated with information collected during the Creative System Modelling (CSM) workshops held in the case study areas. In fact, during the CSM workshops participating local actors were asked to share their opinions on the main issues affecting the area considered, with respect to IWRM and in the context of climate change. This process led to the validation of the IIT, which can then be used in several ways to foster IWRM. An example will be discussed in Chapter 8, where criteria derived from indicators will be used to assess relative effectiveness of IWRM responses to cope with flood risk under the impact of climate change.

6 Conclusions and recommendations

The creation of the IIT enabled the integration of two processes, one local actor/end-user driven, and the other researcher driven. The framework (Theme, Domain, Sub-domain) can thus be seen as the interface between the contributions of local actors and BRAHMATWINN research partners towards the formalization of the problem. Sub-domains represent the level we have decided to deal with in future steps of the project, because they represent the complexity of the system at a level of definition local actors and end-users can deal with.

The IIT was further used in three workshops (Salzburg, October 2008; Kathmandu, November 2008; Kathmandu, November 2009) providing the possibility to local actors to give a final validation of it.

From the results obtained the following recommendation can be made:

1. System analysis and modelling should be integrated into the process of defining integrated indicators, as well as in the workshop discussions with local actors and stakeholders, to enhance their appreciation of the pressing needs. The latter should be addressed when assessing the system dynamics, and when validating their process models to better contribute to decision making for adaptive IWRM and to cope with vulnerabilities.

2. Integrated indicators as analysed in this study should be integrated in the ultimate Integrated Land and Water Resources Management System (ILWRMS) to support decision making on all governance levels, and this is described in more detail in Chapter 9.

Acknowledgements. We would like to acknowledge the following BRAHMATWINN research partners for contributing to the creation of the Integrated Indicator Table: Geoinformatik Department, Friedrich-Schiller Univerity, Jena (Germany); Department for Geography, Ludwig-Maximilians University, Munich (Germany); Institute for Atmospheric and Environmental Sciences, Goethe University Frankfurt (Germany); Centre for Geoinformatics, University of Salzburg, Salzburg (Austria); GeoData Insitute, University of Southampton (UK); Centre for Water Law, Policy and Science, University of Dundee (UK); Department of Limnology and Hydrobotany, University of Vienna (Austria), Department of Geosciences, University of Oslo (Norway); ICIMOD, Kathmandu (Nepal), Royal University of Bhutan, Thimphu (Bhutan); Indian Institute of Technology Roorkee, Roorkee (India). Last but not least we would also like to acknowledge the contribution of other FEEM researchers who worked in previous phases of the BRAHMATWINN project: Jacopo Crimi, Alessandra Sgobbi and Yaella Depietri.

The interdisciplinary BRAHMATWINN EC-project carried out between 2006–2009 by European and Asian research teams in the UDRB and in the UBRB enhanced capacities and supported the implementation of sustainable Integrated Land and Water Resources Management (ILWRM).

References

DESA/DSD: Indicators of Sustainable Development: Framework and Methodologies, Division for Sustainable Development (DSD), edited by: Department of Economic and Social Affairs (DESA), Background Paper No. 3, Commission on Sustainable Development, Ninth Session, 16–27 April 2001, New York, 2001.

EEA: Environmental Indicators: typology and overview, edited by: European Environment Agency (EEA), Technical report no. 25, Copenhagen, 1999.

EEA: EEA core set of indicators, Guide, edited by: European Environmental Agency (EEA), Technical report, 2005.

Giupponi, C., Sgobbi, A., Mysiak, J., Camera, R., and Fassio, A.: NetSyMoD – An Integrated Approach for Water Resources Management, in: Integrated Water Management, edited by: Meire, P., Coenen, M., Lombardo, C., Robba, M., and Sacile, R., Springer, Netherlands, 69–93, 2008.

OECD: OECD Key Environmental Indicators, edited by: Organisation for Economic Development and Co-Operation (OECD), 2007.

www.unece.org/env/europe/monitoring/IandR_en.html, last access: March 2011.

http://www.euro.who.int/en/what-we-do/data-and-evidence/environment-and-health-information-system-enhis/publications, last access: March 2011.

4

Evaluation of HARMONIE in the KNMI Parameterisation Testbed

E. I. F. de Bruijn and W. C. de Rooy

Royal Netherlands Meteorological Institute, De Bilt, The Netherlands

Correspondence to: E. I. F. de Bruijn (cisco.de.bruijn@knmi.nl)

Abstract. HARMONIE, a non-hydrostatic NWP model has a single column version which is used for testing and validation of physical parameterisations. Since January 2010, this single column model (SCM) has been run on a daily basis in the KNMI parameterisation testbed (KPT). In this testbed, the HARMONIE SCM is run with different options and the output is compared with a wide variety of observations and other participating SCMs as well as large-eddy simulations (LES) model output. The evaluation presented here makes use of the advanced observation site Cabauw in the Netherlands, with a focus on shallow convection, turbulence and cloud formation. The examples shown illustrate the potential of the daily monitoring and in-depth evaluation to detect and improve model deficiencies.

1 Introduction

HARMONIE is a non-hydrostatic NWP model that is used for mesoscale predictions. It is developed in cooperation with the HIRLAM and ALADIN consortia. In this paper we focus on HARMONIE cycle 36 with AROME physics (Seity et al., 2012). Due to the high horizontal resolution of $2.5 \times 2.5 \text{ km}^2$, we assume that deep convection is resolved, while shallow convection still needs to be parameterised. For the validation of the physical parameterisations, a single column version of HARMONIE is available. Since January 2010, this single column model (SCM) is run on a daily basis in the KNMI parameterisation testbed (KPT) (see Neggers et al., 2012 and Sect. 2.1). In this testbed the HARMONIE SCM is validated against observations and large- eddy simulations (LES) (see Heus et al., 2012 and Sect. 2.2). The output can also be compared with other SCM's, extracted from different NWP models participating in the KPT. The KPT is suitable and has been successfully used for evaluating model performances for fast processes, such as shallow convection, turbulence and surface processes. As a result, several deficiencies in the model's physical parameterisations have been found and improved. The aim of this article is to illustrate the potential of the daily monitoring in the KPT by showing an evaluation of the HARMONIE SCM focusing on low clouds and mist. This is done by showing two examples in which striking differences occur between two HARMONIE model configurations with different convection and cloud schemes. Observations from Cabauw and LES output are used as reference.

2 Set-up of the system

2.1 The parameterisation testbed

In the KPT, a so-called host model provides data for running a SCM, because time and height dependent geostrophic wind speed and advection fields of temperature, humidity are needed. The driver files, containing this information, are derived from the most recent regional atmospheric climate model (RACMO) (van Meijgaard et al., 2008) forecast. RACMO has a resolution of 0.2 degrees and is initialized with the ECMWF analysis and forced by boundaries from the same model. During the complete 72 h forecast, the SCM output is nudged to the RACMO state with a relaxation time of 6h to prevent it from drifting away. As a result the SCM is still able to develop short living features like clouds, but will also stay close to the background model RACMO on longer timescales.

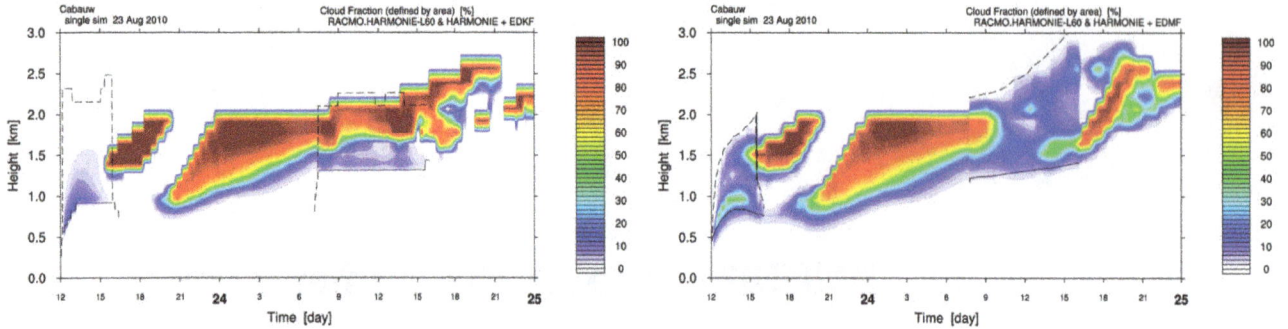

Figure 1. Cloud fraction of EDKF (left), EDMF (right) for 24 August 2010 (analysis 23 August 2010, 12:00 UTC) . The solid line is lifting condensation level and dashed line is the updraft termination height.

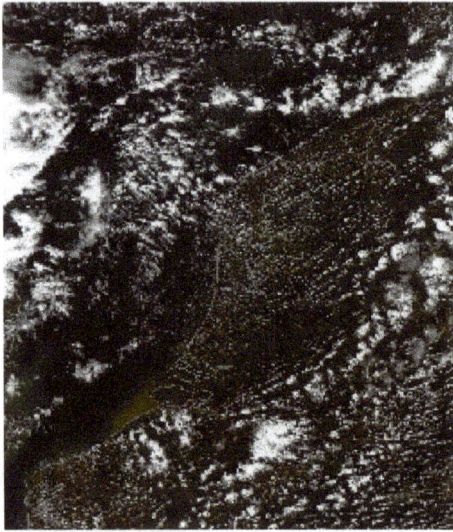

Figure 2. MODIS VIS satellite picture on 24 August 2010, 12:21 UTC.

2.2   Observations and LES

For evaluation we use observations and LES. The Cabauw site in the Netherlands offers a wide variety of quality observations. These observations comprise a 200 m tall tower with sensors for temperature, humidity and wind components and ground-based remote sensing instruments like lidar and cloud radar. Also LES output is used for validation. LES with its very high resolution ($100 \times 100\,\mathrm{m}^2$) in the horizontal and 40 m in the vertical are well capable of resolving convection and for convective situations they can be considered as pseudo observations (see e.g. Siebesma and Cuijpers, 1995). LES produce additional information of the atmospheric state which the instrumentation does not provide, for instance the vertical profiles of cloud cover and mass flux.

2.3   Convection schemes

In this study we focus on the validation of two alternative mass flux shallow convection schemes. The Eddy Diffusivity Kain Fritsch (EDKF) scheme (Pergaud et al., 2009) represents convection with one updraft and lateral mixing between this updraft and the environment is described according to Kain and Fritsch (1990). The alternative convection scheme, noted here as the EDMF scheme uses a dry and wet updraft (Neggers et al., 2009) and lateral mixing is described according to de Rooy and Siebesma (2008). As discussed in de Rooy et al. (2012), it can be expected that the lateral mixing is better represented, i.e. in better correspondence with LES in the EDMF scheme than in the EDKF scheme. Consequently, e.g. the mass flux profile and cloud top height should be estimated more accurately with the EDMF scheme.

2.4   Cloud schemes

The EDKF scheme is combined with a statistical cloud scheme in which variance of the moisture deficit is produced by turbulence only. In addition to the cloud cover resulting from this statistical cloud scheme, an extra term proportional to the updraft fraction from the convection scheme is added. The EDMF scheme uses a full statistical cloud scheme with a variance of the moisture deficit produced by turbulence and convection. Apart from turbulence and convection there can be other sources of variance like gravity waves and mesoscale organization. To account for this, the EDMF scheme applies an additional variance term proportional to the saturation total water specific humidity. As explained in de Rooy et al. (2010), in this way we add the characteristics of a RH-scheme, where cloud cover is simply a function of the relative humidity to a statistical cloud scheme.

3   Results

Figure 1 presents the development of the cloud cover during 24 August 2010, based upon an analysis on 23 August, 12:00 UTC. In EDKF (left) there is no break-up of the stratocumulus cloud deck, whereas in EDMF the cloud

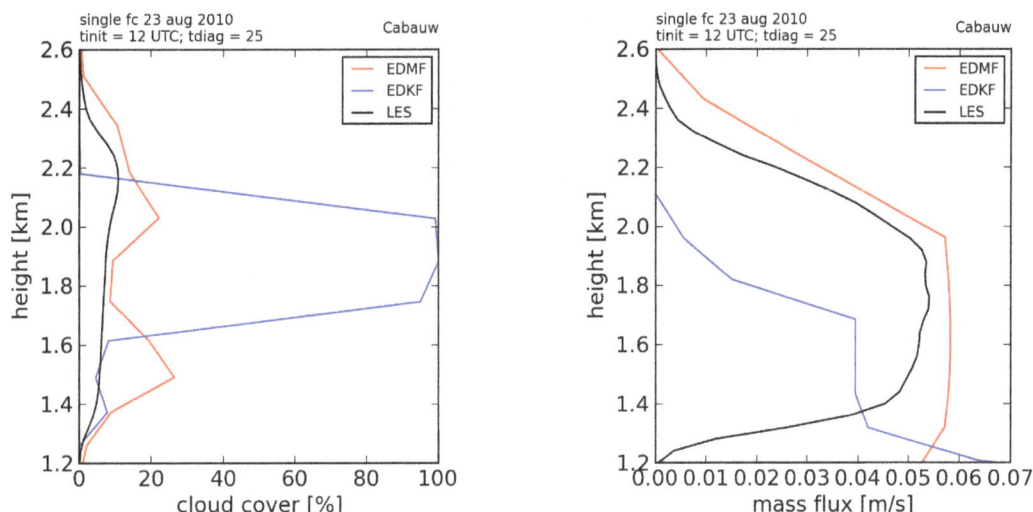

Figure 3. Profiles of cloud fraction (right) and mass flux (left) at $t = +25$ h, validation time for 24 August 2010, 13:00 UTC. Note that the cloud base height is at approximately 1.4 km.

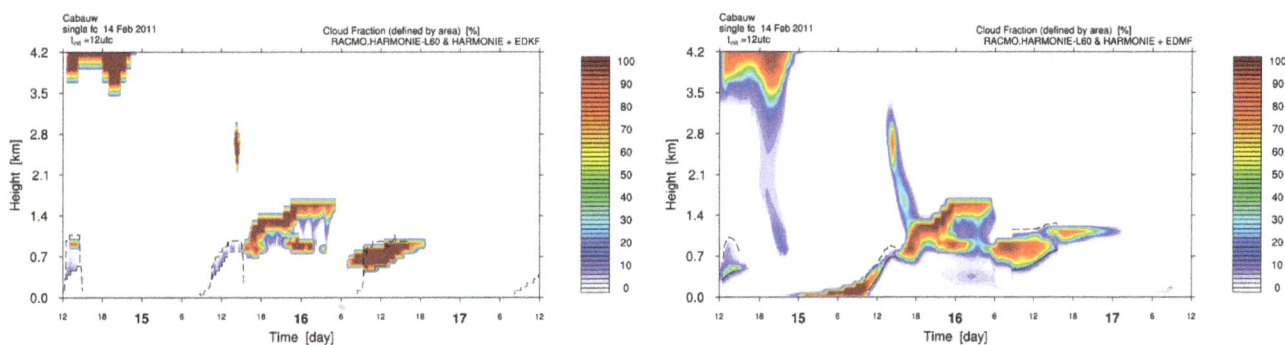

Figure 4. Cloud fraction represented by EDKF (left) and EDMF (right) for 15 February 2011. Note the development of fog during the night with EDMF.

fraction is becoming less after 25 h of forecast time. The cloud base and top are represented by the lifting condensation level (solid line) and updraft termination height (dashed line). These variables are only present when the convection scheme is active. In EDMF, the updraft termination height increases as soon as the cloud deck breaks up. In reality the stratocumulus deck did break up as illustrated by a satellite picture of the Netherlands (see Fig. 2) and the cloud fraction according to LES (see Fig. 3 left panel). Figure 3 reveals the good correspondence of the cloud cover profile of LES and EDMF in contrast with the large overcast with EDKF. If we take a look at the corresponding mass flux profile for this hour (Fig. 3 right panel), we see that the EDKF scheme deposits the moisture in too shallow layer, whereas the EDMF mass flux profile shows a much better correspondence with LES. As a result the EDKF scheme produces a persistent, not observed stratocumulus layer. Due to a better representation of lateral mixing, EDMF shows a better correspondence to the LES mass flux profile as could be expected based on de Rooy et al. (2012). However, the too low cloud top height in

EDKF might also be affected by the missing of the release of energy due to coagulation. As a result of this, the updraft in EDKF does not gain buoyancy from the coagulation process and terminates at a too low level.

During the second case of 15 February 2011, radiation fog was developed in Cabauw. At 00:00 UTC, the reported 10 m windspeed was 1 m s^{-1} and the fog layer had a vertical extent of 400 m. The fog layer slightly increased and started to disappear after 04:00 UTC. The EDKF parameterisation was not able to capture the fog (Fig. 4) while EDMF with a modified cloud scheme was able to give some fog warning. The synoptical observations (Fig. 5) gave a reduced visibility of less than 200 m during the morning hours and back scatter observations from a Lidar device (not shown) in Cabauw confirmed the presence of very low clouds. The improved fog forecast is related to the modification in the statistical cloud scheme. Note that the convection scheme is not active under these stable conditions. In typical fog conditions there is no convection and limited turbulence. Consequently, the variance of the moisture deficit used in the statistical cloud

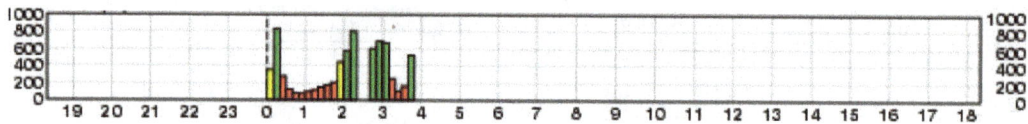

Figure 5. Synoptical visibility observations in [m] in Cabauw during 14–15 February 2011.

scheme can be extremely low. As a result no fog might be produced despite the small moisture deficit. The extra variance term in the EDMF cloud scheme helps to produce fog in these kind of conditions. It should be noted that the fog disappeared in the course of the morning, whereas the EDMF parameterisation was too late in predicting the vanishing of the fog.

4 Conclusions and discussion

Single column models and the KNMI parameterisation testbed have been shown to be useful tools for model evaluation of fast local processes like turbulence and convection. Based on daily monitoring and an in-depth validation against different observations, several deficiencies could be linked to particular physical processes captured inadequately by the corresponding parameterisations. The cases presented here do not stand alone but illustrate typical behavior of the EDKF and EDMF convection and cloud scheme in certain conditions. For example with the EDKF option, the convective transport is often too shallow which results in an accumulation of moisture in a too shallow layer and accordingly a too persistent stratocumulus cloud deck.

Another example deals with insufficient production of variance of moisture deficit in certain circumstances like fog. The modification of the statistical cloud scheme of EDMF turned out to be crucial to capture many observed fog cases. However, in this particular case the simulated fog is too persistent.

The SCM approach has also drawbacks. Due to the coarse prescribed dynamical tendencies, mesoscale circulations can not be captured adequately. The data to force the model should be accurate and preferably not derived from a foreign host model to circumvent the mismatch between the physical parameterisations. Therefore, it is recommended to make tendencies from the 3-D HARMONIE model available for running the SCMs. In a new cycle of HARMONIE planned for 2012, there is an option of extracting fields from the 3-D model, in order to generate a forcing column for the SCM.

Acknowledgements. Roel Neggers from KNMI is kindly acknowledged for making the KPT system available. We are also grateful to the editor and three reviewers for their comments and constructive suggestions which improved the manuscript considerably.

Edited by: G.-J. Steeneveld
Reviewed by: R. Ronda and two anonymous referees

References

de Rooy, W. C. and Siebesma, A. P.: A simple parameterization for detrainment in shallow cumulus, Mon. Weather Rev., 136, 560–576, 2008.

de Rooy, W., de Bruijn, C., Tijm, S., Neggers, R., Siebesma, P., and Barkmeijer, J.: Experiences with Harmonie at KNMI, HIRLAM Newsletter, 56, 21–29, 2010.

de Rooy, W. C., Bechtold, P., Fröhlich, K., Hohenegger, C., Jonker, H., Mironov, D., Siebesma, A. P., Teixeira, J., and Yano, J.-I.: Entrainment and detrainment in cumulus convection: an overview, Q. J. Roy. Meteorol. Soc., in press, 2012.

Heus, T., van Heerwaarden, C. C., Jonker, H. J. J., Pier Siebesma, A., Axelsen, S., van den Dries, K., Geoffroy, O., Moene, A. F., Pino, D., de Roode, S. R., and Vilà-Guerau de Arellano, J.: Formulation of the Dutch Atmospheric Large-Eddy Simulation (DALES) and overview of its applications, Geosci. Model Dev., 3, 415–444, doi:10.5194/gmd-3-415-2010, 2010.

Kain, J. S. and Fritsch, J. M.: A one-dimensional entraining/detraining plume model and its application in convective parameterization, J. Atmos. Sci., 47, 2784–2702, 1990.

Neggers, R. A. J., Kohler, M., and Beljaars, A. C. M.: A dual mass flux framework for boundary layer convection. Part I: Transport, J. Atmos. Sci., 66, 1464–1487, 2009.

Neggers, R. A. J., Siebesma, A. P., and Heus, T.: Continuous single-column model evaluation at a permanent observational supersite, B. Am. Meteor. Soc., in press, 2012.

Pergaud, J., Masson, V., Malardel, S., and Couvreux, F.: A parameterization of dry thermals and dry thermals and shallow cumuli for mesoscale numerical weather prediction, Bound.-Lay. Meteorol., 132, 83–106, 2009.

Seity, Y., Brousseau, P., Malardel, S., Hello, G., Benard, P., Bouttier, F., Lac, C., and Masson, V.: The AROME-France Convective-Scale Operational model, Mon. Weather Rev., 139, 976–991, 2012.

Siebesma, A. P. and Cuijpers, J. W. M.: Evaluation of parametric assumptions for shallow cumulus convection, J. Atmos. Sci., 52, 650–666, 1995.

van Meijgaard, E., van Ulft, L. H., van de Berg, W. J., Bosveld, F. C., van den Hurk, B. J. J. M., Lenderink, G., and Siebesma, A. P.: The KNMI regional atmospheric climate model RACMO, version 2.1. KNMI publication TR-302, 43 pp., 2008.

5

An overview of drought events in the Carpathian Region in 1961–2010

J. Spinoni[1], T. Antofie[1], P. Barbosa[1], Z. Bihari[2], M. Lakatos[2], S. Szalai[3], T. Szentimrey[2], and J. Vogt[1]

[1]JRC-IES, Ispra, Italy
[2]Hungarian Meteorological Service, Budapest, Hungary
[3]Szent Istvan University, Gödöllö, Hungary

Correspondence to: J. Spinoni (jonathan.spinoni@jrc.ec.europa.eu)

Abstract. The Carpathians and their rich biosphere are considered to be highly vulnerable to climate change. Drought is one of the major climate-related damaging natural phenomena and in Europe it has been occurring with increasing frequency, intensity, and duration in the last decades. Due to climate change, land cover changes, and intensive land use, the Carpathian Region is one of the areas at highest drought risk in Europe. In order to analyze the drought events over the last 50 yr in the area, we used a 1961–2010 daily gridded temperature and precipitation dataset. From this, monthly $0.1° \times 0.1°$ grids of four drought indicators (Standardized Precipitation-Evapotranspiration Index (SPEI), Standardized Precipitation Index (SPI), Reconnaissance Drought Indicator (RDI), and Palfai Aridity/Drought Index (PADI)) have been calculated. SPI, SPEI, and RDI have been computed at different time scales (3, 6, and 12 months), whilst PADI has been computed on an annual basis. The dataset used in this paper has been constructed in the framework of the CARPATCLIM project, run by a consortium of institutions from 9 countries (Austria, Croatia, Czech Republic, Hungary, Poland, Romania, Serbia, Slovakia, and Ukraine) with scientific support by the Joint Research Centre (JRC) of the European Commission. Temperature and precipitation station data have been collected, quality-checked, completed, homogenized, and interpolated on the $0.1° \times 0.1°$ grid, and drought indicators have been consequently calculated on the grid itself. Monthly and annual series of the cited indicators are presented, together with high-resolution maps and statistical analysis of their correlation. A list of drought events between 1961 and 2010, based on the agreement of the indicators, is presented. We also discuss three case studies: drought in 1990, 2000, and 2003. The drought indicators have been compared both on spatial and temporal scales: it resulted that SPI, SPEI, and RDI are highly comparable, especially over a 12-month accumulation period. SPEI, which includes PET (Potential Evapo-Transpiration) as RDI does, proved to perform best if drought is caused by heat waves, whilst SPI performed best if drought is mainly driven by a rainfall deficit, because SPEI and RDI can be extreme in dry periods. According to PADI, the Carpathian Region has a sufficient natural water supply on average, with some spots that fall into the "mild dry" class, and this is also confirmed by the FAO-UNEP aridity index and the Köppen-Geiger climate classification.

1 Introduction

The Carpathian Mountains are one of the longest (approximately 1500 km) and most important mountain chains in Europe: they extend over seven countries ranging from the Czech Republic to Serbia, encompassing Slovakia, Poland, Hungary, Ukraine, and Romania. The Carpathians repre- sent a link between North-European taiga and Mediterranean landscapes and they often are a natural barrier for air masses thus, for instance in winter, climate is oceanic to the West from the Carpathians, snowy on the mountain ridge, cold and continental to the East, and Mediterranean to the South-East (Romania). Because of the special orography of the Carpathians, the basin effects are manifold and they cause many different site-specific phenomena, such as a stable boundary layer in winter, rain shadows, or temperature inversions. The Carpathians are extraordinarily rich in flora and endemic plants, as they include the widest primeval forests across all Europe. Moreover, the Carpathians have one of the most valuable biodiversity in Europe, in fact many different bird species and the largest communities of carnivores and predators such as bears and wolves live there.

The Carpathian Region has always been sensitive to hydrological extremes. Examples are the frequent droughts that affected the Great Hungarian Plain, Romania and Serbia from 1983 to 1995. Recent changes in human activities as the development of mass tourism or unsustainable rates of soil exploitation lead to land degradation, a decrease in agricultural production, and an increase in waste and pollution (UNEP, 2007). Deforestation, global warming, and soil erosion processes have been causing floods, droughts, and landslides with a higher frequency in the last 15 yr, especially in the South-Eastern area (e.g. the long drought period in Romania between 2000 and 2003, see e.g. Kozak et al., 2011a).

In the last years a few projects have been developed in order to preserve the unique landscapes, local cultural heritages, and biodiversity of the whole region. Examples are the Carpathian EcoRegion Initiative (Webster et al., 2001), the Carpathian Convention (Kozak et al., 2011a), CarpathCC (Szalai, 2012), and CARPIVIA (http://www.carpivia.eu). Regional studies (Bartholy et al., 2004; Lakatos et al., 2011, for Hungary) and climate change projections by means of scenarios and circulation models (Krüzsely et al., 2011) are quite frequent in the scientific literature related to the Carpathians, as well as small scale or local analyses about floods and/or droughts (Paltineanu et al., 2007; Parajka et al., 2010). On these topics, see also two projects promoted by the European Union: CLAVIER EU Project (http://www.clavier-eu.org) and CECILIA EU Project (http://www.cecilia-eu.org). Until 2010, the Carpathian Region lacked a high-quality climate dataset. To fill this gap and with the financial support of the European Parliament, the European Commission launched and financed the CARPATCLIM Project in late 2010. A consortium of hydro-meteorological institutions from nine countries (Austria, Croatia, Czech Republic, Hungary, Poland, Romania, Serbia, Slovakia, and Ukraine), together with the European Commission's Joint Research Center (JRC, Institute for Environment and Sustainability) set out for the creation of a digital climate atlas of the Carpathian Region (Szalai and Vogt, 2011). This atlas (that will be completed in 2013) is based on a 1961–2010 daily gridded database of 14 meteorological variables: the spatial resolution of the grids is $0.1° \times 0.1°$, the area under examination is 17–27° E (Longitude) and 44–50° N (Latitude), excluding the territories under the political administration of Bosnia-Herzegovina.

After the introduction, the second section of this paper is dedicated to the construction of the dataset: data collection, quality check procedures, homogenization methods, harmonization techniques, and interpolation models are described. Mean temperature (T_M) and precipitation (RR) grids have been used as input for calculating a set of four drought indicators: Standardized Precipitation Index (SPI), Standardized Precipitation-Evapotranspiration Index (SPEI), Reconnaissance Drought Indicator (RDI), Palfai Aridity/Drought Index (PADI), and two climate indicators, Köppen-Geiger climate classification (KG), and FAO-UNEP aridity index. Section 3 provides details on basic features, pros, and cons of these indicators. By means of the listed indicators, computed on different time scales, we performed a detailed study of the drought events of the last 5 decades: a table with the list of relevant drought occurrences and a 1961–2010 monthly drought series for each indicator are presented in Sect. 4. Three case studies (drought in 1990, 2000, and 2003) are then discussed in detail (Sect. 5). A close examination on aridity and shifts in climate classes in the Carpathian Region is presented in Sect. 6. Finally, a summary of the results together with a short overview on some expected outcomes of the CARPATCLIM Project conclude the paper.

2 Data

In most trans-national projects, the main problem is related to different data-sharing policies at country level: local authorities, national or regional meteorological services, independent data providers, etc., usually manage the climate data following different strategies. The philosophy of the CARPATCLIM project lies on the fact that the national members of the consortium retain the property of their data that remain under the custody of the respective owners: no large common database has been created, but the data have been collected and homogenized following shared quality assurance and interpolation methods. Each country, except Austria, collected its own dataset and exchanged data within a belt of 50 Km from the borders with their neighboring countries to enable the cross-border harmonization. For each variable, each member homogenized the records, interpolated them on a national grid, and then the single national products have been merged into harmonized daily grids for the entire Carpathian Region. All the countries used the same homogenization and interpolation methods, in order to avoid producing artificial spatial inhomogeneities. Eventually, daily grids of fourteen variables for the period 1961–2010 have been computed: minimum (T_N), mean (T_M), and maximum temperature (T_X), precipitation (RR), wind speed, wind gust, snow depth, snow water equivalent, relative humidity, air pressure, water vapor pressure, sunshine duration, global solar radiation, and cloud cover. We focus on the creation of T_M and RR grids, because they have been the inputs for calculating the drought indicators.

Project members collected data from various sources such as national meteorological datasets, hand-written annals, and long-term records obtained from regional providers; the

Table 1. Number of mean temperature (T_M) and precipitation (RR) station records used to compute the grids.

PROVIDER	T_M	RR
Czech Rep.	6	23
Croatia	7	26
Hungary	37	176
Poland	9	35
Romania	93	158
Serbia	41	94
Slovakia	26	85
Ukraine	39	130
TOTAL	258	727

station network for RR data (727, see Table 1) is much denser than for T_M data (258). Daily T_M values have been computed as the sum of daily measured T_N and T_X divided by two. In Fig. 1 we show the geographical distribution of the stations.

After the border data exchange, all the 1961–2010 daily records have been quality-checked, completed, and homogenized by means of the Multiple Analysis of Series for Homogenization (MASH, Szentimrey, 1999) software, implemented by the Hungarian Meteorological Service (OMSZ). The original MASH was developed for the homogenization of monthly data series based on hypothesis testing. Later versions have been adapted for the homogenization of daily data series also (Szentimrey, 2008). Depending on the distribution of the examined variable, MASH can be based on an additive or a multiplicative model. In the last version, the following subjects were elaborated for monthly and daily series: series comparison, break-point (change-point) and outlier detection, correction of series, missing data completion, automatic usage of metadata, and a verification procedure to evaluate the homogenization results. The most significant improvements carried out by the current version (MASHv3.03) are connected with the automation of the procedures.

MASH has been chosen because it was recognized as one of the best performing homogenization methods with long-term temperature and precipitation monthly series (Venema et al., 2012). Here we list the basic steps run by MASHv3.03 in the frame of CARPATCLIM: from daily values it calculated monthly series and subsequently estimated monthly inhomogeneities. On the basis of monthly inhomogeneities, it performed a smooth estimation for daily inhomogeneities, then it automatically corrected the daily series, quality checked the homogenized daily data, completed the missing daily values, recalculated monthly series from homogenized daily series, and finally tested the monthly series for homogeneity. All the records shown in Fig. 1 have been completed and homogenized.

After a further border data exchange to ensure the harmonization of the dataset, the station data have been interpolated onto the $0.1° \times 0.1°$ regular grid (see the black rectangle

in Fig. 1). Each country member interpolated T_M and RR series onto national daily grids by means of the Meteorological Interpolation based on Surface Homogenized data basis software (MISH, Szentimrey and Bihari, 2007). The daily national grids have been merged and harmonized into grids for the whole Carpathian Region (17–27° E; 44–50° N) from 1 January 1961 to 31 December 2010. The MISH was developed at OMSZ too: it is a spatial interpolation method with a strong mathematical background that leads to an efficient use of all the valuable meteorological and auxiliary information (Szentimrey et al., 2011). As for MASH, additive (T_M) or multiplicative (RR) interpolation scheme was chosen. In the additive case a regression-Kriging based procedure is performed, while in the multiplicative case everything has been led back to the additive case by a logarithmic transformation. The climate statistical parameters that determine the optimal interpolation parameters by minimizing the expected error are modeled with the help of auxiliary variables. The biggest difference between MISH and the common geo-statistical interpolation methods lies in the application of the meteorological data series for modeling: in geo-statistics (e.g. Cressie, 1991), the sample for modeling is usually based on a single realization in time of the predictor, whilst MISH takes into account the whole data series, i.e. a sample in time and space as well. The auxiliary variables applied in the realization of gridded climatologies for the CARPATCLIM project are: spatial distance, elevation, and the so called AURELHLY (Benichou and Le Breton, 1987) principal components. After the automatic modeling procedure, the gridding interpolation was automatically performed by MISHv1.03.

3 Drought indicators

Though no universal definition exists, the word drought usually refers to a temporal, albeit prolonged shortfall in precipitation as compared to the climatological normal for a defined period of time. Drought is a slowly developing phenomenon with widespread impacts over extended regions and we usually distinguish between meteorological, agricultural, hydrological, and socio-economic drought, depending on the impacts of the rainfall deficit. Due to this complexity, many indicators have been proposed to evaluate the occurrence, duration, and intensity of a drought (e.g. EEA, 2008). We selected four meteorological drought indicators (SPI, SPEI, RDI, and PADI) based on their wide use or regional relevance. The first three are statistical indicators that can be used to detect droughts on a long-term interval, whilst PADI may be used on annual basis as the deviation from the normal value. We based our study on four indicators to provide the user with a more objective determination of drought phenomena in the Carpathian region in the period 1961–2010.

SPI, first introduced by McKee et al. (1993), is a statistical indicator that compares the precipitation during a period

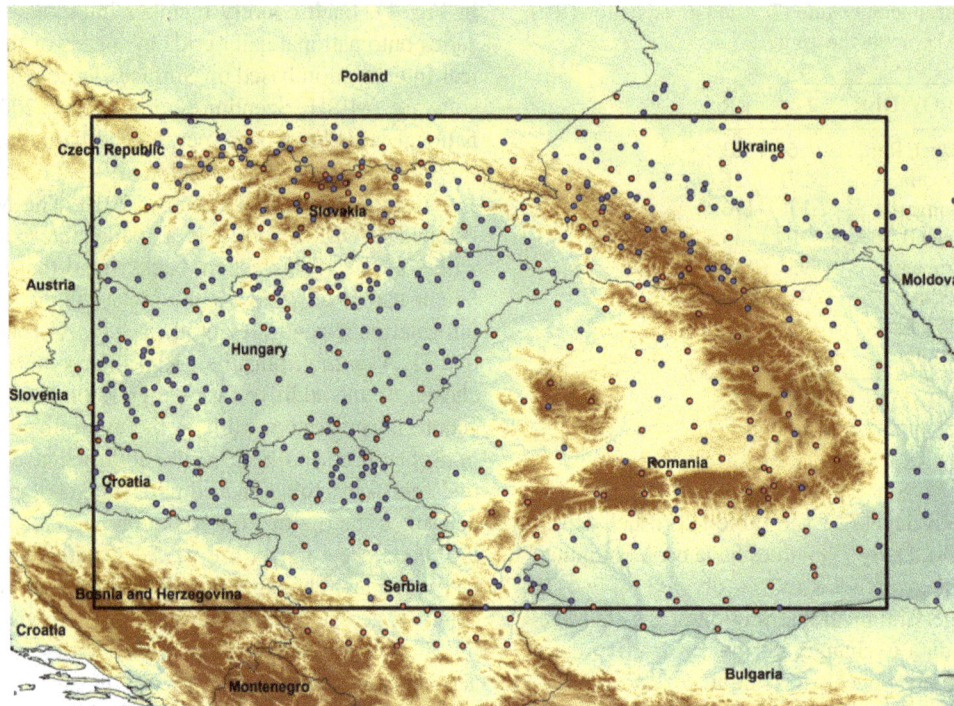

Figure 1. Stations with both RR and T_M data (red dots) and station with RR data only (blue dots). The black rectangle encloses the Carpathian Area (44–50° N, 17–27° E). Bosnia-Herzegovina must not be considered.

Table 2. Classification used for SPI, SPEI, and RDI (left), and for PADI (right). This table also represents the legend for Fig. 3–6.

SPI-SPEI-RDI	Class	PADI	Class
≥ 2.0	Extreme Wet	< 4	Normal
$1.5 \leq\ < 2.0$	Very Wet	$4 \leq\ < 6$	Sub-Humid
$1.0 \leq\ < 1.5$	Wet	$6 \leq\ < 8$	Mild Dry
$-1.0 <\ < 1.0$	Normal	$8 \leq\ < 10$	Dry
$-1.5 <\ \leq -1.0$	Dry	$10 \leq\ < 15$	Very Dry
$-2.0 <\ \leq -1.5$	Very Dry	$15 \leq\ < 30$	Heavy Dry
≤ -2.0	Extreme Dry	> 30	Extreme Dry

of n months versus the long term rainfall distribution at the same location and for the same period of time; it may also be used to determine the drought onset and duration. In this paper, we deal with SPI-3, SPI-6, and SPI-12 (3, 6, and 12 months of rainfall accumulation as input variables); SPI has been calculated on the basis of a Gamma distribution, chosen for it best fits precipitation sums (Thom, 1966). Because the Gamma function is not defined in $x = 0$ (no rainfall), the cumulative probability distribution must be transformed into a standardized distribution with mean 0 and standard deviation 1. Practically, SPI values are the number of deviations left (dry), or right (wet) from 0 (see Table 2 for the classification); the magnitude of the departure from the mean gives us a probabilistic measure of a wet or dry event: following

Guttman (1999), a drought occurs anytime the SPI is continuously negative and reaches −1 or less; the drought ends when SPI becomes positive, though some consequences of drought may be left over for months. SPI needs only RR data, it is multi-scalar in time and space, it is good for comparing indices between different locations, and the frequencies of extreme events are comparable. Since the fitting of the data to the theoretical distribution is an approximation, the choice of the distribution itself can introduce a bias. Depending on the fit between empirical and theoretical distribution, dry regions can be misrepresented.

SPEI was introduced by Vicente-Serrano et al. (2010): its theoretical background is very similar to SPI, but instead of the accumulated rainfall, it is based on the accumulated difference between rainfall and potential evapo-transpiration (PET). As for the calculation of SPI, a Gamma distribution can be used (namely, a shifted version), but a log-logistic distribution or a Paerson-III similarly perform. We chose the shifted Gamma distribution as to compare SPI and SPEI in the best way. We computed SPEI-3, SPEI-6, and SPEI-12. SPEI uses the same classification as SPI (see Table 2). The use of SPEI has two main advantages: it has a better connection to soil water balance than SPI and it considers also temperature (used to compute PET), which is very important in a climate change environment. We used an improved version of the Thornthwaite's model (original: see Thornthwaite, 1948; improved: see Willmott et al., 1985; van der Schrier et al., 2011) that needs only T_M and Latitude as

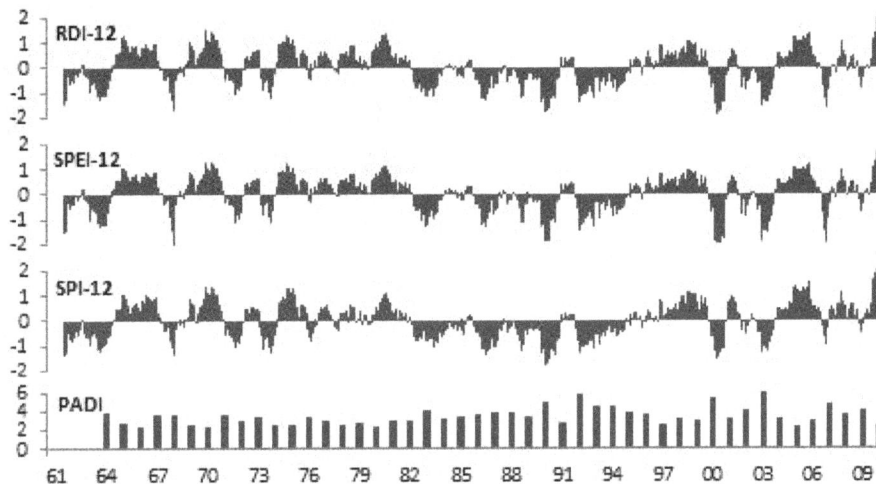

Figure 2. Monthly 1961-2010 drought series of RDI-12, SPEI-12, RDI-12, and annual 1964–2010 series of PADI.

inputs, and it is proven to perform close to the more enhanced Penman-Monteith's model (Allen et al., 2006) if applied to drought indicators (van der Schrier et al., 2011). Due to the fact that RR values can be much higher than PET, precipitation is the basic driver of SPEI, however SPEI may overestimate droughts especially during prolonged heat waves or in very dry areas if large differences between actual and potential evapo-transpiration hold. Be aware of the fact that the subtraction of RR and PET may not precisely follow a Gamma distribution.

RDI (Tsakiris and Vangelis, 2005) is based on the cumulative ratio of precipitation and PET: it is simple, universal and as SPEI, and it includes PET. Because it is based on a ratio, it may be extreme in very arid or wet periods. RDI is a monthly indicator and we computed it as RDI-3, RDI-6, and RDI-12; unlike SPI and SPEI we chose a log-normal distribution for RDI, as suggested by Tsakiris and Vangelis (2005); RDI follows the same classification used for SPI and SPEI (Table 2). When it is computed as RDI-12 for December, it can be compared to FAO-UNEP aridity Index (UNEP, 1992).

The Palfai Aridity Index (*PAI*) was proposed by Palfai (1990) for drought monitoring in Hungary and it has been applied mostly in Eastern European countries. It has been reviewed and modified in the *PADI* in order to be applied to other regions (Kozak et al., 2011b). PADI is an annual indicator ($°C\,mm^{-1}$) based on monthly T_M and RR, annual means and cumulates, and precipitation sum of the previous 3 yr (see Table 2 for the classes); it is based on three correction factors related to temperature, precipitation, and groundwater availability. PADI is not a standardized indicator, so the classes are not objective (Kozak et al., 2011b). It is not suitable for assessing the occurrence of droughts in real time, and we did not differentiate between plains and mountains in the correction factor related to groundwater availability because it is an approximation not calibrated on real data.

The cited indicators have been used to reconstruct 1961–2010 drought series for the Carpathian Region (Sect. 4) and also to analyze three drought events (Sect. 5). Details about the equations used to compute the drought indicators can be found in the cited literature and in Spinoni et al. (2013).

4 Drought events in 1961–2010

For each grid point, we computed annual values of PADI from 1964 to 2010 (PADI needs monthly precipitation data of the previous 3 yr, so we could compute it from 1964 only), and monthly SPI, SPEI, and RDI at 3, 6, and 12-month timescales from 1961 to 2010. It clearly emerges that SPI, SPEI, and RDI, if computed at the same time scale, are highly correlated. We compared the indicators over each grid point and then we averaged the correlation coefficient (r) over the whole grid. For 3-month indicators, r is highest for SPEI-SPI (0.95), followed by RDI-SPEI (0.91), and SPI-RDI (0.82); for 6-month indicators r is highest for RDI-SPEI (0.97), followed by SPEI-SPI (0.95), and SPI-RDI (0.91); for 12-month indicators, r is highest for RDI-SPEI (0.99), followed by SPI-RDI (0.96), and SPEI-SPI (0.94).

In order to calculate drought series for the Carpathian Region, we averaged, for each month or year, the values of the considered indicator over the entire gridded area. In the future, we will construct drought series for climatic sub-regions of the Carpathians, dealing in particular with the differences amongst the plain regions West and East to the Carpathian chain, and also versus the mountain area itself. In Fig. 2 we show the 1961–2010 series for SPI-12 (top), SPEI-12 (center), and RDI-12 (bottom), followed by the annual series of PADI from 1964 to 2010. All the monthly indicators agree: the worst droughts occurred in 1990, 2000, and 2003; less intense or prolonged droughts took place in 1964, 1970, 1973/74, 1983, 1987, 1992, and 2007. Over the whole period the mean value of PADI is 3.51, the highest positive

Table 3. Most relevant drought events from 1961 to 2010 in the Carpathians.

	Drought	Duration			Intensity			Peak			PADI
N	Period	3-m	6-m	12-m	3-m	6-m	12-m	3-m	6-m	12-m	Anom
1	Winter 61/62	4	5	5	3	4	3	Oct-61	Jan-62	Dec-61	–
2	Winter 63/64	8	8	10	2	2	6	Jan-64	Nov-63	Feb-64	0.28
3	First half 68	4	3	6	2	2	6	Jun-68	Jun-68	Jun-68	0.16
4	Early 72	4	5	6	3	2	3	Mar-72	Mar-72	Mar-72	−0.50
5	Spring 74	4	5	5	3	4	1	Apr-74	Apr-74	Apr-74	−0.85
6	Winter 86/87	4	5	11	3	4	4	Nov-86	Feb-87	Feb-87	0.41
7	Early 89	3	3	5	2	1	2	Mar-89	Mar-89	Mar-89	−0.05
8	1990	9	12	13	3	7	11	Feb-90	Mar-90	Aug-90	1.57
9	Late 92	3	9	14	2	2	4	Sep-92	Aug-92	Aug-92	2.44
10	2000	9	10	10	6	8	8	Dec-00	Oct-00	Dec-00	2.00
11	Spring 02	6	6	–	4	3	–	Feb-02	May-02	–	0.54
12	Late 03	5	9	10	4	5	7	Jun-03	Aug-03	Jan-04	2.50
13	Early 07	4	8	5	3	3	3	Dec-06	Feb-07	Jun-07	1.36

anomalous years are 1990 (5.07), 1992 (5.94), 2000 (5.51), and 2003 (6.00).

Following McKee et al. (1993) and Guttman (1999), a "drought" takes place when the indicator is "constantly negative and more negative than −1 for at least 1 month before it turns back to positive values". On the basis of this definition, from 1961 to 2010, RDI-3 detects 19 events, SPEI-3 17 events, and SPI-3 15 events; RDI-6 detects 14 events, SPEI-6 13 events, and SPI-6 12 events; RDI-12 detects 12 events, SPEI-12 11 events, and SPI-12 9 events only. In Table 3 we list the "drought events" that have been detected by at least two indicators out of SPI, SPEI, and RDI. The numbers shown in Table 3 are based on a "3-month accumulation drought series" that has been calculated as the average between SPI-3, RDI-3, and SPEI-3 series over the whole grid (*3-m* in Table 3). We did the same for 6-month and 12-month indicators (*6-m* and *12-m* in Table 3). For each event, *Duration* stands for the number of months from the first month where the indicator becomes lower than −1 to the last month with a negative value before the indicator turns back positive. *Intensity* stands for the number of months in which the drought indicator is lower than −1. *Peak* refers to the month with the lowest value in the "drought period". PADI's annual anomaly is positive when the year is "drier" than the normal value of 1961–2010, negative when it is "wetter". Table 3 provides a complete overview on the droughts in the Carpathian Region over the last 5 decades: at a first look, the drought frequency is increasing, in fact in the 00s four events have been detected, whilst 3 events have occurred in the 60s, and 2 in the 70s, 80s, and 90s. A previous study, carried by Snizell et al. (1998), reported as well a statistically significant increase in drought frequency in Hungary, especially in the late 80s and in the 90s. However, the increase is just based on a simple observation of the fact that the number of droughts is slightly increasing in the last decades as compared to ear-

lier decades. An analysis based on statistical significance of such an increase has not been carried out because it would have been based on the arbitrariness of defining a "drought" event. Out of the 13 events listed, 3 of them can be considered exceptional: the drought in 1990 (the longest one), the drought in 2000 (the most intense), and the drought in the second half of 2003, which followed the heat wave of summer 2003. It seems that in the last 15–20 yr the droughts have also been longer and more intense than in the past and this is probably due to the temperature rise in the Carpathians because of climate change. A deeper analysis on the temporal evolution of drought frequency, intensity, and duration in the Carpathian area will be performed in the future.

5 Case studies: the droughts of 1990, 2000, and 2003

The drought of 1990 (see Vermes and Mihalyfy, 1995) was a long and particularly intense one, especially in February, March, and autumn. If we look at Table 4 we notice that, from a meteorological point of view, it started between January (RDI-3) and February (SPEI-3, SPI-3) and ended in October. For 6-month indicators it started in February, peaked in March, and softened progressively till it ended in spring 1991. Also 12-month indicators see a long drought that started in June, lasted approximately one year, and ended up in spring 1991. The temporal correlation between indicators is evident and the indicators suggest that the drought in 1990 was most intense in February and March, and stroke also in autumn on a hydrological perspective. It was the longest one in 1961–2010, especially in the western side of the Carpathians. We also present the maps of SPI-3, SPEI-3, and RDI-3 for the peak month February 1990 (Fig. 3): for this particular month we notice that SPI-3 and SPEI-3 show very similar spatial patterns, and RDI-3 differs from the other drought indicators as it detects a more intense drought in all the regions

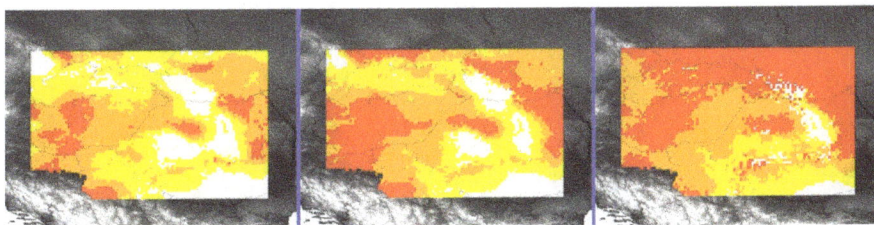

Figure 3. Left to right: drought maps of SPI-3, SPEI-3, and RDI-3 related to February 1990. See Table 2 for colour-scale.

Figure 4. Spatio-temporal evolution of drought from October 2000 (top images), to November 2000 (center) and December 2000 (bottom). From left column to right, SPI-6, SPEI-6, and RDI-6. See Table 2 for colour-scale.

North, North-East and East to the Carpathian Chain. This is due to the highly positive T_M anomalies in winter 1989/90 (up to 3.2 °C compared to the long-term T_M in winters from 1961 to 1990), so RDI is extreme due to low values of PET in the ratio between RR and PET.

The drought of 2000 hit the whole Pannonian Basin, an area located within the natural borders of the Alps, the Carpathian Mountains, the Dinarides, and the Balkan Mountains (Szalai et al., 2000). It was particularly heavy in Romania, where it caused more than 500 million dollars of economic damage (EM-DAT, the International Disasters Database, see http://www.emdat.be). All 3-month and 6-month indicators agree on the fact that the drought started in June 2000, peaked in the last months of 2000, and ended between February 2001 (SPI-3, SPEI-3) and May 2001 (RDI-6). The 12-month indicators detect the drought with a five to six months delay (see Table 5), so the drought-involved period is shifted onwards in time. This case study remarks that it is important to deal with drought indicators with dif-

ferent accumulation periods in order to account for the various features of a drought event. The main driver was the rainfall deficit, but also the temperatures in the second half of 2000 were higher than the normal values and forced the drought to be intense: in this case the drought shows very similar spatial features if evaluated with any of the three indicators, though in Central Hungary, Serbia and Romania the values of SPEI-6 are the lowest (i.e. more intense drought), especially in October and November. The spatial correlation between drought indicators is very high, as we see in Fig. 4, where the evolution of RDI-6, SPEI-6, and SPI-6 from October to December 2000 is presented. In the "peak months" the regions most involved were Hungary, Romania and Serbia, whilst the regions North to the Carpathian Mountains do not seem to be involved by this drought event.

The drought that followed the European spring and summer positive temperature anomalies *of 2003* is probably the most known of the last 15 yr: it caused huge damages in agricultural production, especially in Central and Eastern Europe

Figure 5. Left to right: SPI-3, SPEI-3, and RDI-3. Top to bottom: April 2003, June 2003, August 2003, October 2003. See Table 2 for colour-scale.

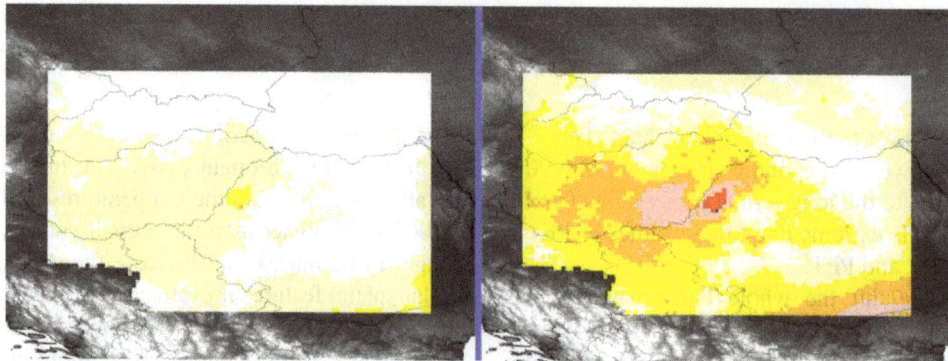

Figure 6. Palfai Aridity and Drought Index: mean value in 1964–2010 (left), 2003 (right). See Table 2 for colour-scale.

(Rebetez et al., 2006). The 2003 drought was caused by a severe lack of summer precipitation and extremely high temperatures (up to 4 °C above 1961–1990 mean values in Central Europe) as the heat wave lasted from April (or May) to September. Compared to the other drought phenomena described above, it was very intense but "limited" to 4–6 months as it was mainly concentrated between May and September 2003: in fact, all the indicators reached values

lower than −1.50 in these months only. SPI detects it one month in advance than RDI and SPEI, but if analyzed with SPI only, the drought event is not found to be as intense as SPEI and RDI suggest, in particular from June to September (see SPEI and RDI for 3 and 6 months in Table 6). In October, T_M and RR turned back to almost normal values, so the drought conditions disappeared faster than the in the first two case studies described in this paper. Moreover, because the

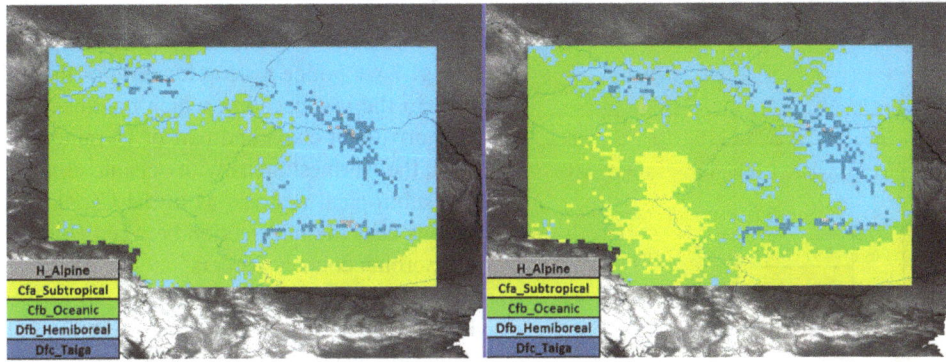

Figure 7. Köppen-Geiger climate maps of the Carpathians for 1961–1990 (left) and 1981–2010 (right).

Table 4. The drought of 1990 for SPI, SPEI, and RDI. The bold boundaries enclose the drought event.

M	Y	SPI3	SPEI3	RDI3	SPI6	SPEI6	RDI6	SPI12	SPEI12	RDI12
12	89	-0.92	-0.77	-0.70	-0.68	-0.46	-0.61	-0.41	-0.32	-0.47
1	90	-1.40	-1.17	-0.19	-0.63	-0.55	-0.54	-0.35	-0.26	-0.37
2	90	-1.40	-1.62	-2.00	-1.27	-1.36	-1.31	-0.26	-0.23	-0.34
3	90	-1.12	-1.82	-1.74	-1.52	-1.96	-2.00	-0.34	-0.34	-0.45
4	90	-0.29	-0.86	-1.06	-1.10	-1.42	-1.58	-0.44	-0.28	-0.42
5	90	-0.65	-0.86	-0.86	-1.30	-1.82	-1.65	-0.80	-0.75	-0.82
6	90	-0.52	-0.22	-0.36	-1.10	-1.19	-1.28	-1.31	-1.32	-1.35
7	90	-1.06	-0.71	-0.91	-1.02	-1.07	-1.11	-1.17	-1.20	-1.20
8	90	-1.27	-0.97	-1.19	-1.31	-1.20	-1.33	-1.78	-2.09	-1.82
9	90	-0.95	-0.54	-0.83	-1.01	-0.56	-0.83	-1.73	-1.96	-1.71
10	90	-0.49	-0.27	-0.43	-1.08	-0.74	-0.97	-1.64	-1.94	-1.65
11	90	-0.01	0.10	0.05	-0.96	-0.70	-0.88	-1.57	-1.86	-1.64
12	90	0.20	0.12	-0.07	-0.60	-0.35	-0.50	-1.15	-1.07	-1.23
1	91	0.34	-0.44	-0.67	-0.69	-0.46	-0.56	-1.18	-1.10	-1.20
2	91	0.46	-0.35	0.31	0.30	-0.09	-0.02	-1.22	-1.02	-1.14
3	91	-1.22	-1.16	-1.08	-0.67	-0.78	-0.95	-1.22	-0.95	-1.07
4	91	0.73	-0.59	-0.62	-0.79	-0.78	-0.77	-1.41	-1.14	-1.21
5	91	0.30	0.58	0.63	0.07	0.23	0.44	-0.84	-0.34	-0.46
6	91	0.33	0.64	0.64	-0.37	0.00	-0.05	-0.73	-0.28	-0.41
7	91	0.92	0.92	0.96	0.44	0.52	0.51	-0.12	0.12	0.05
8	91	0.54	0.45	0.45	0.54	0.58	0.58	0.26	0.42	0.41

Table 5. The drought of 2000 for SPI, SPEI, and RDI. The bold boundaries enclose the drought event.

M	Y	SPI3	SPEI3	RDI3	SPI6	SPEI6	RDI6	SPI12	SPEI12	RDI12
5	00	-0.32	-0.56	-0.60	0.06	-0.08	-0.32	0.59	0.28	0.20
6	00	-1.55	-1.97	-1.76	-1.09	-1.37	-1.33	-0.06	-0.21	-0.33
7	00	-1.18	-1.30	-1.23	-0.83	-1.11	-0.98	-0.21	-0.26	-0.33
8	00	-1.37	-1.60	-1.49	-1.19	-1.56	-1.37	-0.63	-0.77	-0.82
9	00	-0.55	-0.42	-0.54	-1.34	-1.71	-1.49	-0.61	-0.60	-0.67
10	00	-1.53	-1.74	-1.80	-1.73	-2.38	-1.91	-0.97	-1.25	-1.12
11	00	-1.03	-1.20	-1.21	-1.68	-2.32	-1.94	-1.22	-1.89	-1.51
12	00	-1.49	-2.37	-2.25	-1.18	-1.63	-1.40	-1.54	-2.43	-1.91
1	01	-0.46	-0.85	-1.50	-1.50	-2.15	-1.82	-1.47	-2.32	-1.77
2	01	-0.06	-0.17	-0.47	-0.94	-1.18	-1.24	-1.48	-2.37	-1.79
3	01	0.80	0.53	-0.38	0.38	-0.87	-1.40	-1.30	-2.27	-1.69
4	01	0.98	0.73	0.40	0.43	0.11	-0.50	-1.11	-1.74	-1.38
5	01	0.50	0.30	0.22	0.28	0.13	-0.98	-1.12	-1.79	-1.39
6	01	0.40	0.48	0.45	0.77	0.65	0.63	-0.24	-0.27	-0.39
7	01	0.60	0.55	0.55	0.94	0.74	0.74	-0.06	-0.21	-0.34
8	01	0.68	0.52	0.55	0.74	0.49	0.51	0.13	-0.03	-0.16
9	01	1.18	0.87	0.94	1.05	0.85	0.91	0.71	0.44	0.36

anomalous temperature and rainfall regimes lasted approximately 6 months and were forerun and followed by quasi-normal conditions, the 12-month indicators never peaked to values lower than −2.0. As we see in Fig. 5, all the 3-month indicators agree: in the Carpathians, the first clear drought signals appeared in April, the drought peaked twice (June and August), and it vanished almost completely in October. Drought shifted from West to East and the geographical pat-

terns are very similar in all the indicators, nevertheless for SPEI the drought was a bit more intense and widespread in August. As for the drought of 2000, the Carpathians acted as a natural barrier for the drought: the regions North and East to the mountain barrier experienced a less intense drought than the regions West and South to it. In the Carpathians, elevation plays a leading role; in fact it is also worth noticing that, according to RDI-3, while drought was starting in the westernmost areas of the Carpathian region, in April 2003, the mountaintops of the Carpathian Chain experienced wet conditions.

The whole 2003 was extremely dry in Europe, including our study area, which is shown in Fig. 6: the differences amongst the mean value of PADI in 1964–2010 (left)

Table 6. The drought that followed the summer heat wave of 2003 as seen by for SPI, SPEI, and RDI. The bold boundaries enclose the drought event.

M	Y	SPI3	SPEI3	RDI3	SPI6	SPEI6	RDI6	SPI12	SPEI12	RDI12
3	03	-0.42	-0.12	0.37	-0.13	0.06	0.33	0.05	0.02	-0.09
4	03	-1.39	-0.69	-0.49	-0.90	-0.59	-0.13	-0.01	-0.01	-0.10
5	03	-1.50	-1.69	-1.57	-1.05	-0.97	-0.98	-0.10	-0.20	-0.27
6	03	-1.77	-2.58	-2.05	-1.72	-2.21	-1.90	-0.47	-0.65	-0.69
7	03	-0.94	-1.94	-1.31	-1.47	-2.09	-1.68	-0.46	-0.51	-0.59
8	03	-1.33	-2.25	-1.63	-1.79	-2.84	-2.10	-1.03	-1.24	-1.22
9	03	-0.49	-0.65	-0.64	-1.42	-2.27	-1.73	-1.29	-1.89	-1.54
10	03	0.08	0.15	0.10	-0.58	-1.01	-0.86	-1.08	-1.50	-1.29
11	03	0.54	0.66	0.68	-0.59	-0.84	-0.79	-1.12	-1.47	-1.31
12	03	0.39	0.54	0.73	-0.18	-0.18	-0.26	-1.17	-1.45	-1.40
1	04	0.46	-0.57	-0.79	-0.29	-0.23	-0.29	-1.20	-1.54	-1.36
2	04	0.40	0.44	0.58	0.65	0.78	0.95	-0.94	-1.17	-1.12
3	04	0.77	0.72	0.22	0.75	0.82	0.92	-0.69	-0.88	-0.93
4	04	0.50	0.42	0.23	0.00	-0.09	-0.32	-0.50	-0.85	-0.88
5	04	-0.03	0.12	0.07	0.16	0.30	0.31	-0.39	-0.45	-0.53
6	04	0.21	-0.01	-0.11	0.23	0.33	0.30	0.01	0.08	0.01

and PADI in 2003 (right) are remarkable. According to the long-term mean PADI, the Carpathians are in "normal conditions", except of Hungary and South-Eastern corner (Romania), where "sub-humid" conditions can be found (see Table 2 for the complete classification). In 2003, the "sub-humid" areas turned to be mild-dry, the South-Eastern corner was "dry", and the areas along the country border between Hungary and Romania were very or heavy dry, confirming that 2003 was anomalous in the Carpathians.

6 Shifts in climatic regions

Following the Köppen-Geiger's (KG) climate classification (Köppen, 1936; Geiger, 1961), the Carpathian Mountains are a climatic barrier between oceanic (South and West) and continental (North and East) climate. In Fig. 7 we show the KG maps related to 1961–1990 and 1981–2010, computed using monthly and average T_M and RR values over the two 30-yr periods. Out of the 31 KG climate classes, only 5 are present in the Carpathian Region: alpine (H: alpine or ET: tundra) on the mountain peaks, two oceanic subclasses (Cfa: humid subtropical and Cfb: mild oceanic), and two continental subclasses (Dfb: continental hemi-boreal and Dfc: taiga). No desert, steppe or dry Mediterranean classes can be found. However, the subtropical (Cfa) is very similar to a semi-arid class. From 1961–1990 to 1981–2010 the alpine climate slightly decreased (0.32 to 0.17 %), subtropical strongly increased (4.94 to 16.86 %), oceanic increased

(46.25 to 55.20 %), whilst hemi-boreal strongly decreased (45.33 to 24.63 %), and taiga is constant (3.12 to 3.14 %). In Fig. 7 we notice a steep increase of the mild oceanic climate in the last decades, which confines the continental to the Carpathian chain and part of Ukraine only; on the other hand, the warmest areas with oceanic climate became semiarid subtropical regions, as in the Csongrad and Vojovdina regions near the intersection between Hungarian, Romanian and Serbian country borders.

The FAO-UNEP aridity index (UNEP, 1992) is probably the most widespread indicator which quantifies the aridity of an area according to climate factors: it is calculated as the ratio between annual precipitation and evapotranspiration (ET). We used PET instead of ET, because of simplicity and because we are interested in climate features only, not in biological or agronomic factors. If the ratio is higher than 0.75, there is no desertification risk, if it is lower than 0.03 the area is desert-like. In between, we call it "hyper-arid" if the index is ≥ 0.03 but < 0.05, "arid" if ≥ 0.05 but < 0.2, "semi-arid" if ≥ 0.2 but < 0.5, "dry or sub-humid" if ≥ 0.5 but > 0.65, "mild dry/humid" if the ratio lies between 0.65 and 0.75. In the Carpathian region, no area falls in a "dry/arid/desert" category in 1961–1990, neither in 1981–2010. Only a small increase in the "mild dry/humid" class can be seen: from 2.34 % in 1961–1990 to 4.39 % in 1981–2010, mainly due to 2 dry years, 2000 and 2003. Regions that turned into "mild dry/humid" class are located in the south-eastern Romanian corner of the Carpathian region. Going eastwards, the Eastern Romania is facing a desertification threat, especially in the Danube Delta and on the Black Sea coast, two areas out of the Carpathian Region (Spinoni et al., 2012). Similar results can be found using other aridity indices, as Crowther's (Bove et al., 2005), De Martonne's (De Martonne, 1926), and Bagnouls-Gaussen's (Kosmas et al., 1999): the Carpathians are not under an increased desertification risk due to natural factors. However, in spite of the seemingly favourable mean conditions, episodic droughts can cause very serious damage.

7 Summary and conclusions

A 1961–2010 daily gridded dataset of T_M and RR, collected by the CARPATCLIM consortium, has been used as the basis for computing four drought indicators (RDI, SPEI, SPI, and PADI) and two climate indicators (KG and FAO-UNEP) over the Carpathian Region.

In this paper we dealt with drought events in the last 5 decades in the Carpathian Region: the most intense droughts took place in 1990, 2000, and 2003, followed by other 10 notable events. On the other hand, 2005 and 2010 were the wettest years. We discussed in detail the three most important drought events: all the indicators agreed on the temporal structure and geographical patterns of the droughts. SPI, SPEI, and RDI proved to be highly correlated if computed at the same accumulation scale (3, 6 or 12 months). PADI confirmed the a-normality of 1990, 2000, and 2003 yr. We do not recommend using PADI as a standalone indicator, because it is not able to capture the monthly evolution of a drought. In general, the drought frequency is slightly increasing: in fact, during in the last decade (2001–2010), 4 drought events occurred out of the 13 detected between 1961 and 2010. Anyway, this rise is not confirmed by significance tests, but we think that the increasing tendency is an important indication based on the fact that in the recent decades more drought events took place than in the earlier decades.

The Carpathian Mountains are an orographic border between mild oceanic (South and West) and continental (North and East) climates. In the last 20 yr, a shift from oceanic to continental climate can be seen, especially in the Romanian part of the Carpathians and on the country borders between Serbia and Hungary. Using the KG climate classification, no desert, steppe or arid areas are present in the area under examination; furthermore, using the FAO-UNEP aridity's index, it is clear that the Carpathians cannot be considered an arid area.

In the future, we plan comparing the results obtained by means of the CARPATCLIM dataset with an independent dataset collected by JRC (Spinoni et al., 2013). The results obtained in this paper and the future comparisons will be part of the European Drought Observatory, a web-portal developed by the DESERT Action of the Climate Risk Management Unit of Institute for Environment and sustainability of JRC (http://edo.jrc.ec.europa.eu/).

Acknowledgements. We wish to thank all the members of the CARPATCLIM consortium, who held a fundamental role in data collection and calculation of the gridded variables, and actively discussed about the methodologies used in this paper. Strictly in alphabetical order, we acknowledge Igor Antolovic, Ingeborg Auer, Oliver Bochnicek, Sorin Cheval, Alexandru Dumitrescu, Natalia Gnatiuk, Johann Hiebl, Peter Kajaba, Piotr Kilar, Tamas Kovacs, Gabriela Ivanakova, Danuta Limanowka, Monica Matei, Dragan Mihic, Janja Milkovic, Yurii Nabyvanets, Pavol Nejedlik, Akos Nemeth, Predrag Petrovic, Robert Pyrc, Radim Tolasz, Tatjana Savic, Oleg Skrynyk, Pavel Štastný, Petr Štěpánek, Pavel Zahradníček.

In the end, we would like to thank the three anonymous referees that improved this paper with their precious suggestions and remarks.

Edited by: I. Auer
Reviewed by: three anonymous referees

References

Allen, R. A., Pruitt, W. O., Wright, J. L., Howell, T. A., Ventura, F., Snyder, R., Itenfisu, D., Steduto, P., Berengena, J., Baselga Yrisarry, J., Smith, M., Pereira, J. S., Raes, D., Perrier, A., Alves, I., Walter, I., and Elliott, R.: A recommendation on standardized surface resistance for hourly calculation of reference ET_0 by the FAO56 Penman-Monteith method, Agr. Water Manage., 81, 1–22, 2006.

Bartholy, J., Pongcraz, R., and Molnar, Z.: Classification and analysis of pat climate information based on historical documentary sources for the Carpathian basin, Int. J. Climate, 24, 1759–1776, 2004.

Benichou, P. and Le Breton, O.: Prise en compte de la topographie pour la cartographie des champs pluviometriques statistiques, La Meteorologie, 19, 23–34, 1987.

Bove, B., Brindisi, P., Glisci, C., Pacifico, G., and Summa, M. L.: Indicatori Climatici di Desertificazione in Basilicata, Forest, 2, 74–84, 2005.

Cressie, N.: Statistics for Spatial Data, Wiley, New York, 900 pp., 1991.

De Martonne, E.: Une nouvelle function climatologique: L'indice d'aridite, La Meteorologie, 2, 449–458, 1926.

EEA: Drought Management Plan Report. Including Agricultural, Drought Indicators and Climate Change Aspects, Technical Report 2008, 023, 132 pp., 2008.

Geiger, R.: Köppen-Geiger/Klima der Erde (Wandkarte 1:16 Mill.), Überarbeitete Neuausgabe von Geiger, R., Klett-Perthes, Gotha, 1961.

Guttman, N. B.: Accepting the Standardized Precipitation Index: a calculation algorithm, J. Am. Water Resour. As., 35, 311–322, 1999.

Köppen, W.: Das geographisca System der Klimateologie, edited by: Köppen, W. and Geiger, G., 1. C., Gebr., Borntraeger, 1–44, 1936.

Kosmas, C., Kirkby, M., and Geeson, N.: The MEDALUS Project. Mediterranean Desertification and Land Use, Manual on key indicators of Desertification and mapping environmentally sensitive areas to desertification, Eur. Commission, Brussels, Project Report, 265 pp., 1999.

Kozak, J., Björnsen Gurung, A., and Ostapowicz, K.: Research Agenda for the Carpathians: 2010–2015, 43 pp., http://mri.scnatweb.ch/download-document?gid=1204, Krakow, 2011a.

Kozak, P., Palfai, I., Herceg, A., and Fiala, K.: Palfai Drought Index (PADI) – Expansion of Applicability of Hungarian PAI for South East Europe (SEE) Region, 27th Conference of the Danubian Countries, Budapest (HUN), 2nd Symposium, 16–17 June 2011b.

Krüzselyi, I., Bartholy, J., Horányi, A., Pieczka, I., Pongrácz, R., Szabó, P., Szépszó, G., and Torma, Cs.: The future climate characteristics of the Carpathian Basin based on a regional climate model mini-ensemble, Adv. Sci. Res., 6, 69–73, doi:10.5194/asr-6-69-2011, 2011.

Lakatos, M., Szentimrey, T., and Bihari, Z.: Application of gridded daily data series for calculation of extreme temperature and precipitation indices in Hungary, Időjárás, 115, 99–109, 2011.

McKee, T. B., Doesken, N. J., and Kleist, J.: The relationship of drought frequency and duration to time scales, Proc. of 8th Conf. on Appl. Climatol., Anaheim, California, Am. Met. Soc., 179–184, 17–22 January 1993.

Palfai, I.: Description and forecasting of droughts in Hungary, Proc. of 14th Congress on Irrigation and Drainage (ICID), Rio de Janeiro, 1, Cap. 1-C, 151–158, 1990.

Paltineanu, C., Mihailescu, F., Seceleanu, I., Dragota, C., and Vasenciuc, F.: Ariditatea, seceta, evapotranspiratia si cerintele de apa ala culturilor agricole in Romania, Ovidius University Press,

Constanta, 319 pp., ISBN 978-973-614-412-7, 2007.

Parajka, J., Kohnova, S., Balint, G., Barbuc, M., Borga, M., Claps, P., Dumitrescu, A., Gaume, E., Hlavcova, K., Merz, R., Pfaundler, M., Stancalie, G., Szolgay, J., and Blöschl, G.: Seasonal characteristics of flood regimes across the Alpine-Carpathian range, J. Hydrol., 394, 78–89, 2010.

Rebetez, M., Mayer, H., Dupont, O., Schindler D., Gartner, K., Kropp, J., P., and Menzel, A.: Heat and drought 2003 in Europe: a climate synthesis, Ann. Forest Sci., 63, 569–577, 2006.

Snizell, C., Bussay, A., and Szentimrey, T.: Drought tendencies in Hungary, Int. J. Climatol., 18, 1479–1491, 1998.

Spinoni, J., Antofie, T., Barbosa, P., De Jager, A., Klein Tank, A., Micale, F., Naumann, G., Sepulcre-Canto, G., Singleton, A., van der Schrier, G., and Vogt, J.: 2001–11 high-resolution drought climatologies for Europe, EMS Annual Meeting Abstracts, Vol. 9, Lodz, 10–14 September 2012.

Spinoni, J.: 1951–2011 European Drought and Climate Atlas. Testing a wide set of indicators for the European Drought Observatory, EUR 25235 EN, Luxembourg (LUX), Publications Office of the EU, in press, 2013.

Szalai, S.: EU Projects in the Carpathian Region: CARPATCLIM and CarpathCC, Carpathian Convention Working Group on sustainable Forest Management, Donji Milanovac, Serbia, 23–24 April 2012.

Szalai, S. and Vogt, J.: CARPATCLIM – high resolution gridded database of the Carpathian Region and calculation of drought indices as a contribution to the European Drought Observatory, WRCP Conference, Denver, USA, T185A, 24–28 October 2011.

Szalai, S., Szinell, C., and Zoboki J.: Drought Monitoring in Hungary, Proceedings of an Expert Group Meteeng, Lisbon, Portugal, AGM-2, WMO/TD No. 1037, 5–7 September 2000.

Szentimrey, T.: Multiple Analysis of Series for Homogenization (MASH), Proceedings of the 2nd Seminar for Homogenization of Surface Climatological Data, Budapest, Hungary; WMO, WCDMP-No. 41, 27–46, 1999.

Szentimrey, T.: Development of MASH homogenization procedure for daily data, Proceedings of the Fifth Seminar for Homogenization and Quality Control in Climatological Databases, Budapest, Hungary, 2006; WCDMP-No. 68, WMO-TD NO. 1434, 116–125, 2008.

Szentimrey, T. and Bihari, Z.: Mathematical background of the spatial interpolation methods and the software MISH (Meteorological Interpolation based on Surface Homogenized Data Basis), Proceedings from the Conference on Spatial Interpolation in Climatology and Meteorology, Budapest, Hungary, 2004, COST Action 719, COST Office, 17–27, 2007.

Szentimrey, T., Bihari, Z., Lakatos, M., and Szalai, S.: Mathematical, methodological questions concerning the spatial interpolation of climate elements, Proceedings from the Second Conference on Spatial Interpolation in Climatology and Meteorology, Budapest, Hungary, 2009, Időjárás 115, 1–11, 2011.

Thom, H. C. S.: Some Methods of Climatological Analysis. WMO Technical note 81, Secretariat of the WMO, Geneva, Switzerland, 53 pp., 1966.

Thornthwaite, C. W.: An Approach toward a Rational Classification of Climate, Geogr. Rev., 38, 55–94, 1948.

Tsakiris, G. and Vangelis, H.: Establishing a Drought Index incorporating Evapotranspiration, Eur. Wat., 9/10, 11–13, 2005.

UNEP: World Atlas of Desertification. Edward Arnold, London, 1992.

UNEP: Carpathians Environment Outlook 2007, Published by the United Nations Environment Program, ISBN 978-92-807-2870-5, J. No: DEW/0999/GE, 2007.

Van der Schrier, G., Jones, P. D., and Briffa, K. R.: The sensitivity of the PDSI to the Thornthwaite and Penman-Monteith parameterizations for potential evapotranspiration, J. Geophys. Res., 116, D03106, doi:10.1029/2010JD015001, 2011.

Venema, V. K. C., Mestre, O., Aguilar, E., Auer, I., Guijarro, J. A., Domonkos, P., Vertacnik, G., Szentimrey, T., Stepanek, P., Zahradnicek, P., Viarre, J., Müller-Westermeier, G., Lakatos, M., Williams, C. N., Menne, M. J., Lindau, R., Rasol, D., Rustemeier, E., Kolokythas, K., Marinova, T., Andresen, L., Acquaotta, F., Fratianni, S., Cheval, S., Klancar, M., Brunetti, M., Gruber, C., Prohom Duran, M., Likso, T., Esteban, P., and Brandsma, T.: Benchmarking homogenization algorithms for monthly data, Clim. Past, 8, 89–115, doi:10.5194/cp-8-89-2012, 2012.

Vermes, L. and Mihalyfy, A.: Proceedings of the International Workshop on Drought in the Carpathians' Region. Ed. MTESZ, 352 pp., ISBN 9638012714, 9789638012715, 1995.

Vicente-Serrano, S. M., Begueria, S., and Lopez-Moreno, J. I.: A Multiscalar Drought Index Sensitive to Global Warming: The Standardized Precipitation Evapotranspiration Index, J. Climate, 23, 1696–1718, 2010.

Webster, R., Holt, S., and Avis, C.: The status of the Carpathians. A report developed as a part of The Carpathian Ecoregion Initiative, 67 pp., 2001.

Willmott, C. J., Rowe, C. M., and Mintz, Y.: Climatology of the terrestrial seasonal water cycle, J. Climatol., 5, 589–606, 1985.

6

A modification of the mixed form of Richards equation and its application in vertically inhomogeneous soils

F. Kalinka[1,2] and B. Ahrens[1]

[1]Goethe-University Frankfurt am Main, Institute for Atmospheric and Environmental Sciences, Germany
[2]LOEWE BiK-F research centre Frankfurt am Main, Germany

Abstract. Recently, new soil data maps were developed, which include vertical soil properties like soil type. Implementing those into a multilayer Soil-Vegetation-Atmosphere-Transfer (SVAT) scheme, discontinuities in the water content occur at the interface between dissimilar soils. Therefore, care must be taken in solving the Richards equation for calculating vertical soil water fluxes. We solve a modified form of the mixed (soil water and soil matric potential based) Richards equation by subtracting the equilibrium state of soil matrix potential ψ_E from the hydraulic potential ψ_h. The sensitivity of the modified equation is tested under idealized conditions. The paper will show that the modified equation can handle with discontinuities in soil water content at the interface of layered soils.

1 Introduction

The Richards equation (RE) (Richards, 1931) is commonly used in many Land Surface Models (LSMs) to describe vertical soil water flow in unsaturated zones. It is a combination of mass conservation and Darcy's law. There are three different expressions for it, (1) by soil water content ("θ-based"), (2) by hydraulic potential ("ψ-based") and (3) by both, hydraulic potential and soil water content ("mixed form") as the dependent variables (e.g., Hillel, 1980; Bohne, 2005). The "ψ-based" RE is often used by soil scientists and has the advantage to be continuous in the dependent variable at soil interfaces, but the numerical solution shows errors in mass balance for any iteration method (Picard, Newton-Raphson, etc.) (Celia et al., 1990). The "θ-based" RE is favored by climate modelers because of its excellent mass balance, but has the disadvantages that it cannot handle water flows in saturated zones and that discontinuities in the dependent variable occur across soil layer boundaries (Hills et al., 1989). Using the "mixed form" of RE with a finite element method for numerical solution, Celia et al. (1990) showed, that mass balance problems were consistently low in a homogeneous soil and that discontinuity problems are better-natured than in the "θ-based" RE.

Implementing a soil data map that include information about vertical soil properties like soil type, care must be taken in solving RE to model unsaturated soil water flow in layered soils, because discontinuities in the water content occur at the interface between dissimilar soils. Many efforts have been done to handle with this problem (Hills et al., 1989; Romano et al., 1998; Schaudt and Morrill, 2002; Matthews et al., 2004; Barontini et al., 2007). But to a greater or lesser extent, all methods and results show problems in mass conservation. Zeng and Decker (2008) applied a modified mixed form of RE for unsaturated homogeneous soils, where mass conservation is nearly given. We expected, that this combination will also lead to better results in modeling water flows in inhomogeneous soils and herefore, implemented this formulation into the LSM model TERRA-ML, which is included in the non-hydrostatic mesoscale weather forecast model COSMO (http://www.cosmo-model.org/) as well as in the regional climate model COSMO-CLM (http://www.clm-community.eu/).

In the Theory section we give a short overview over TERRA-ML and discuss the applied modifications. In the Result section we show the application of the modified RE in a sand-loam-sand and a loam-sand-loam soil column under idealized conditions (no infiltration, no gravitational drainage and initialization with the hydrostatic equilibrium state). Finally the results are summarized and discussed within the Conclusion section.

2 Theory

2.1 Model description

TERRA-ML is the actually implemented LSM of the non-hydrostatic mesoscale weather forecast model COSMO and of the regional climate model COSMO-CLM. Only a brief description of the model can be given here but more details can be found in Schrodin and Heise (2001), Doms et al. (2005) and Heise et al. (2006). Combining conservation of mass with Darcy's law leads to the mixed form of RE, describing one-dimensional water flow in soils

$$\frac{\partial \theta}{\partial t} = \frac{\partial}{\partial z}\left[K\frac{\partial(\psi_m + \psi_z)}{\partial z}\right] - S \qquad (1)$$

where θ [m m^{-1}] is the volumetric soil moisture, t is time, z [m] is the vertical coordinate (positive upward), ψ_m [m] is the matric potential, ψ_z [m] is the gravitational potential, K [m s^{-1}] is the hydraulic conductivity and S is a sink term (e.g. transpiration loss in the root zone). Taking the hydraulic diffusivity (e.g. Kabat et al., 1997)

$$D_\theta = -K_\theta\left(\frac{\partial \psi_m}{\partial \theta}\right) \qquad (2)$$

into account, leads to the θ-based RE for soil water fluxes used in TERRA-ML

$$\frac{\partial \theta}{\partial t} = \frac{\partial}{\partial z}\left[-D\frac{\partial \theta}{\partial z} + K\right] - S. \qquad (3)$$

The hydraulic conductivity K and the hydraulic diffusivity D both depend on soil moisture θ and soil properties. Both functions are parameterized according to Rijtema (1969) using exponential functions.

Soil matric potential ψ_m for unfrozen soil is defined according to Clapp and Hornberger (1978) at the node depth z_i of each layer i as

$$\psi_{m,i} = \psi_{sat,i}\left(\frac{\theta_i}{\theta_{sat,i}}\right)^{-B_i} \qquad (4)$$

where the air entry potential at saturation ψ_{sat} is an exponential function of percentages of sand, and the pore size distribution index B is a linear function of percentages of clay (Cosby and Hornberger, 1984).

Celia et al. (1990) showed, that best simulation results with respect to mass conservation are obtained using the mixed form of RE. Additionally, using an inhomogeneous soil type distribution, Eq. (3) will be discontinuous in θ at the interface between two different soil horizons, so that $\partial\theta/\partial z$ is not solvable using small soil layer thicknesses. However, the hydrological potential ψ is continuous, that is the reason why we test Eq. (1) as possible alternative to Eq. (3). Furthermore Grasselt et al. (2008) showed that, using a linear function of K, rather than the operationally used exponential function used in TERRA, also leads to results closer to observations. That is why we have implemented an empirical linear function of hydraulic conductivity according to Clapp and Hornberger (1978).

Table 1. Soil type dependent hydrological parameters

Soil-type	% sand	% clay	ψ_{sat} [mm]	B
sand	90	5	−51.29	3.705
loam	40	20	−229.09	6.09

2.2 Equilibrium state

For idealized case studies in numerical weather prediction as well as in climate simulations, it is common to do model-runs in equilibrium state as initialisation. In our case the hydro-static equilibrium state of Eq. (1) is theoretically reached, if no fluxes occur between the soil layers ($\partial\theta_E/\partial t = 0$), that is

$$\psi_{m,E} + \psi_z = \psi_{sat}\left(\frac{\theta_E}{\theta_{sat}}\right)^{-B} + \psi_z = C = \psi_{sat} + z_{WTD} \qquad (5)$$

as a special solution under assumption of a constant ψ_h in saturated zones ($C = \psi_{sat} + z_{WTD}$). Here z_{WTD} is the water table depth (WTD) and the subscript E in θ_E and $\psi_{m,E}$ denotes the equilibrium state. The layer averaged $\bar{\theta}_{E,i}$ can then be derived by integrating over soil layers.

2.3 Modified Richards equation

Because the derivative of C with depth vanishes, we can rewrite Eq. (1) to

$$\frac{\partial \theta}{\partial t} = \frac{\partial}{\partial z}\left[K\frac{\partial(\psi_m + \psi_z - C)}{\partial z}\right] - S. \qquad (6)$$

The method of subtracting the equilibrium state has the theoretical advantage, that sharp bends in the hydraulic potential ψ_h, arising at the interface between two dissimilar soils because of different soil type properties, will be minimized in the numerical solution. Furthermore, small discontinuities in ψ_m, occurring at the boundary between layered soils as a result of the numerical solution of Eq. (4) in combination with the soil type dependent hydrological parameters given in Table 1, are also reduced.

2.4 Model setup

For all simulations we used the mixed form of RE (Eq. 1) first and the modified form (Eq. 6) afterwards. Here, the operationally used "θ-based" RE (Eq. 3) is not used anymore because of its disadvantage that it can not simulate soil water fluxes across soil layer boundaries, as discussed in Sect. 2.1. In general, horizontal fluxes are not considered in TERRA-ML, we therefore simulated only one column. The following model setup has been used:

– specific soil type properties B, ψ_{sat}, % sand and % clay were used as given in Table 1,

– time stepping $\Delta t = 60$ s,

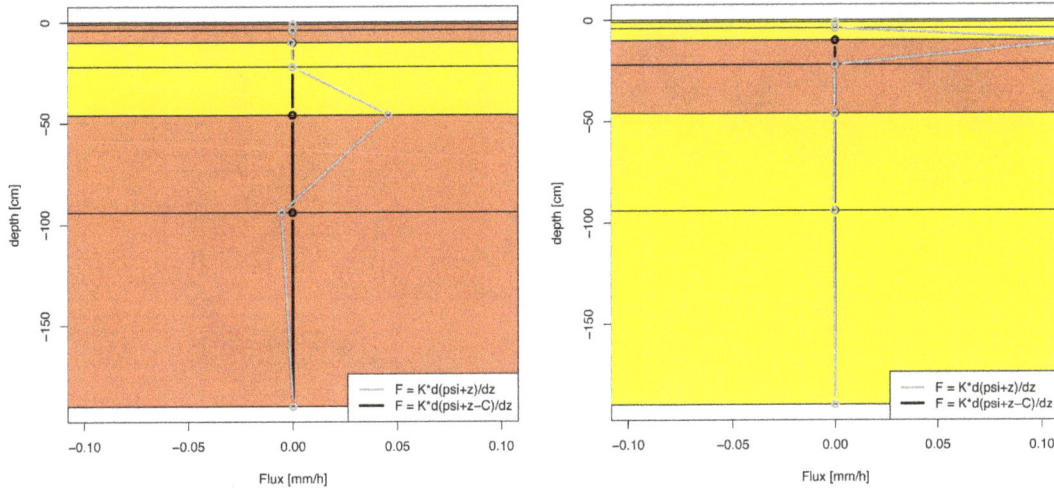

Figure 1. (left) Simulated mean water fluxes [mm h^{-1}] in a loam-sand-loam soil column after an integration period of 30 days with a WTD of 3 m: the grey line shows water fluxes of the old formulation (Eq. 1), the black line of the new formulation (Eq. 6). (right) Same as (left) exept with a sand-loam-sand soil column.

- simulation time = 30 days,

- in a first simulation the soil was divided into seven layers until 250 cm depth, albeit layer thicknesses increase with depth (operationally applied in the COSMO model) and in a further simulation in 25 layers with thicknesses of 10 cm each,

- consideration of a coarse soil in between fine soils (sand-loam-sand) and in a further simulation a fine soil in between coarse soils (loam-sand-loam),

- prescription of different WTDs in the model domain (0.5 m, 1 m and 2 m) and below model domain (3 m and 5 m),

- obtaining $C = \psi_{sat} + z_{WTD}$ with z_{WTD} = depth of WTD,

- neglecting the sink Term S in both, Eq. (1) and Eq. (6),

- initialization of soil moisture with its equilibrium state,

- no infiltration as upper boundary condition and

- no gravitational drainage as lower boundary condition.

The last four assumptions ensure a closed system, where no fluxes should be simulated in a perfect model.

3 Results

3.1 Varying layer thickness

Applying the model setup described above with the old formulation of RE (Eq. 1), using a WTD of 3 m and a seven layer soil distribution, fluxes up to 0.05 mm per hour occur especially from the sand column into the loam column in both, a loam-sand-loam distribution and a sand-loam-sand distribution (grey lines in Fig. 1). These fluxes can not be removed by using smaller soil layer thicknesses to minimize ∂z and therefore to improve results of numerical solution. Figure 2 shows fluxes occurring while using homogeneous soil layer thicknesses of 10 cm. Fluxes from the sand layer into the loam layer do not get smaller, they rather enlarge up to 0.6 mm/h in a loam-sand-loam distribution and up to 0.3 mm/h in a sand-loam-sand column (grey lines).

Using the modified formulation of RE (Eq. 6) instead of Eq. (1), no fluxes occur so that mass is perfectly conserved, independent of soil layer thickness and soil type distribution (black lines in Figs. 1 and 2, pink line in Fig. 2).

3.2 Influence of water table depth

We have implemented different WTDs to show its influence on soil water fluxes in an inhomogeneous soil type distribution. If the WTD is below the model domain (3 m and 5 m), fluxes only occur at the interfaces between different soil horizons (Fig. 2). If the WTD is in the model domain (0.5, 1 and 2 m), layers with depth below or at the WTD become saturated, so that $\theta_E = \theta_{sat}$ and $\psi_E = \psi_{sat}$. Figure 3 shows simulated fluxes for different WTDs. It is conspicuous, that below and at saturated layers, huge fluxes occur up to 230 mm h^{-1}. Again they especially occur at the interfaces between different soil horizons as a result of the influence of different soil properties and this shows that the mixed form of RE is not adequate in simulating water flows in saturated soils. Testing the sensitivity of the modified RE to the WTD, results again indicate perfect mass conservation. The black, red and the dark green lines in Fig. 3 show, that no fluxes occur, independent on WTD and soil type distribution. This demonstrates the efficiency of the modified RE.

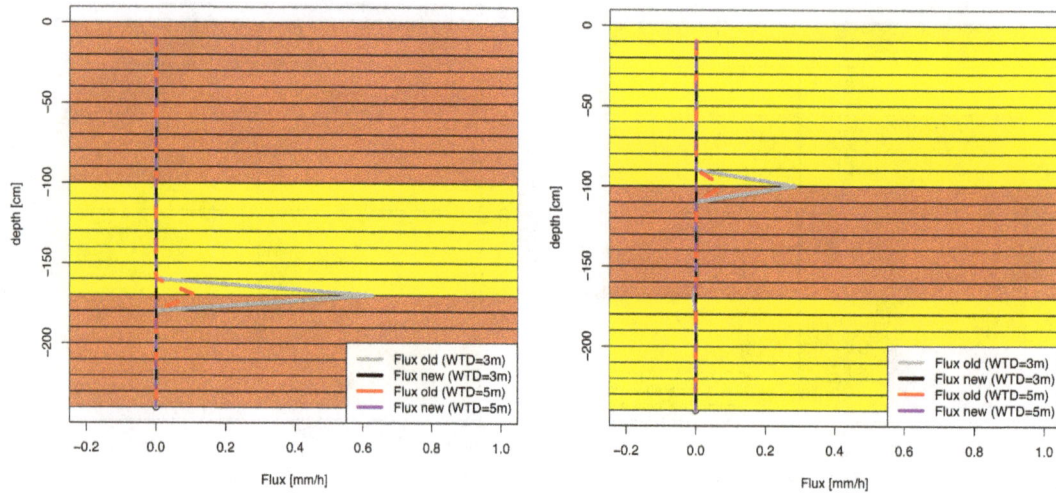

Figure 2. (left) Simulated mean water fluxes [mm h^{-1}] in a loam-sand-loam soil column after an integration period of 30 days with homogeneous soil layer thicknesses of 10 cm and with different WTDs at 3 m and 5 m in a loam-sand-loam soil column: the grey and red lines show fluxes in mm h^{-1} of the old formulation (Eq. 1), the black and purple lines of the new formulation of RE (Eq. 6). (right) Same as (left) except using a sand-loam-sand soil type distribution.

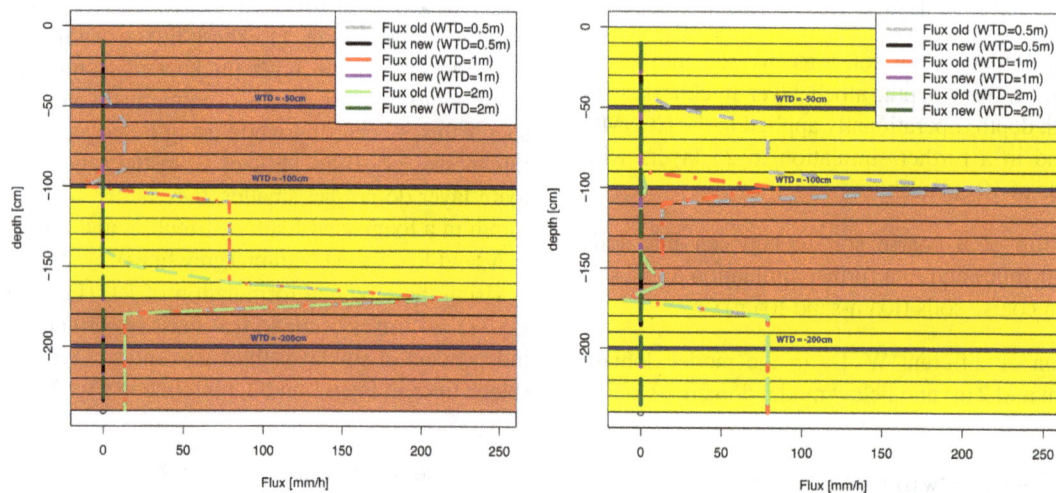

Figure 3. (left) Same as Fig. 2a, except using WTDs within the model domain at 0.5, 1 and 2 m (indicated by the blue lines). Results of the new formulation of RE (black, purple and dark green lines) overlap each other. (right) Same as (left) except using a sand-loam-sand soil type distribution.

4   Conclusions

This study shows that the "θ-based" and the "mixed form" of RE are not adequate in simulating water transports in non-homogeneous unsaturated soils. Therefore, we tested a modified form of the "mixed" RE after Zeng and Decker (2008), where the equilibrium state is subtracted of the hydraulic potential. Results show almost perfect mass conservation, independent of soil layer thicknesses used for the numerical solution, of the WTD (if it is below or in the model domain) and of what soil type distribution is adopted. However, all simulations were done under idealized test cases (neglecting the sink term S, prescribing a water table depth, initializing soil moisture with its equilibrium state, no infiltration as upper boundary condition and no gravitational drainage). Further simulations under more realistic conditions still have to be done.

Acknowledgements. This work was supported by the Hessian Centre on Climate Change within the research project area INKLIM-A. The authors also acknowledge funding from the Hessian initiative for the development of scientic and economic excellence (LOEWE) at the Biodiversity and Climate Research Centre (BiK-F), Frankfurt/Main.

Edited by: H. Formayer
Reviewed by: two anonymous referees

References

Barontini, S., Ranzi, R., and Bacchi, B.: Water dynamics in a gradually nonhomogeneous soil described by the linearized Richards equation, Water Resour. Res., 43, W08411, doi:10.1029/2006WR005126, 2007.

Bohne, K.: An introduction into applied soil hydrology, Catena Verlag GMBH, Reiskirchen, Germany, 2005.

Celia, M. A., Bouloutas, E. T., and Zarba, R. L.: A general mass-conservative numerical solution for the unsaturated flow equation, Water Resour. Res., 26, 1483–1496, 1990.

Clapp, R. B. and Hornberger, G. M.: Empirical equations for some soil hydraulic properties, Water Resour. Res., 14, 601–604, 1978.

Cosby, B. J. and Hornberger, G. M.: A statistical exploration of the relationships of soil moisture characteristics to the physical properties of soils, Water Resour. Res., 20, 682–690, 1984.

Doms, G., Förstner, J., Heise, E., Herzog, H.-J., Raschendorfer, M., Schrodin, R., Reinhardt, T., and Vogel, G.: A description of the non hydrostatic regional model LM Part II: Physical parameterization, Deutscher Wetterdienst, Offenbach, Germany, 2005.

Grasselt, R., Schüttemeyer, D., Warrach-Sagi, K., Ament, F., and Simmer, C.:: Validation of TERRA-ML with discharge measurements, Meteorolog. Z., 17, 763–773, 2008.

Heise, E., Ritter, B., and Schrodin, R.: Operational implementation of the Multilayer Soil Model, Consortium for Small-Scale Modelling (COSMO), Technical Report No. 9, Deutscher Wetterdienst, Offenbach, Germany, 2006.

Hillel, D.: Fundamentals of soil physics, Academic, San Diego, California, USA, 1980.

Hills, R. G., Porro, I., Hudson, D. B., and Wierenga, P. J.: Modeling one-dimensional infiltration into very dry soils – 1. Model development and evaluation, Water Resour. Res., 25, 1259–1269, 1989.

Kabat, P., Hutjes, R. W. A., and Feddes, R. A.: The scaling characteristics of soil parameters: From plot scale heterogeneity to subgrid parameterization, J. Hydrol., 190, 363–396, 1997.

Matthews, C. J., Cook, F. J., Knight, J. H., and Braddock, R. D.: Handling the water content discontinuity at the interface between layered soils within a numerical scheme, 3rd Australian New Zealand Soils Conference, Sydney, Australia, 5–9 December 2004, S15/1791, 2004.

Richards, L. A.: Capillary conduction of liquids through porous mediums, J. Appl. Phys., 1, 318–333, 1931.

Rijtema, P. E.: Soil moisture forecasting, Technical Report Nota 513, Instituut voor Cultuurtechniek en Waterhuishouding, Wageningen, 18 pp., 1969.

Romano, N., Brunone, B., and Santini, A.: Numerical analysis of one-dimensional unsaturated flow in layered soils, Adv. Water Res., 21, 315–324, 1998.

Schaudt, K. J. and Morrill, J. C.: On the treatment of heterogeneous interfaces in soil, Geophys. Res. Lett, 29, 1395, doi:10.1029/2001GL014376, 2002.

Schrodin, R. and Heise, E.: The Multi-Layer version of the DWD soil model TERRA-ML, Consortium for Small-Scale Modelling (COSMO), Technical Report No. 2, Deutscher Wetterdienst, Offenbach, Germany, 2001.

Zeng, X. and Decker, M.: Improving the numerical solution of soil moisture-based Richards equation for land models with a deep or shallow water table, J. Hydrometeor., 10, 308–319, 2008.

7

Diurnal course analysis of the WRF-simulated and observation-based planetary boundary layer height

H. Breuer[1], F. Ács[1], Á. Horváth[2], P. Németh[3], and K. Rajkai[4]

[1]Eötvös Loránd University, Department of Meteorology, 1117, Pázmány P. s. 1/a, Budapest, Hungary
[2]Hungarian Meteorological Service, 8600 Vitorlás utca 17, Siófok, Hungary
[3]Hungarian Meteorological Service, Marcell György Observatory, P.O. Box 39, 1675 Budapest, Hungary
[4]Institute for Soil Sciences and Agricultural Chemistry, Centre for Agricultural Research, Hungarian Academy of Sciences, 1022 Herman Ottó 15, Budapest, Hungary

Correspondence to: H. Breuer (bhajni@nimbus.elte.hu)

Abstract. Weather Research and Forecasting (WRF) single-column model simulations were performed in the late summer of 2012 in order to analyse the diurnal changes of the planetary boundary layer (PBL). Five PBL schemes were tested with the WRF. From the radiometer and wind-profiler measurements at one station, derived PBL heights were also compared to the simulations. The weather conditions during the measurement period proved to be dry; the soil moisture was below wilting point 85 percent of the time. Results show that (1) simulation-based PBL heights are overestimated by about 500–1000 m with respect to the observation-based PBL heights, and (2) PBL height deviations between different observation-based methods (around 700 m in the midday) are comparable with PBL height deviations between different model schemes used in the WRF single-column model. The causes of the deviations are also discussed. It is shown that in the estimation of the PBL height the relevance of the atmospheric profiles could be as important as the relevance of the estimation principles.

1 Introduction

In the last decade the importance of planetary boundary layer (PBL) modelling has increased since high-resolution models require proper description of turbulence. Up until today several PBL schemes have been implemented into the single-column models (e.g. Holt and Raman, 1988; Cuxart et al., 2006; Svensson et al., 2011) or in numerical weather (e.g. Steeneveld et al., 2008; Shin and Hong, 2011; Xie et al., 2012) or climate (e.g. Engeln and Teixeira, 2013) prediction systems. In spite of growing interest, studies focusing on the daily cycle of PBL height are rare (e.g. Hernández-Ceballos et al., 2012). This study intends to bridge this gap giving a detailed analysis of the daily cycle of PBL height in a hot summer period during 2012 in the Carpathian Basin. Observed and Weather Research and Forecasting (WRF)-simulated atmospheric profiles together with different PBL height estimating methods are used and compared in evaluating the diurnal course.

2 Model

The simulations were carried out with the WRF 3.4.1 (Skamarock et al., 2008) single-column model (SCM). The SCM was used in a 1 km domain with 60 levels. Simulation time was 48 h, with a 5 s time step, from which the last 24 h were analysed. This model setup was performed for each day during the analysis. The Noah LSM (Chen and Dudhia, 2001) was used as the land surface scheme with four soil layers. The following main physical parameterisations were used: RRTM (rapid radiative transfer model) (Mlawer et al., 1997) for radiation transfer, WSM (WRF Single Moment) five-class for cloud microphysics (Hong et al., 2004), and the cumulus convection was calculated explicitly. The atmosphere of the SCM model was initialised at 00:00 UTC by radiosounding measurements. Advection is calculated as the model uses a 3×3 grid, but only the middle grid represents the results. Measurement-driven or 3-D-model-driven advection forcing was turned off. In the model the wind component

Figure 1. Location of the upper air observatory and the soil measurement sites.

tendencies and advection calculation are based on Ghan et al. (2000) following the work of Randall and Cripe (1999). As such, an upstream advection scheme is used in the model, in which the tendencies are determined with an advective timescale defined as the ratio of horizontal domain size and average measured/initial wind speed. Without outside wind forcing, this advective time scale is responsible for the wind and advection tendency. The required soil temperature was taken from the Global Forecast System. At the same time measurements were used for soil moisture as no measurements were available there.

3 Measurements

Radiometer and wind-profiler measurements were also used in evaluating the diurnal course of PBL height. The lowest measurement height of the wind profiler is around 154 m. Data are available at every next 220 m until a height of 3–4 km. The radiometer measurements are set to every 50 m until a height of 500 m. From there to a height of 2000 m, the measurement step is doubled, and from 2000 to 10 000 m the step is 250 m. Radiosonde measurements at 00:00 UTC for initialising the SCM were also conducted at the observatory operated by the Hungarian Meteorological Service. Soil moisture measurements were conducted at depths of 10–40 and 40–70 cm in the vicinity of the observatory at five sites (Fig. 1). At the sites marked by a circle, the soil texture is sand, and at the others it is loam. All of the sites had different cultivations: oat, alfalfa, maize, sunflower and grass (maximum leaf area index: 3, 2, 3.5, 2.5 and $2\,\mathrm{m^2\,m^{-2}}$, respectively). All measurements refer to the period 6 July–8 October 2012. The area around observatory was exceptionally dry until mid-September; the soil moisture was usually below wilting point except on about 10 days when local showers occurred at various sites.

4 PBL schemes

4.1 WRF single-column model

Using the SCM, five fundamentally different schemes were tested (Table 1). The Yonsei University (YSU, Hong et al., 2006; Hong, 2010) model and the Asymmetric Convective Model 2 (ACM2, Pleim, 2007) are mainly non-local mixing schemes, but the latter changes the calculations in stable conditions to local mixing. In both models, the same bulk Richardson number formalism is used, but the approaches are somewhat different. The critical Ri number in the case of YSU is 0.25 under stable and 0 under unstable conditions, while in the ACM2 it is defined as 0.25. Also in case of YSU the whole atmospheric profile is searched through for the critical values, while in case of ACM2 in unstable conditions the bulk Ri method is only used over the entrainment layer. The Mellor–Yamada–Janjić (MYJ, Janjić, 1990, 2002), the quasi-normal scale elimination (QNSE, Sukoriansky et al., 2005) and the Bougeault–Lacarrère (BouLac, Bougeault and Lacarrère, 1989) schemes predict the turbulent kinetic energy (TKE) in every model level and step and have a 1.5-order closure theory in the treatment of turbulence. The QNSE is based on the MYJ scheme but has improved mixing in stable conditions. In the MYJ and QNSE models the PBL height is defined where the turbulence disappears. This is determined from TKE, where the critical TKE drops below $0.202\,\mathrm{m^2\,s^{-2}}$. In the BouLac scheme, a more measurement-oriented approach, the parcel method is used to define the PBL height.

4.2 PBL height estimation from measurements

The estimation of PBL height has many forms depending on the measurements available (Seibert et al., 1997). Five methods were applied to radiometric and wind-profiler measurements in order to check variability of the estimations (Table 2). Methods applied based on the calculation of potential temperature (Θ) are André and Mahrt (1982) – hereinafter PTMG – and, naturally, the parcel method. Based on the turbulent nature of the mixing layer, the bulk Richardson

Table 1. Main physical features of the PBL parameterisations used in the WRF-SCM model (YSU – Yonsei University, MYJ – Mellor–Yamada–Janjić, QNSE – quasi-normal scale elimination, ACM2 – Asymmetric Convective Model 2, BouLac – Bougeault–Lacarrère).

Scheme abbreviations	Mixing	Order of closure	PBL height determination	Critical value
YSU	non-local	1	bulk Ri	bulk $Ri_{cr} = 0.25$ – stable bulk $Ri_{cr} = 0$ – unstable
MYJ	local	1.5	TKE	$TKE_{cr} = 0.202\,\mathrm{m^2\,s^{-2}}$
QNSE	local	1.5	TKE	$TKE_{cr} = 0.202\,\mathrm{m^2\,s^{-2}}$
ACM2	non-local in unstable, local in stable conditions	1	bulk Ri	bulk $Ri_{cr} = 0.25$
BouLac	local	1.5	parcel method	–

Table 2. Basic characteristics of the PBL height estimations applied to the measurements (PTMG – maximum gradient of potential temperature, MW – Matyasovszky and Weidinger (1998), RI – critical bulk Richardson number, SNR – maximum of signal-to-noise ratio).

Scheme abbreviations	PBL height determination	Critical value
PTMG	potential temperature	maximum gradient
Parcel method	potential temperature	$\Theta = \Theta_{surface}$
MW	virtual temperature (T_v)	$\overline{\mathrm{grad}T_v} = 0.0095\,\mathrm{K\,m^{-1}}$ grad T_v at PBL top $= 0.008\,\mathrm{K\,m^{-1}}$
RI	bulk Ri number	bulk $Ri_{cr} = 0.25$
SNR-lability	signal-to-noise ratio	maximum of SNR

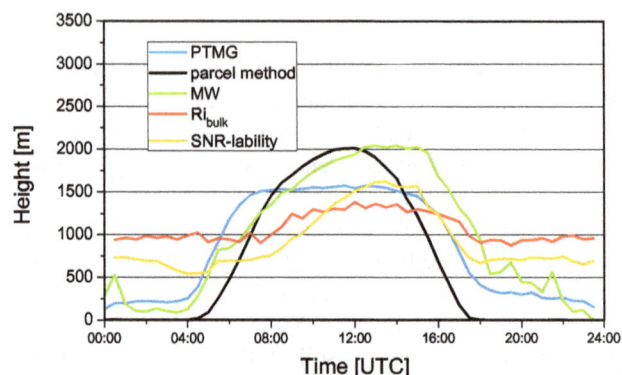

Figure 2. Average (July–September, 2012) diurnal course of PBL height derived from wind-profiler and radiometer measurements.

number (Ri_{bulk}) was also estimated for defining PBL height, where the critical value was chosen as 0.25. Apart from these methods, a virtual temperature-based method (Matyasovszky and Weidinger, 1998, hereinafter: MW method) was also chosen. In this case the PBL height is defined where the average gradient of virtual temperature in the PBL is equal to $0.0095\,\mathrm{K\,m^{-1}}$ and the gradient at the PBL top is equal to $0.008\,\mathrm{K\,m^{-1}}$. The constants were statistically derived from radiosounding measurements. The wind-profiler measurements allowed the use of height-corrected (Lee and Kawai, 2011) signal-to-noise ratio (SNR) to define the top of the PBL as the maximum of SNR (Angevine et al., 1994). This method was modified when stable stratification was found with radiometric measurements; the maximum was searched for only as long as the SNR increased with height.

5 Results

5.1 Averaged diurnal PBL heights

Considering the 3-month averages of diurnal course, methods incorporating potential temperature gradient and RI show a plateau-like behaviour (Fig. 2). In those cases the maximum PBL height changes between 1200 and 1500 m for several hours with minimal changes. The gradual increment and sudden change are obtained from stratification-dependent SNR and the MW (Matyasovszky and Weidinger, 1998) method. The highest average PBL height from measurements is around 2000 m in the case of the MW and parcel method. While the parcel method gives a bell-shaped curve with a maximum at 12:00 UTC, with the MW and SNR methods this is found at around 14:00 UTC. Between the two most used methods (RI and parcel), the difference is about 600 m on average. The greatest increasing rate in PBL height evolution is found between 04:00 and 05:00 UTC with the PTMG and MW method, followed by the parcel method at around 06:00 UTC, the RI and the SNR method at 08:00 UTC.

The PBL height estimations obtained by WRF-SCM (Fig. 3) can be divided into two groups: one group is formed by MYJ and QNSE while the other by YSU, ACM2 and BouLac. MYJ and QNSE estimate the PBL height about

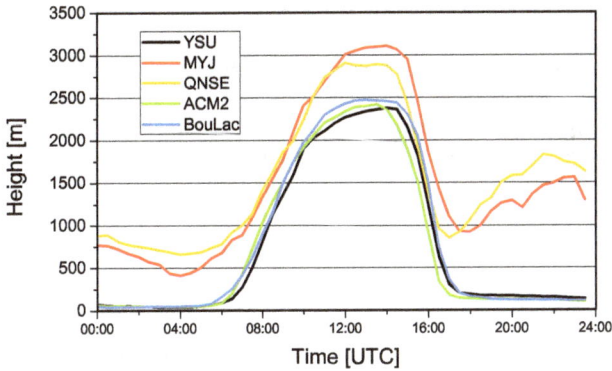

Figure 3. Average (July–September, 2012) diurnal course of PBL height using the WRF-SCM model.

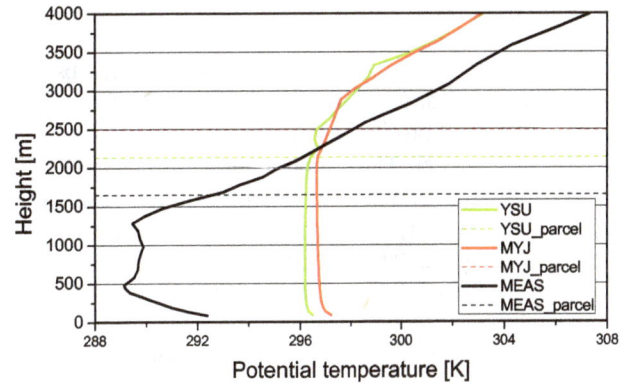

Figure 4. Measured and simulated potential temperature profile and parcel method estimated PBL heights (dashed lines) on 17 July 2012 at 12:00 UTC.

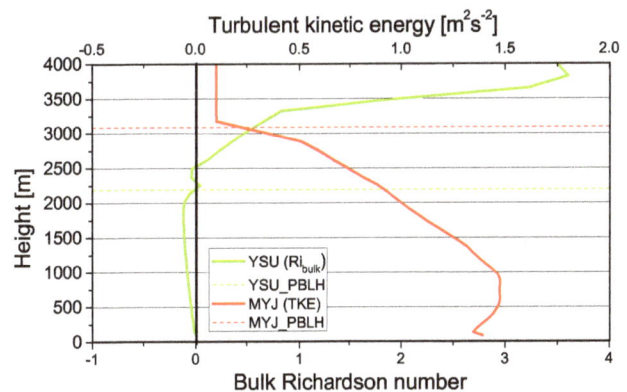

Figure 5. Simulated turbulent kinetic energy (MYJ) and calculated bulk Richardson number (YSU) profile and simulated PBL heights (dashed lines) on 17 July 2012 at 12:00 UTC.

500 m higher than the other schemes during daytime. MYJ and QNSE are also somewhat different: these differences reach about 200 m around 14:00 UTC, and they are still greater between 14:30 and 16:30 UTC. Note that the differences between the two groups are greater in the nighttime than in the daytime period.

5.2 The importance of atmospheric profiles and the estimation principles

The scatter of PBL height diurnal courses, irrespective of simulations or observations used, is considerably high: around noon between 1300 and 3000 m, in the midnight between few tens of metres and about 1500 m. These enormous variations are caused by both the differences in the estimation principles and the differences in the atmospheric profiles used. The relevance of atmospheric profiles will be demonstrated in the comparison of measurement/MYJ/YSU potential temperature profiles (Fig. 4), while the relevance of the estimation principles used in different methods will be shown comparing the TKE/Ri profiles (Fig. 5). Considering the profile of Θ it can be said that the simulations were warmer at the surface with about 4 K, which is not surprising given the conditions of the simulations. However the measured profile shows about a 3.5 K decrease in the surface layer which is about 300 m thick, while in simulation the decrease is only 0.5 K in the same distance. Furthermore, while in the simulations, the profile barely changes in the mixed layer, the measurements show considerable fluctuation. The Θ profile shows a stable stratification from around 1300, 2500 and 2900 m in case of measurement and the simulations (YSU, MYJ), respectively. The PBL height estimated with the parcel method is 1650, 2130 and 2488 m, respectively. However the model simulations put the PBL height to 2188 and 3087 m in case of YSU and MYJ because the determination principle is different (Fig. 5). In Fig. 5, the calculated bulk Ri number and TKE profiles can also be seen for the YSU and MYJ simulation. Up until about 2300 m the Ri profile shows a weak turbulence where it reaches $Ri_{cr} = 0$. Even though a

small mixing can be found in the next 200 m, the Ri shows constant stable stratification from 2500 m. In the MYJ simulations the TKE is maximal at 1000 m and decreasing from there until about 3000 m. In studies where seasonal or annual averages are considered, usually the average YSU PBL heights are higher than the MYJ ones (e.g. Hu et al., 2010; Coniglio et al., 2013). In the CASES-99 campaign for stably stratified, mostly cloudless days, the results are the same (Svensson et al., 2011) with SCM models, but the results are on the contrary when an unstable day is chosen (Shin and Hong, 2011) in a full WRF system. In our simulations most of the days were cloudless and unstably stratified. On days when the atmosphere was stably stratified, the YSU simulations gave higher PBL heights than the MYJ. Therefore we suppose that the differences are a result of different atmospheric stratification.

6   Conclusions

PBL height diurnal variations were estimated at the observatory of the Hungarian Meteorological Service from radiometer and wind-profiler measurements as well as by the WRF-SCM model. Three-month diurnal averages were calculated over the summer drought period. The main results are as follows: (1) around noon, the simulation-based PBL heights (range: 2400–3000 m) were always higher than the observation-based PBL heights (range: 1300–2000 m), which can be a result of the difference between the measured and the simulated temperatures. (2) Around noon, the scatter of the PBL heights obtained by observation-based methods (Fig. 2, about 700 m) is comparable with the scatter of the PBL heights obtained by simulation-based methods (Fig. 3, 500–700 m). (3) The enormous scatter of PBL height diurnal courses is generated by the differences between the estimation principles and by the differences between the atmospheric profiles estimated by simulation or observation tools. The results of the analysis suggest that the atmospheric profile differences could be as important as the estimation principle differences.

Acknowledgements. The project is supported by the Hungarian Scientific Research Found (OTKA K-81432). The authors also thank István Aszalos for his help with the radiometer data.

Edited by: G.-J. Steeneveld
Reviewed by: three anonymous referees

References

André, J. C. and Mahrt, L.: The nocturnal surface inversion and influence of clear-air radiative cooling, J. Atmos. Sci., 39, 864–878, 1982.

Angevine, W. M., White, A. B., and Avery, S. K.: Boundary layer depth and entrainment zone characterization with a boundary-layer profiler, Bound.-Layer Meteorol., 68, 375–385, 1994.

Bougeault, P. and Lacarrère, P.: Parameterization of orography-induced turbulence in a mesobeta-scale model, Mon. Weather Rev., 117, 1872–1890, 1989.

Chen, F. and Dudhia, J.: Coupling an advanced land surface-hydrology model with the Penn State-NCAR MM5 modeling system. Part I: Model implementation and sensitivity, Mon. Weather Rev., 129, 569–585, 2001.

Coniglio, M. C., Correia, J., Marsh, P. T., and Kong, F.: Verification of convection-allowing WRF model forecasts of the planetary boundary layer using sounding observations, Weather Forecast., 28, 842–862, 2013.

Cuxart, J., Holtslag, A. A. M., Beare, R. J., Bazile, E., Beljaars, A., Cheng, A., Conangla, L., Ek, M., Freedman, F., Hamdi, R., Kerstein, A., Kitagawa, H., Lenderink, G., Lewellen, D., Mailhot, J., Mauritsen, T., Perov, V., Schayes, G., Steeneveld, G.-J., Svensson, G., Taylor, P., Weng, W., Wunsch, S., and Xu, K.-M.: Single-column model intercomparison for a stably stratified atmospheric boundary layer, Bound.-Layer Meteorol., 118, 273–303, 2006.

Ghan, S., Randall, D., Xu, K.-M., Cederwall, R., Cripe, D., Hack, J., Iacobellis, S., Klein, S., Krueger, S., Lohmann, U., Pedretti, J., Robock, A., Rotstayn, L., Somerville, R., Stenchikov, G., Sud, Y., Walker, G., Xie, S., Yio, J., and Zhang, M.: A comparison of single column model simulations of summertime midlatitude continental convection, J. Geophys. Res., 105, 2091–2124, 2000.

Hernández-Ceballos, M. A., Adame, J. A., Bolivar, J. P., and de la Morena, B. A.: The performance of different boundary-layer parameterisations in meteorological modelling in a southwestern coastal area of the Iberian Peninsula, ISRN Meteorol., Vol. 2012, 983080, 13 pp., 2012.

Holt, T. and Raman, S.: A review and comparative evaluation of multilevel boundary layer parameterizations for first-order and turbulent kinetic energy closure models, Rev. Geophys., 26, 761–780, 1988.

Hong, S.-Y.: A new stable boundary-layer mixing scheme and its impact on the simulated East Asian summer monsoon, Q. J. Roy. Meteorol. Soc., 136, 1481–1496, 2010.

Hong, S.-Y., Dudhia, J., and Chen, S. H.: A revised approach to ice microphysical processes for the bulk parameterization of clouds and precipitation, Mon. Weather Rev., 132, 103–120, 2004.

Hong, S.-Y., Noh, Y., and Dudhia, J.: A new vertical diffusion package with an explicit treatment of entrainment processes, Mon. Weather Rev., 134, 2318–2341, 2006.

Hu, X.-M., Nielsen-Gammon, J. W., and Zhang, F. Q.: Evaluation of three planetary boundary layer schemes in the WRF Model, J. Appl. Meteorol. Clim., 49, 1831–1844, 2010.

Janjić, Z. I.: The step–mountain coordinate–physical package, Mon. Weather Rev., 118, 1429–1443, 1990.

Janjić, Z. I.: Nonsingular implementation of the Mellor-Yamada Level 2.5 Scheme in the NCEP Meso model, NCEP Off. Note 437, 61 pp., NCEP, Camp Springs, Md, 2002.

Lee, S.-J. and Kawai, H.: Mixing depth estimation from operational JMA and KMA wind-profiler data and its preliminary applications: Examples from four selected sites, J. Meteorol. Soc. Japan, 89, 15–28, 2011.

Matyasovszky, I. and Weidinger, T.: Charaterizig air pollution potential over Budapest using macrocirculation types, Időjárás, 102, 219–237, 1998.

Mlawer, E. J., Taubman, S. J., Brown, P. D., Iacono, M. J., and Clough, S. A.: Radiative transfer for inhomogeneous atmosphere: RRTM, a validated correlated-k model for the longwave, J. Geophys. Res., 102, 16663–16682, 1997.

Pleim, J. E.: A combined local and nonlocal closure model for the atmospheric boundary layer. Part I: Model description and testing, J. Appl. Meteorol. Clim., 46, 1383–1395, 2007.

Randall, D. A., and Cripe, D. G.: Alternative methods for specification of observed forcing in single-column models and cloud system models, J. Geophys. Res., 104, 24527–24545, 1999.

Seibert, P., Beyrich, F., Gryning, S. E., Joffre, S., Rasmussen, A., and Tercier, P.: Mixing height determination for dispersion modelling. In: COST Action 710 Harmonization of the pre-processing of meteorological data for atmospheric dispersion models, Final Report EUR 18195 EN, Report of Working Group 2, 121 pp., 1997.

Shin, H. H. and Hong, S. Y.: Intercomparison of planetary boundary-layer parametrizations in the WRF model for a single day from CASES-99, Bound-Layer Meteorol, 139, 261–281, 2011.

Skamarock, W. C., Klemp, J. B., Dudhia, J., Gill, D. O., Barker, D. M., Duda, M., Huang, X.-Y., Wang, W., and Power, J. G.: A description of the Advanced Research WRF Version 3. NCAR Technical Note, NCAR/Tech Notes-475+STR, 125 pp., 2008.

Steeneveld, G. J., Mauritsen, T., de Bruijn, E. I. F., de Arellano, J. V. G., Svensson, G., and Holtslag, A. A. M.: Evaluation of limited-area models for the representation of the diurnal cycle and contrasting nights in CASES-99, J. Appl. Meteorol. Clim., 47, 869–887, 2008.

Sukoriansky, S., Galperin, B., and Perov, V.: Application of a new spectral theory of stable stratified turbulence to the atmospheric boundary layer over sea ice, Bound.-Layer Meteorol., 117, 231–257, 2005.

Svensson, G., Holtslag, A. A. M., Kumar, V., Mauritsen, T., Steeneveld, G. J., Angevine, W. M., Bazile, E., Beljaars, A., de Bruijn, E. I. F., Cheng, A., Conangla, L., Cuxart, J., Falk, M. J., Larson, V. E., Mailhot, J., Masson, V., Park, S., Pleim, J., and Söderberg, S.: Evaluation of the diurnal cycle in the atmospheric boundary layer over land as represented by a variety of single column models – the second GABLS experiment, Bound.-Layer Meteorol., 140, 177–206, 2011.

von Engeln, A. and Teixeira, J.: A planetary boundary layer height climatology derived from ECMWF reanalysis data, J. Climate, 26, 6575–6590, 2013.

Xie, B., Fung, J. C. H., Chan, A., and Lau, A.: Evaluation of nonlocal and local planetary boundary layer schemes in the WRF model, J. Geophys. Res., 117, D12103, doi:10.1029/2011JD017080, 2012.

8

Sensitivity of the RMI's MAGIC/Heliosat-2 method to relevant input data

C. Demain, M. Journée, and C. Bertrand

Royal Meteorological Institute of Belgium, Brussels, Belgium

Correspondence to: M. Journée (michel.journee@meteo.be)

Abstract. Appropriate information on solar resources is very important for a variety of technological areas. Based on the potential of retrieving global horizontal irradiance from satellite data, an enhanced version of the Heliosat-2 method has been implemented at the Royal Meteorological Institute of Belgium to estimate surface solar irradiance over Belgium from Meteosat Second Generation at the SEVIRI spatial and temporal resolution. In this contribution, sensitivity of our retrieval scheme to surface albedo, atmospheric aerosol and water vapor contents is investigated. Results indicate that while the use of real-time information instead of climatological values can help to reduce to some extent the RMS error between satellite-retrieved and ground-measured solar irradiance, only the correction of the satellite-derived data with in situ measurements allows to significantly reduce the overall model bias.

1 Introduction

There are several methods for converting satellite images into surface solar irradiance (SSI). The Heliosat-2 method (Rigollier et al., 2004) is a well-known method of inverse type. The principle of the method is that a difference in global radiation perceived by the sensor aboard a satellite is only due to a change in the apparent albedo, which is itself due to an increase of the radiation emitted by the atmosphere towards the sensor (i.e., Cano et al., 1986; Raschke et al., 1987). A key parameter is the cloud index (also denoted as effective cloud albedo), n, determined by the magnitude of change between what is observed by the sensor and what should be observed under a very clear sky. To evaluate the all-sky SSI, a clear-sky model is coupled with the retrieved cloud index which acts as a proxy for cloud transmittance. Inputs to the Heliosat-2 method are not the visible satellite images in digital counts as in the original version of the method (Cano et al., 1986) but images of radiances/reflectances:

$$n^t(i, j) = \frac{\rho^t(i, j) - \rho_{cs}^t(i, j)}{\rho_{max}^t(i, j) - \rho_{cs}^t(i, j)} \quad (1)$$

where $n^t(i, j)$ is the cloud index at time t for the pixel (i, j); $\rho^t(i, j)$ is the reflectance or apparent albedo observed by the sensor at time t; $\rho_{max}^t(i, j)$ is the apparent albedo of the bright-

est cloud at time t; $\rho_{cs}^t(i, j)$ is the apparent ground albedo under clear-sky condition at time t. With calibrated radiances as input, Heliosat-2 offers the opportunity to replace some of the empirical parameters in the scheme with known physical quantities from external sources.

A modified version of the Heliosat-2 calculation scheme has been implemented at the Royal Meteorological Institute of Belgium (RMI) to retrieve SSI values from Meteosat Second Generation (MSG; Schmetz et al., 2002) satellite over Belgium. Section 2 presents the modifications/adaptations made to the method at RMI to exploit the enhanced capabilities of MSG in the SSI retrieval. Performance of our algorithm is discussed in Sect. 3. Sensitivity of the retrieval scheme to clear-sky model input data is presented in Sect. 4. Final remarks and conclusion are given in Sect. 5.

2 Description of the MAGIC/Heliosat-2 method implemented at RMI

The method takes advantage of the enhanced capabilities of the Spinning Enhanced Visible and Infrared Imager (SEVIRI) on board of the MSG platform through an improved scene identification and applies the Modified Lambert Beer function (Müller et al., 2004) within an eigen-vector hybrid look-up table (LUT) approach (Müller et al., 2009) for the clear-sky irradiance computation. It is based on the LibRad-Tran (Mayer and Kylling, 2005) radiative transfer model (RTM) and enables the use of extended information about the atmospheric state. The source code of the clear-sky model (Mesoscale Atmospheric Global Irradiance Code – MAGIC) is available under gnu-public license at http://sourceforge.net/projects/gnu-magic/. It is worth pointing out that even if the SEVIRI sensor comprises a high spatial resolution broadband visible channel (HRV, High Resolution Visible) and that the Heliosat method was originally conceived for working with broadband images, the algorithm as in Posselt et al. (2011) is applied to the visible narrow-band channels of SEVIRI (centered at 0.6 µm (VIS06) and 0.8 µm (VIS08), respectively). Indeed, while the MSG HRV channel has a higher spatial sampling distance than the SEVIRI spectral channels (i.e., 1 km vs. 3 km at the subsatellite point, respectively), Journée et al. (2012) have shown that for a mid-latitude region with a rather flat orography like Belgium, the MSG-based daily SSI retrieval is much more sensible to the temporal resolution than to the spatial resolution of the satellite images. Therefore, only working with SEVIRI spectral images does not require to deal with the georeferencing of the MSG HRV images (e.g., the original HRV geolocation performed by the EUMETSAT ground segment is only accurate up to ±3 HRV pixels) which allows to save CPU time without a noticeable loss of precision. The retrieval process runs over a spatial domain ranging from 48.0° N to 54.0° N and from 2.0° E to 7.5° E within the MSG field-of-view. In this domain, the SEVIRI spatial sampling distance degrades to about 6 km in the north–south direction and 3.3 km in the east–west direction.

The schematic view of the procedure is shown in Fig. 1. First, cloud mask, snow mask and cloud phase are derived over the domain in real time for each 15-min MSG time slot with the EUMETSAT Satellite Application Facility on Support to Nowcasting and Very Short Range Forecasting (NWC SAF) software (Derrien and Le Gléau, 2005) using the MSG SEVIRI spectral information and 24 h numerical weather forecasts from the European Centre for Medium-Range Weather Forecasts (ECMWF) two times a day. In parallel, the SEVIRI VIS06 and VIS08 data are converted from counts to reflectance.

Second, at the end of each day, the VIS06 and VIS08 reflectances and the NWC SAF algorithm products related to the 96 MSG time slots of the elapsed day are used to determine the effective cloud albedo (or cloud index), n. For each pixel in the image and each daytime MSG time slot, clear-sky reflectances in the VIS06 and VIS08 spectral bands, ρ_{clr}, are determined from a trailing window of 60 days MSG SEVIRI reflectances at 0.6 µm and 0.8 µm, respectively, according to Ipe et al. (2003). Overcast visible reflectances, ρ_{max}, at 0.6 and 0.8 µm are determined by a LUT approach using the cloud phase information provided by the NWC SAF software. It relies on the LibRadTran (Mayer and Kylling, 2005)

simulated outgoing radiances in the SEVIRI 0.6 and 0.8 µm spectral bands. RTM computations were performed assuming a cloud optical depth of 128 and a pure cloud thermodynamic phase (i.e., water cloud or ice cloud). Because of the low reflectance of a majority of natural land surfaces at 0.6 µm and the reduced influence of the vegetation seasonal cycle on the reflectance signature at this wavelength (Asner, 1998), the algorithm is applied to the VIS06 channel over land surface and to the VIS08 channel over water surface. Based on the NWC SAF cloud and snow masks the computed cloud index at 0.6 and 0.8 µm are either corrected or not corrected for possible cloud shadow or snow contamination.

Third, SSI is derived for each pixel and MSG time slot by the combination of the satellite clearness index $k(n)$ (a decreasing function of n, Hammer et al., 2003) and SSI_{clr}, the corresponding clear-sky surface irradiance calculated by MAGIC:

$$SSI = k(n) \cdot SSI_{clr} \tag{2}$$

Note that for effective cloud albedo values between 0 and 0.8, SSI is the clear-sky irradiance which is not reflected back to space by clouds:

$$SSI = (1 - n) \cdot SSI_{clr} \tag{3}$$

Finally, the MSG retrieved all-sky SSI are integrated over the entire diurnal cycle and merged with the corresponding daily global solar irradiation recorded within the Belgian ground radiometric network operated by RMI.

3 Performance of the RMI MAGIC/Heliosat-2 retrieval scheme

The key factor determining the short-term variability of the all-sky irradiance is cloudiness. While the Heliosat-2 method is normally capable of reproducing most of the irradiance variability caused by clouds (i.e., Hammer et al., 2003; Rigollier et al., 2004; Lefèvre et al., 2007; Moradi et al., 2009), current satellite instruments are insufficient to accurately sense small broken clouds. The method then interprets the scene as a large thin cloud rather than a patchwork of small thick clouds, so that the modeled irradiance varies less and at a lower frequency than what terrestrial instruments would record. Uncertainty in the modeled instantaneous irradiance under intermittent cloudiness is therefore higher than under cloudless or overcast conditions. At low solar elevations, uncertainty in the cloud index generally increases because of complex 3-D reflections effects off the sides of clouds, and parallax effects in the case of high clouds. Additionally, a high surface albedo (e.g., snow-covered areas) lowers the contrast of the signal recorded by the satellite and introduces error in the cloud index. The major challenge here is to correctly attribute any high radiance value retrieved from satellite imagery to either a cloud scene or

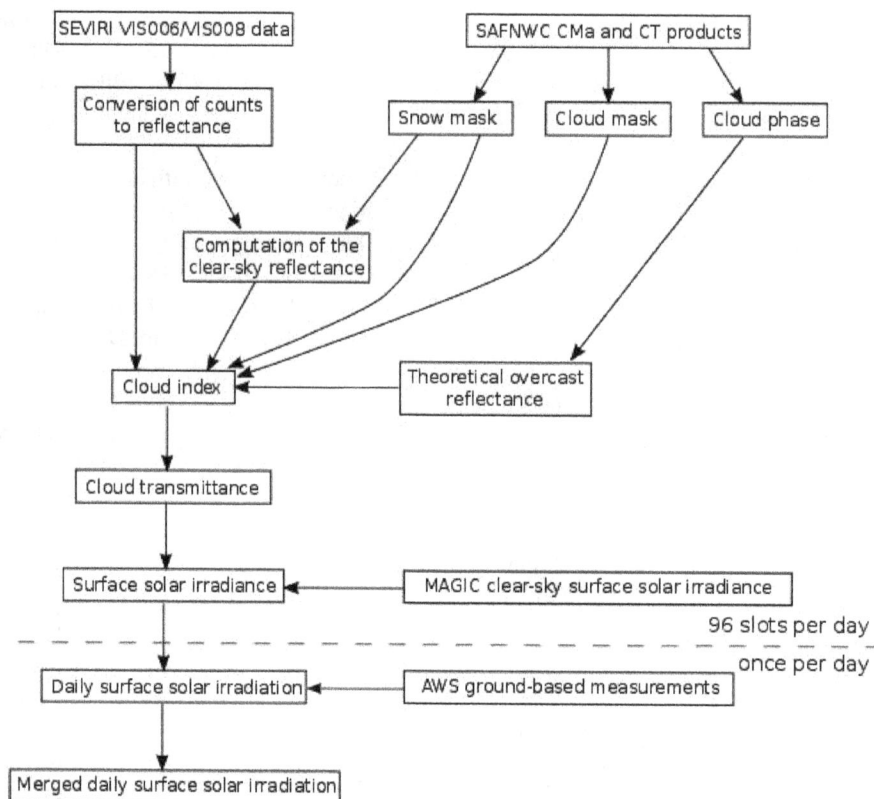

Figure 1. Schematic view of the MAGIC/Heliosat-2 process implemented at RMI.

a high-albedo surface. Uncertainties are thus expected to be higher during winter months.

Measurements from 11 radiometric stations operated by RMI were used to evaluate the performance of our algorithm. Global horizontal irradiance measurements are made with a 5 secondes time step and then integrated on a 10-min basis in the RMI data warehouse where they undergo a set of semi-automatic quality assessment procedures (Journée and Bertrand, 2011). Daily SSI values are obtained by simple summation for the 10-min integrated ground-based measurements and by trapezoidal integration over the diurnal cycle for the MSG-based instantaneous values. The merged daily SSI estimations are obtained using the kriging with external drift geostatistical approach as described in Journée and Bertrand (2010). The retrieved daily all-sky SSI are evaluated first for non-merged data through a direct comparison between satellite derived and ground-measured data and second by leave-one-out cross-validation (CV) for satellite-ground merged estimations on the basis of one year of quality controlled data (i.e., 2011). Two statistical error indices are considered: the mean bias error (MBE) and root mean square error (RMSE).

Table 1 indicates that overall for the year 2011 the satellite-retrieved daily SSI data using climatological input data for the MAGIC clear-sky model (i.e., CLIM(SA,AOD,WVC)) present an overall positive bias of 3.52 % (i.e., a MBE of 119.02 Wh m^{-2}). The bias is maximum in winter and minimum in autumn (i.e., MBE of 7.67 % (67.75 Wh m^{-2}) and of 2.48 % (62.62 Wh m^{-2}) for DJF and SON, respectively). The reported spring and summer biases are in the order of 3.07 % (142.09 Wh m^{-2}) and 3.83 % (167.22 Wh m^{-2}) for MAM and JJA, respectively. The bias originates in the use of RTM simulated overcast visible reflectances, ρ_{max}, in the cloud index, n, computation (see Eq. 1) instead of actual measured overcast reflectances. Simulated overcast visible reflectances tend to overestimate the actual ones (i.e., RTMs provide an upper bound on ρ_{max}, which often exceeds the actual ρ_{max}) which leads to reduce the magnitude of the effective cloud albedo, and subsequently to an increase in the retrieved SSI. While the observation angles of geostationary sensors remain invariable, measurements depend on varying illumination angles and changes during the course of the day/year. The larger bias at wintertime relies on an enhanced overestimation in simulated ρ_{max} at high solar zenith angles (low elevation). Fortunately, correcting the satellite-derived daily SSI values by using in situ measurements reduces the overall model bias to 0.03 % (MBE$_{cv}$ of 1.01 Wh m^{-2}) and its seasonal magnitude (e.g., MBE$_{cv}$ of −0.67 Wh m^{-2} (−0.08 %) for DJF). Finally, thanks to the enhanced SEVIRI scene identification the method does not exhibit a particular weakness over snowed land surface.

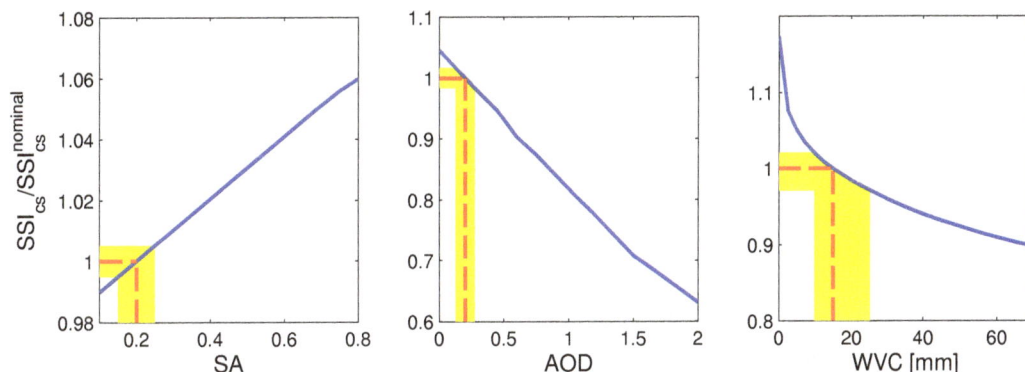

Figure 2. Sensitivity of the MAGIC computed clear-sky SSI with respect to the surface albedo (SA), the aerosol optical depth (AOD) and the atmospheric water vapor content (WVC). Values are normalized with respect to the daily integrated clear-sky SSI computed over Belgium with the nominal values for the 3 parameters (i.e., SA = 0.2, AOD = 0.2 and WVC = 15 mm).

4   Sensitivity of the retrieval scheme to the MAGIC clear-sky model input

To estimate the all-sky irradiance at each time step, the clear-sky SSI is coupled with the cloud index, which acts as a proxy for cloud transmittance. The clear-sky radiative model is at the core of this process, as it calculates solar irradiance for any given sun position, state of the atmosphere, and terrain conditions. The performance of the clear-sky model depends on the physical soundness of its parameterizations, and its ability to deal properly not only with typical cases, but also with extreme values of the main inputs, such as high load of aerosols, extreme humidity conditions, high site elevation, low solar elevation, unusual surface reflectance patterns, or shading effects. The sun geometry is a deterministic parameter, which can be evaluated with satisfactory accuracy. The modeling of how topography affects the direct and diffuse radiation components is also deterministic, but more complex. The site's elevation determines the atmospheric pressure and the related extinction process, whereas the surrounding terrain determines the possible shading of the sun and/or parts of the sky. Accounting for the terrain complicates calculations and requires a spatial resolution at least an order of magnitude better than the nominal resolution of the satellite image (e.g., Cebecauer et al., 2011). Clear-sky atmospheric conditions are variable, both spatially and temporally. These variations are related to the changing concentrations of the main radiatively active atmospheric constituents, namely aerosols, water vapor and ozone. In the current operational scheme, climatological input data are used by the MAGIC clear-sky code. The drawback of such an approach is that long-term averages artificially remove the natural daily fluctuations and the failure to represent extreme events. Moreover, because aerosols and water vapor are highly variable over space and time a relative coarse spatio-temporal resolution of aerosol or water vapor data prevents a detailed rendering of local small-scale patterns and could be a source of bias between estimated and measured SSI.

Figure 2 displays the sensitivity of the MAGIC computed clear-sky SSI with respect to the surface albedo (SA), the aerosol optical depth (AOD) and the atmospheric water vapor content (WVC). Basically, one year of simulations based on a 15-min time step were carried out by running the MAGIC code with SA, AOD, and WVC values covering a large range of variation. In these simulations, each parameter was varied over the largest physical range while the other two were kept fixed to their nominal values for Belgium (annual mean climatological values at Uccle). The results are illustrated in Fig. 2 for the annual mean of the daily integrated clear-sky SSI. Note that values in this figure are normalized with respect to the daily integrated clear-sky SSI computed with the nominal values for the 3 parameters (i.e., SA = 0.2, AOD = 0.2, WVC = 15 mm). Also provided on the panels in Fig. 2 (yellow strip) is the typical variability range for the 3 parameters in Belgium (i.e., SA ranges from 0.15 to 0.25, AOD from 0.125 to 0.275 and WVC from 10 to 25) as retrieved from climatology. Clearly, the largest impact in the simulated MAGIC clear-sky SSI over Belgium is expected to come from the integrated WVC input (variation of ±3.2 % in the simulated annual mean clear-sky daily SSI) and the impact on SSI will be much higher for low atmospheric WVC. By contrast, variation in the simulated annual mean clear-sky daily SSI over Belgium equals ±0.5 % and ±2.1 % for the SA and AOD input values, respectively. Note that for SA values, a large sensitivity up to 5 % could appear in wintertime in the presence of snowed land surface. It is worth pointing out that if a pixel is detected as snowed by the NWC SAF snow mask, a snow albedo is considered in the retrieval process instead of the pixel default albedo value.

To further assess the potential impact of SA, AOD and WVC on the MSG retrieved all-sky daily SSI values, one year (i.e., 2011) of SSI retrievals were performed over the RMI's MAGIC/Heliosat-2 domain by considering for each of the 3 input parameters both the MAGIC default climatological values and time series derived from in situ data, satellite

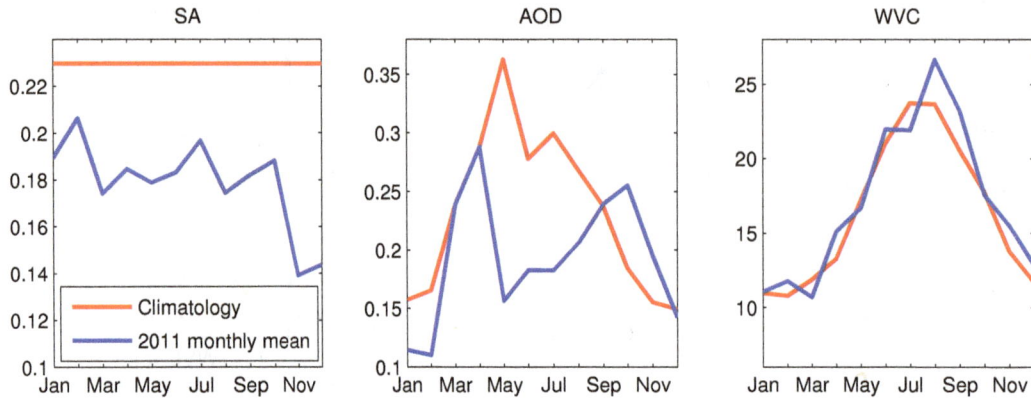

Figure 3. Comparison between climatological and observed/estimated 2011 monthly mean values for surface albedo, aerosol optical depth and water vapor content (in mm) for Uccle (Brussels).

observations or model-based forecasts. Basically, beside the MAGIC default climatological surface albedo data set (i.e., annual mean surface albedo at $1 \times 1°$ spatial resolution from Clouds and the Earth's Radiant Energy System Clouds and Surface and Atmospheric Radiation Budget data product, Rutan and al., 2009) daily surface albedo values produced by the EUMETSAT Land Surface Analysis Satellite Application Facility at the MSG/SEVIRI spatial resolution (Geiger et al., 2008) for the year 2011 were considered. Note that in both cases the MAGIC model accounts for the solar zenith angle dependency of SA. Similarly, a monthly mean AOD data set was generated at the MSG/SEVIRI spatial resolution over Belgium by aggregation of the 10×10 km MODIS Collection 5 aerosols optical depth retrievals over land (Levy et al., 2007) and by combination with AERONET level 2.0 monthly mean AOD values from the 6 AERONET sites located within the domain to supplement the MAGIC AOD default values (i.e., monthly mean AOD at $1 \times 1°$ spatial resolution from Kinne et al., 2006 median model). Finally, 3-hourly water vapor profiles from the $0.25 \times 0.25°$ ECMWF model forecasts updated twice a day and interpolated at each MSG time slot were considered as an additional source of WVC information to the MAGIC WVC default values (i.e., long-term monthly mean values at $0.25 \times 0.25°$ spatial resolution taken from the ECMWF ERA-40/Interim reanalysis data set). Comparison between MAGIC default SA, AOD and WVC values for Uccle (Brussels) and the corresponding 2011 monthly mean estimations is provided in Fig. 3. Note that because ozone has only a small impact on SSI (e.g., Cebecauer et al., 2011), a climatological value is assumed sufficient and no further analysis of the potential impact of this parameter on the retrieval scheme performance was considered here.

Table 1 summarizes in terms of error indices the results obtained for different combinations of the 3 ancillary parameter data sources in the SSI retrieval process against the ground-based measurements. Because SA values from the LSA SAF are lower than the default MAGIC values,

use of the LSA SAF albedo product in the retrieval process allows to reduce the overall bias by 1.14 % (e.g., from 119.02 Wh m^{-2} for CLIM(SA,AOD,WVC) to 80.46 Wh m^{-2} for SA+CLIM(AOD,WVC)). While the bias is reduced for each season, the better improvement appears in winter (DJF) where the MBE decreases from 67.75 Wh m^{-2} (7.67 %) for CLIM(SA,AOD,WVC) to 52.49 Wh m^{-2} (or 5.94 %) for SA+CLIM(AOD,WVC). Because the estimated 2011 monthly mean AOD values are lower than the MAGIC default values excepted in autumn (see upper right panel in Fig. 3), use of the 2011 AOD values degrades the overall performance of the method (i.e., MBE of 159.69 Wh m^{-2} and RMSE of 320.53 Wh m^{-2} for the AOD+CLIM(SA+WVC) retrievals vs. MBE of 119.02 Wh m^{-2} and RMSE of 291.05 Wh m^{-2} for the CLIM(SA,AOD,WVC) retrievals, respectively). The largest score reduction occurs at summertime where the MBE for JJA reaches up to 234.19 Wh m^{-2} (or 5.35 %) vs. a MBE of 167.22 Wh m^{-2} or 3.84 % for CLIM(SA,AOD,WVC). Note that on the contrary a slight improvement (RMS reduction of 0.01 %) is reported for SON. Regarding the WVC, because estimations from the ECMWF model forecasts are lower than the MAGIC default values, use of the ECMWF model forecasts tend to increase the overall statistical error indices as indicated in Table 1. However, the situation differing from one season to another, combining both the LSA SAF albedo values and the WVC values from the ECMWF model forecasts in the retrieval process (i.e., SA+WVC+CLIM(AOD) in Table 1) produces the best score in terms of RMSE reduction (273.72 Wh m^{-2} or 8.08 % vs. 291.05 Wh m^{-2} or 8.60 % for CLIM(SA,AOD,WVC)). By contrast, accounting for the WVC values from the ECMWF model forecast and the estimated 2011 monthly mean AOD values generate the worst SSI retrieval in terms of error indices with an overall RMS of 178.99 Wh m^{-2} and a RMSE of 326.93 Wh m^{-2}. Finally, the various statistics provided in Table 1 indicate that the largest improvement in the estimated daily all-sky SSI values over Belgium in terms of bias reduction and RMSE decrease

Table 1. MBE and RMSE of (A) non-merged and (B) merged daily-integrated SSI using the MAGIC default climatology and dynamical inputs (i.e., SA, AOD and WVC) versus global solar radiation measurements at 11 stations in Belgium for the year of 2011.

(A) Non-merged daily integrated SSI				
MAGIC input	MBE ($Wh\,m^{-2}$)	relative MBE (%)	RMSE ($Wh\,m^{-2}$)	relative RMSE (%)
CLIM(SA,AOD,WVC)	119.02	3.52	291.05	8.60
SA+CLIM(AOD,WVC)	80.46	2.38	276.99	8.18
AOD+CLIM(SA,WVC)	159.69	4.72	320.53	9.47
WVC+CLIM(SA,AOD)	138.32	4.09	292.56	8.64
SA+WVC+CLIM(AOD)	99.50	2.94	273.72	8.08
SA+AOD+CLIM(WVC)	120.75	3.57	300.40	8.88
AOD+WVC+CLIM(SA)	178.99	5.29	326.93	9.66
SA+WVC+AOD	139.79	4.13	302.54	8.94

(B) Merged daily integrated SSI				
MAGIC input	MBE_{cv} ($Wh\,m^{-2}$)	relative MBE_{cv} (%)	$RMSE_{cv}$ ($Wh\,m^{-2}$)	relative $RMSE_{cv}$ (%)
CLIM(SA,AOD,WVC)	1.01	0.03 %	244.41	7.22 %
SA+CLIM(AOD,WVC)	−0.06	< 0.01 %	247.71	7.32 %
AOD+CLIM(SA,WVC)	1.74	0.05 %	244.45	7.22 %
WVC+CLIM(SA,AOD)	1.42	0.04 %	243.20	7.19 %
SA+WVC+CLIM(AOD)	0.21	0.01 %	246.60	7.29 %
SA+AOD+CLIM(WVC)	0.74	0.02 %	247.66	7.32 %
AOD+WVC+CLIM(SA)	2.08	0.06 %	243.43	7.19 %
SA+WVC+AOD	0.98	0.03 %	246.58	7.29 %

is obtained by merging both satellite retrievals and ground-based measurements. Influence of the MAGIC input data sources appears as quite negligible in the merged daily all-sky SSI products. As an example, the reported $RMSE_{cv}$ values in Table 1 differ by less than 5 $Wh\,m^{-2}$ or 0.13 % between the 8 different merged daily products.

5 Conclusions

This study aimed at assessing the sensitivity of the MAGIC/Heliosat-2 method over Belgium to the surface albedo, atmospheric aerosol optical depth and integrated water vapor content values required as input data by the MAGIC clear-sky model. Our results indicate that irrespective of the ancillary data sources for these 3 parameters, our retrieval scheme is positively biased over Belgium. Because simulated overcast visible reflectances tend to overestimate the actual ones our cloud index computation leads to reduce the magnitude of the effective cloud albedo, and subsequently to an increase in the retrieved SSI. Use of observed/estimated values during the retrieval process instead of climatological means can help to reduce to some extent the bias and RMS error between satellite retrieved and ground measured all-sky

SSI. The largest errors reduction is obtained when considering both the actual surface albedo and integrated water vapor content values in the retrieval process. Because the estimated 2011 aerosol optical depth values were on average larger than the default MAGIC values, the overall bias of the method was increased when performing the SSI retrievals with these estimations. However, due to limitations in the AOD retrieval process over land surface (e.g., Hsu et al., 2004) and the limited number of AOD measurements available over the domain, it is likely that our estimated 2011 AOD monthly mean time series are only a crude estimation of the actual 2011 AOD values over Belgium. As an example, the summer AOD background value over Belgium appears well underestimated in our reconstruction which therefore tends to reinforce the positive bias of our retrieval scheme compared to the ground observations and subsequently to a worse performance of the model. Finally, whatever the sources of the MAGIC input data may be, the most accurate mapping of the daily all-sky SSI over Belgium is obtained once ground measurements and satellite-based estimation are merged into a single product. Correcting the satellite-derived data by using in situ measurements reduces the overall model bias. Therefore, the influence of the MAGIC input data sources is largely masked in the merged daily integrated all-sky SSI product, which subsequently appears to be nearly independent of the MAGIC input data sources used during the SSI satellite retrieval process.

Acknowledgements. The authors are grateful to Richard Mueller (German Meteorological Service, Germany) for providing us the MAGIC clear-sky model. We are also grateful to the all of the investigators and their staff who participate in the AErosol RObotic NETwork for establishing and maintaining the sites used in this investigation. This work was supported by the Belgian Science Policy Office (BELSPO) through the ESA/PRODEX program PRODEX-9 contract No. 4000102777 "Surface Solar Radiation".

Edited by: G.-J. Steeneveld
Reviewed by: E. L. A. Wolters, J. Polo, and one anonymous referee

References

Asner, G. P.: Biophysical and Biochemical Sources of variability in Canopy Reflectance, Remote Sens. Environ., 64, 234–253, 1998.

Cano, D., Monget, J. M., Albuisson, M., Guillar, H., Regas, N., and Wald, L.: A method for the determination of the global solar radiation from meteorological satellite data, Sol. Energy, 56, 207–212, 1986.

Cebecauer, T., Suri, M., and Gueymard, C.: Uncertainty sources in satellite-derived direct normal irradiance: how can prediction accuracy be improved globally?, Proceedings of the SolarPACES Conference, Granada, Spain, 20–23 September 2011.

Derrien, M. and Le Gléau, H.: MSG/SEVIRI cloud mask and type from SAFNWC, Int. J. Remote Sens., 26, 4707–4732, 2005.

Geiger, B., Carrer, D., Franchistéguy, L., Roujean, J.-L., and Meurey, C.: Land surface albedo derived on a daily basis from Meteosat Second Generation observations, IEEE T. Geosci. Remote Sens., 46, 3841–3856, 2008.

Hammer, A., Heinemann, D., Hoyer, C., Kuhlemann, R., Lorenz, E., Müller, R., and Beyer, H.: Solar energy assessment using remote sensing technologies, Remote Sens. Environ., 86, 423–432, 2003.

Hsu, N. C., Tsay, S. C., King, M. D., and Herman, J. R.: Aerosol Properties over Bright Reflecting Source Regions, IEEE T. Geosci. Remote Sens., 42, 557–569, 2004.

Ipe, A., Clerbaux, N., Bertrand, C., Dewitte, S., and Gonzalez, L.: Pixel-scale composite top-of-the-atmosphere clear-sky reflectances for Meteosat-7 visible data, J. Geophys. Res., 108, 148–227, 2003.

Journée, M. and Bertrand, C.: Improving the spatio-temporal distribution of surface solar radiation data by merging ground and satellite measurements, Remote Sens. Environ., 114, 2692–2704, 2010.

Journée, M. and Bertrand, C.: Quality control of solar radiation data within the RMIB solar measurements network. Sol. Energy, 85, 72–86, 2011.

Journée, M., Stöckli, R., and Bertrand, C.: Sensitivity to spatio-temporal resolution of satellite-derived daily surface solar irradiation, Remote Sens. Lett., 3, 315–324, 2012.

Kinne, S., Schulz, M., Textor, C., Guibert, S., Balkanski, Y., Bauer, S. E., Berntsen, T., Berglen, T. F., Boucher, O., Chin, M., Collins, W., Dentener, F., Diehl, T., Easter, R., Feichter, J., Fillmore, D., Ghan, S., Ginoux, P., Gong, S., Grini, A., Hendricks, J., Herzog, M., Horowitz, L., Isaksen, I., Iversen, T., Kirkevåg, A., Kloster, S., Koch, D., Kristjansson, J. E., Krol, M., Lauer, A., Lamarque, J. F., Lesins, G., Liu, X., Lohmann, U., Montanaro, V., Myhre, G., Penner, J., Pitari, G., Reddy, S., Seland, O., Stier, P., Takemura, T., and Tie, X.: An AeroCom initial assessment – optical properties in aerosol component modules of global models, Atmos. Chem. Phys., 6, 1815–1834, doi:10.5194/acp-6-1815-2006, 2006.

Lefèvre, M., Diabaté, L., and Wald, L.: Using reduced data sets ISCCP-B2 from the Meteosat satellites to assess surface solar irradiance, Sol. Energy, 81, 240–253, 2007.

Levy, R. C., Remer, L. A., Mattoo, S., Vermote, E. F., and Kaufman, Y. J.: Second-generation operational algorithm: Retrieval of aerosol properties over land from inversion of Moderate Resolution Imaging Spectroradiometer spectral reflectance, J. Geophys. Res., 112, 13211–13221, 2007.

Moradi, I., Müller, R., Alijani, B., and Kamali, G.A.: Evaluation of the Heliosat-II method using daily irradiation data for four stations in Iran, Sol. Energy, 83, 150–156, 2009.

Mayer, B. and Kylling, A.: Technical note: The libRadtran software package for radiative transfer calculations – description and examples of use, Atmos. Chem. Phys., 5, 1855–1877, doi:10.5194/acp-5-1855-2005, 2005.

Müller, R., Dagestad, K-F., Ineichen, P., Schroedter Homscheidt, M., Cros, S., Dumortier, D., Kuhlemann, R., Olseth, J., Piernavieja, G., Reise, Ch., Wald, L., and Heinnemann, D.: Rethinking satellite based solar irradiance modelling – The SOLIS clear sky module, Remote Sens. Environ., 91, 160–174, 2004.

Müller, R., Matsoukas, C., Gratzki, A., Behr, H. D., and Hollmann, R.: The CM-SAF operational scheme for the satellite based retrieval of solar surface irradiance – a LUT based eigenvector hybrid approach, Remote Sens. Environ., 113, 1012–1024, 2009.

Posselt, R., Müller, R., Stöckli, R., and Trentmann, J.: Spatial and temporal Homogeneity of Solar surface Irradiance across satellite generations, Remote Sens., 3, 1029–1046, 2011.

Raschke, E., Gratzki, A., and Rieland, M.: Estimates of global radiation at the ground from reduced data sets of International Satellite Cloud Climatology Project, J. Climate, 7, 205–213, 1987.

Rigollier, C., Lefèvre, M., and Wald, L.: The method Heliosat-2 for deriving shortwave solar radiation from satellite images, Sol. Energy, 77, 159–169, 2004.

Rutan, D., Rose, F., Roman, M., Manalo-Smith, N., Schaaf, C., and Charlock, T.: Development and assessment of broadband surface albedo from Couds and Earth's Radiant Energy System Clouds and Radiation Swath data product, J. Geophys. Res., 114, D08125, doi:10.1029/2008JD010669, 2009.

Schmetz, J., Pili, P., Tjemkes, S., Just, D., Kerkmann, J., Rota, S., and Ratier, A.: An Introduction to Meteosat Second Generation (MSG), B. Am. Meteorol. Soc., 83, 977–992, 2002.

9

Yearly changes in surface solar radiation in New Caledonia

P. Blanc[1], C. Coulaud[2], and L. Wald[1]

[1]MINES ParisTech, PSL Research University, Centre Observation, Impacts, Energy, BP 204,
06905 Sophia Antipolis CEDEX, France
[2]ADEME, Valbonne, France

Correspondence to: L. Wald (lucien.wald@mines-paristech.fr)

Abstract. New Caledonia experiences a decrease in surface solar irradiation since 2004. It is of order of 4 % of the mean yearly irradiation over the 10 years period: 2004–2013, and amounts to $-9\,\mathrm{W\,m^{-2}}$. The preeminent roles of the changes in cloud cover and to a lesser extent, those in aerosol optical depth on the decrease in yearly irradiation are evidenced. The study highlights the role of data sets offering a worldwide coverage in understanding changes in solar radiation and planning large solar energy plants such as the ICOADS (International Comprehensive Ocean-Atmosphere Data Set) of the NOAA and MACC (Monitoring Atmosphere Composition and Climate) data sets combined with the McClear model.

1 Introduction

New Caledonia, a large island in tropical Pacific Ocean (Fig. 1), experiences a very sunny weather. Cloudiness is low and on average 60 % of the solar radiation available at the top of atmosphere reaches the ground. Hence, solar radiation is an option for energy production.

The solar radiation reaching the ground on a horizontal surface, known as the surface solar irradiance (SSI) is made of the direct component that is the radiation coming from the direction of the sun and the diffuse component that is the radiation coming from all other directions (Fig. 2). The sum of the direct and diffuse is called the global radiation. Of interest to concentrating solar technologies (CST) that concentrate sun rays to produce electricity is the direct solar radiation received on a plane normal to the sun rays (DNI).

A preliminary study, not publicly available, has been performed to assess the potentials of electricity production by concentrating Fresnel mirrors using local measurements of DNI and other meteorological data. The yearly sum of DNI has been estimated at $2\,\mathrm{MWh\,m^{-2}}$ at Noumea, the major city. If $190\,000\,\mathrm{m^2}$ of mirrors were installed, they could produce 36 GWh per year. Coupled to a coal fired power station, they could save up to 19 kt of CO_2 per year. A similar quantity

of energy per year would be produced by approximately the same number of PV panels.

This preliminary study has noted that the SSI is varying from years to years. Such variations have an impact on the electricity production and hence on investment and should be taken into account.

The present article deals with the changes of SSI with months and years from 1998 up to 2013 and discusses possible causes.

2 Data

Measurements of global SSI and DNI were collected from Meteo-France for the stations of Koumac and Noumea (Fig. 1). Global SSI and DNI are measured at Koumac while Noumea provides global SSI only. The values delivered by Meteo-France are monthly irradiation, from 1998 to 2013. The clearness index K_T, also called atmospheric transmissivity or transmittance, is computed by dividing the observed monthly irradiation by the monthly irradiation received by a horizontal plane located at the top of atmosphere. K_T allows to disentangling the seasonal variations of the atmosphere transparency from those due to the orbit of the Earth around the Sun.

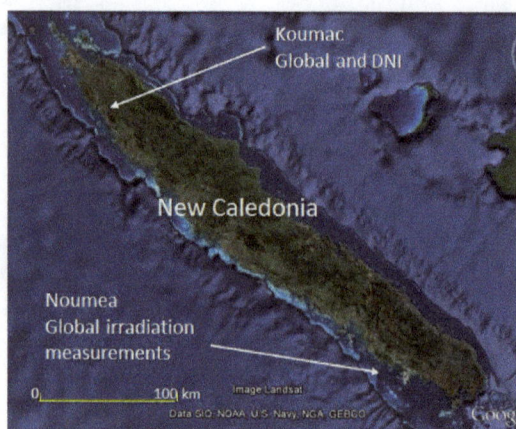

Figure 1. Map of New Caledonia. Location of Noumea and Koumac where Meteo-France measures solar radiation. Source: Google Earth.

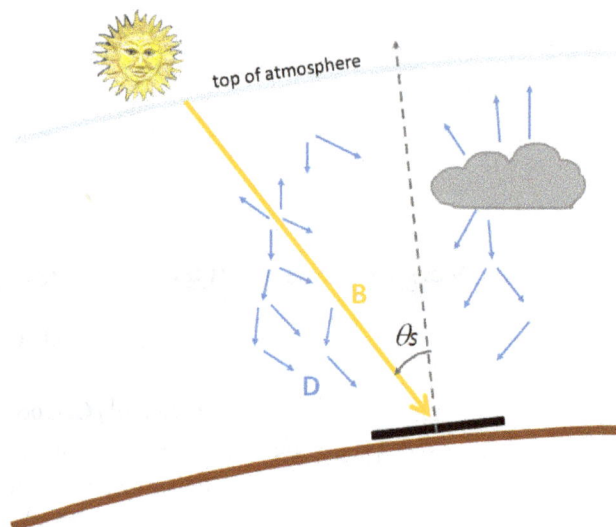

Figure 2. The direct component B is the radiation that comes from the direction of the sun, indicated by the solar zenith angle θ_S. Global = direct B + diffuse D. DNI is the direct for normal incidence.

Re-analyses are another means to assess the SSI. They offer worldwide and multi-decadal coverage. Several authors (see a review in Boilley and Wald, 2015) have compared SSI estimated by re-analyses to coincident ground measurements. They found a tendency of a majority of reanalyses to overestimate the SSI. Reanalyses often predict clear sky conditions while actual conditions are cloudy. This overestimation of occurrence of clear sky conditions leads to an overestimation of the SSI. The opposite is also true though less pronounced: actual clear sky conditions are predicted as cloudy. For these reasons, reanalyses are not used in this study.

Proper processing of satellite images is another means to assess the SSI. Several databases offer daily irradiation over the area under study, such as the NASA SSE data set available at http://eosweb.larc.nasa.gov/sse/. However, none of these data sets can be used here because they are not appropriate for this study like NASA SSE which is limited to the period 1983–2005, or they are not available freely such as the data sets from the 3Tier or GeoModel Solar companies.

The European-funded projects MACC (Monitoring Atmosphere Composition and Climate) have created a set of global aerosol properties together with physically consistent total column content in water vapour and ozone (Kaiser et al., 2012). These data are available on a grid of approximately 100 km every 3 h. Once collected from the MACC web site (http://www.gmes-atmosphere.eu), these data were interpolated in space and aggregated to yield monthly means of aerosol optical depth at 550 nm, and total column contents in water vapour and ozone for Koumac and Noumea, from 2004 to 2013. The McClear model (Lefèvre et al., 2013) uses the MACC data sets as inputs and provide monthly irradiation of global SSI and DNI that should be observed if the sky were clear i.e. cloud-free, from 2004 to 2013. These quantities are called clear-sky SSI and DNI.

As discussed above, cloud coverage from meteorological analyses is not very accurate. Since New Caledonia is

an island of limited extension, it was thought that cloud coverage may be provided by the International Comprehensive Ocean-Atmosphere Data Set (ICOADS). ICOADS is a global ocean marine meteorological and surface ocean dataset formed by merging many national and international data sources that contain measurements and visual observations from ships (merchant, navy, research), moored and drifting buoys, coastal stations, and other marine platforms. Cloud coverage is available in okta on a grid of 2° in size. The KNMI Climate Explorer (climexp.knmi.nl) offers access to these data and processing capabilities. Cloud coverage has been collected from this site as monthly values averaged over the oceanic region in the vicinity of New Caledonia defined by latitude ranging from −22 to −18°, and longitude from 156 to 164°.

The Pacific Ocean is subject to the El Niño Southern Oscillation (ENSO) which is characterized by variations in sea surface temperature in the eastern tropical part and in air surface pressure and cloudiness in the western tropical part. El Niño is the warm oceanic phase in the eastern Pacific and is coupled with high surface air pressure in the western part. The cold oceanic phase, La Niña, in the eastern Pacific accompanies low surface air pressure in the western part. New Caledonia is concerned by the ENSO. The Oceanic Niño Index (ONI) is one of the indicators of the strength of the ENSO events. It is available each month as a running average over three months at http://www.cpc.ncep.noaa.gov/products/analysis_monitoring/ensostuff/ensoyears.shtml. Since ENSO has an effect on cloudiness, the ONI has been included in the study to investigate its relationship with irradiation.

Table 1. Correlation coefficient between SSI or K_T and other variables on a monthly basis.

	Monthly irradiation	Clearness index K_T
Clear sky SSI	0.94	0.41
Clear-sky K_T	0.64	0.23
Cloud cover	−0.29	−0.45
ONI	−0.03	0.04

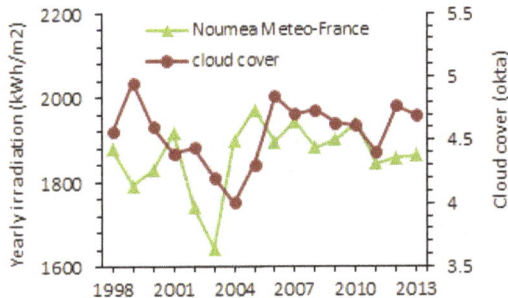

Figure 4. Yearly K_T observed at Noumea, clear-sky K_T from McClear multiplied by 0.75 and ICOADS-derived cloud cover from 2004 to 2013.

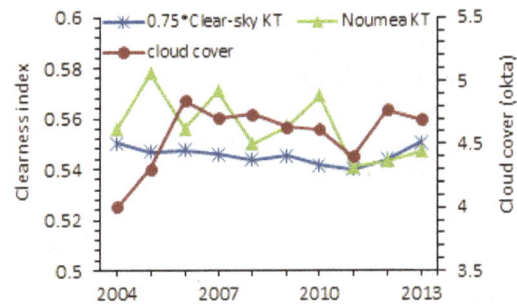

Figure 3. Yearly irradiation observed at Noumea and ICOADS-derived cloud cover from 1998 to 2013.

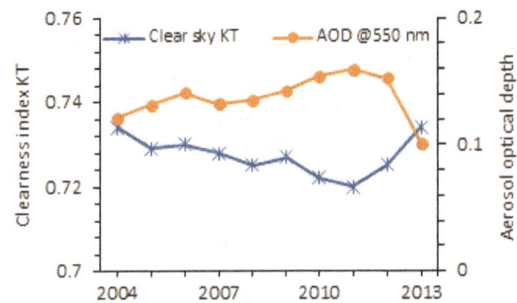

Figure 5. Yearly clear-sky K_T at Noumea from McClear and MACC aerosol optical depth at 550 nm from 2004 to 2013.

3   Results

Figure 3 exhibits the yearly SSI observed at Noumea from 1998 to 2013 as well as the cloud cover from ICOADS.

The SSI fluctuates from year to year and ranges between 1640 and 1970 kWh m^{-2}. The relative change from one year to the previous one ranges from −5 to 7 %. These values are similar to those observed in many places around the world.

One would expect an anti-correlation between the SSI and the cloud cover. A decrease of the SSI from one year to another should correspond to an increase of the cloud cover and vice-versa. This is well observed here for the whole period with a few exceptions: transition 2002–2003, 2004–2005, 2010–2011, and 2011–2012. One may observe that the magnitude of year-to-year change in SSI does not correspond always to the same magnitude of change in cloud cover. Other causes than cloud cover intervene in the change in SSI.

A very large correlation coefficient is found between the monthly SSI and the clear-sky SSI: 0.94 (Table 1). A smaller correlation was expected because of the influence of the cloud cover on the SSI. It is concluded that the seasonal changes on SSI due to changes in the Earth orbit during a year have a large influence on this correlation. This is illustrated by the much less correlation coefficient: 0.23, observed between K_T and the clear-sky K_T. This seasonal influence is seen also in the correlation coefficient between the SSI and the cloud cover which is small: −0.29. If the seasonal effects are removed by analysing K_T instead of the SSI, the correlation is more pronounced: −0.45. The larger the cloud cover,

the smaller K_T. Table 1 shows that there is no correlation with the ONI index.

To better illustrate the causes of changes in yearly SSI, Fig. 4 exhibits the yearly K_T observed at Noumea from 2004 to 2013, together with the clear-sky K_T and the cloud cover. The clear-sky K_T has been multiplied by 0.75 to match better K_T and help in the visual interpretation.

One may see that as a whole, K_T and the clear-sky K_T are following the same trend: decreasing from 2004 to 2011 and since then, a slight increase. The trend in K_T is subject to fluctuations that are mostly caused by cloud cover. Like in Fig. 3, an increase of the cloud cover yields a decrease in K_T and vice-versa, except for 2004–2005 and 2010–2011.

Figure 5 exhibits the yearly clear-sky K_T and the aerosol optical depth at 550 nm. Unsurprisingly, there is a clear link between these variables. As a whole, an increase in aerosol optical depth yields a decrease of the clear-sky K_T. The correlation coefficient is −0.89.

4   Discussion

Noumea experiences year-to-year fluctuation in SSI. The strong negative correlation coefficient: −0.45 between cloud cover and K_T clearly shows the preeminent role of the cloud cover on the SSI.

Cloud cover cannot be the sole cause of the changes in solar radiation. Another reason is an increase in aerosol load. Strong correlations are found between the SSI and K_T, and the clear-sky SSI and K_T. The role of aerosols is evidenced in Fig. 5 where a change in aerosol optical depth yields an opposite change in K_T and further in SSI. Besides their scattering and absorbing effects on solar radiation, aerosols may act as cloud condensation nuclei, thereby increasing cloud reflectivity and lifetime, hence the cloud cover effects on the SSI. The role of aerosols is far from simple; this complexity may contribute to explain the exceptions discussed in Fig. 3.

The relative change in SSI over a period of 10 years can be computed by taking the difference between the SSI of the last and first years, and dividing this difference by the mean value of the SSI over the 10 years (Mueller et al., 2014). This operation cannot be performed in our case with reliable results because the period of measurements is too short: 17 years only. Figure 3 shows that the yearly SSI in 2003 is much smaller than for the other years. This extreme value has an influence on most of the period since it will belong to all moving windows of 10 years, starting from 1998 till 2012. If this extreme is removed, the resulting relative changes would be very different.

Only a visual analysis can be performed. The yearly SSI is less variable after 2003. Since 2004, the SSI has decreased as a whole. The change C over a decade is quantified by calculating the difference between the mean SSI computed for each limit of the interval 2004–2013, i.e. 2004 and 2005, and 2012 and 2013, in the following way where I is the yearly SSI:

$$C = [(I(2012) + I(2013)) - (I(2004) + I(2005))]/2.$$

It is found that C is equal to $-75\,\mathrm{kWh\,m^{-2}}$, i.e. 4 % of the yearly SSI averaged over 2004–2013. When converted in irradiance, C is equal to $-9\,\mathrm{W\,m^{-2}}$. This value is similar to the typical change reported by Wild (2012) for India since 2000: $-10\,\mathrm{W\,m^{-2}}$.

5 Conclusions

This study demonstrates that New Caledonia experiences a decrease in surface solar irradiation since 2004. The decrease over the 10 years period: 2004–2013, is of order of 4 % of the mean yearly irradiation for this period. This change amounts to $-9\,\mathrm{W\,m^{-2}}$ and is very large with respect to climatic changes as reported by Wild (2012). The available data set for the study is limited and hence our conclusions are fairly questionable.

The study has demonstrated the preeminent role of the cloud cover on the decrease of the yearly SSI and has found that it cannot be the sole cause of the changes in solar radiation. Changes in aerosol optical depth play a role in changes in SSI though more moderate.

This study highlights the role of data sets offering a worldwide coverage in understanding changes in solar radiation and planning large solar energy plants. The ICOADS (International Comprehensive Ocean-Atmosphere Data Set) of the NOAA has helped in revealing the role of the cloud cover. The MACC (Monitoring Atmosphere Composition and Climate) data sets combined with the McClear model reveal changes in the optical depth of the aerosols that yield opposite changes in the irradiation under clear sky conditions.

Acknowledgements. The research leading to these results has received funding from the ADEME, research grant no. 1105C0028. The authors are grateful to the anonymous reviewers whose comments helped in the clarity of this article.

Edited by: S.-E. Gryning
Reviewed by: two anonymous referees

References

Boilley, A. and Wald, L.: Comparison between meteorological reanalyses from ERA-Interim and MERRA and measurements of daily solar irradiation at surface, Renew. Energy, 75, 135–143, doi:10.1016/j.renene.2014.09.042, 2015.

Kaiser, J. W., Peuch, V.-H., Benedetti, A., Boucher, O., Engelen, R. J., Holzer-Popp, T., Morcrette, J.-J., Wooster, M. J., and the MACC-II Management Board: The pre-operational GMES Atmospheric Service in MACC-II and its potential usage of Sentinel-3 observations, ESA Special Publication SP-708, Proceedings of the 3rd MERIS/(A)ATSR and OCLI-SLSTR (Sentinel-3) Preparatory Workshop, held in ESA-ESRIN, 15–19 October 2012, Frascati, Italy, 2012.

Lefèvre, M., Oumbe, A., Blanc, P., Espinar, B., Gschwind, B., Qu, Z., Wald, L., Schroedter-Homscheidt, M., Hoyer-Klick, C., Arola, A., Benedetti, A., Kaiser, J. W., and Morcrette, J.-J.: McClear: a new model estimating downwelling solar radiation at ground level in clear-sky conditions, Atmos. Meas. Tech., 6, 2403–2418, doi:10.5194/amt-6-2403-2013, 2013.

Mueller, B., Wild, M., Driesse, A., and Behrens, K.: Rethinking solar resource assessments in the context of global dimming and brightening, Solar Energy, 99, 272–282, doi:10.1016/j.solener.2013.11.013, 2014.

Wild, M.: Enlightening global dimming and brightening, B. Am. Meteorol. Soc., 93, 27–37, doi:10.1175/BAMS-D-11-00074.1, 2012.

10

Development of responses based on IPCC and "what-if?" IWRM scenarios

V. Giannini[1,2], L. Ceccato[1], C. Hutton[3], A. A. Allan[4], S. Kienberger[5], W.-A. Flügel[6], and C. Giupponi[1,2]

[1]Ca' Foscari University of Venice, Italy
[2]Fondazione Eni Enrico Mattei, Venice, Italy
[3]GeoData Institute, Southampton, UK
[4]Centre for Water Law, Policy and Science, University of Dundee, UK
[5]Centre for Geoinformatics, University of Salzburg, Salzburg, Austria
[6]Department of Geoinformatics, Hydrology and Modelling, Friedrich-Schiller University Jena, Germany

Abstract. This work illustrates the findings of a participatory research process aimed at identifying responses for sustainable water management in a climate change perspective, in two river basins in Europe and Asia. The chapter describes the methodology implemented through local participatory workshops, aimed at eliciting and evaluating possible responses to flood risk, which were then assessed with respect to the existing governance framework. Socio-economic vulnerability was also investigated developing an indicator, whose future trend was analysed with reference to IPCC scenarios. The main outcome of such activities consists in the identification of Integrated Water Resource Management Strategies (IWRMS) based upon the issues and preferences elicited from local experts. The mDSS decision support tool was used to facilitate transparent and robust management of the information collected and communication of the outputs.

1 Introduction

The BRAHMATWINN research project has planned a participatory process to integrate scientific and stakeholders' knowledge to deal with water management, climate change, and alpine mountain regions in Europe and Asia. Two parallel streams of research have been developed. On the one hand, research activities in the various disciplinary fields, such as climatology, hydrology, sociology, economics, and governance, relevant for integrated water resources management (IWRM) and the development of adaptation responses. On the other hand, a series of local workshops in the Upper Danube River Basin (UDRB) and the Upper Brahmaputra River Basin (UBRB), have been developed. The first outcome of this integrated and iterative process – the Integrated Indicator Table (IIT) – was described in Chapter 6.

Local actors' (LA) knowledge should be used in social and ecosystem management, in order to integrate scientific with local knowledge. Thus the participation of local actors can contribute significantly to the achievement of project outcomes that are better suited to fulfil local needs (de La Vega-Leinert et al., 2008), increasing the impacts of research efforts. Participatory processes enable sharing information between scientists and stakeholders, creating new opinions, addressing problems, combining expertise, in order to reach agreements and compromise solutions taking into account all interests at stake (Reed, 2008; Renn, 2006).

Besides the relevance given to public participation in IWRM, the necessity of utilizing also more effective tools to support decision making processes has emerged, giving more importance to information and communication technologies (ICT), such as Decision Support System (DSS) tools (Mysiak et al., 2005). In a DSS a conceptual model can be formalized through a joint effort integrating knowledge from disciplinary and local experts, bridging the gap between "hard science" and qualitative assessments (Sgobbi and Giupponi, 2007).

Future socio-economic vulnerability scenarios following the IPCC SRES projections A1, A2, B1, B2 (IPCC, 2000) for the time steps 2000, 2020 and 2050 have been modelled, which are based on the present day vulnerability modelling (Hutton et al., 2011; Kienberger et al., 2009b). The scenario modelling has been carried out in the Salzach River basin and in the Assam NE-India case studies, under the same scenario conditions and following a joint methodology. A condensed vulnerability index, consisting of proxy variables, has been identified and its indicators projected using a correlation with future GDP and population scenarios.

In this chapter we illustrate some of the methods and findings relative to the analysis of the effectiveness of the responses identified to cope with climate change. We present these as a methodological and operational proposal for the management of decision processes in a participatory context during the development of Integrated Water Resources Management (IWRM) options adapting to likely climate change impacts. The feasibility of these responses and strategies are then validated with reference both to the existing governance frameworks in place, and to projected future governance characteristics inferred from the IPCC SRES Scenarios (IPCC, 2000).

2 Role within the integrated project

The participative activities presented in this chapter, as well as those that were carried out in earlier phases of the project, made it possible to maintain open communication with local actors, allowing the project consortium to acquire local knowledge and orient research activities towards needs. They also provided a means of carrying out the twinning of the two river basins, shedding light on commonalities and distinct features. As far as the results of the two workshops discussed in this chapter are concerned, the phases of climate change scenarios presentation and brainstorming set the foundations for the DSS Design, and enabled the setting up of the activities on a common and shared framework, i.e. the features of each river basin. These phases also contributed to raise awareness about climate change dynamics, and to the state-of-the-art downscale modelling approaches. The phases of *DSS Design* and *Analysis of Options* carried out by means of the mDSS software raised great interest among the participants, who were thus involved in the project activities, exposed to preliminary results, and contributed to orient the final phases of the project. Several participants appreciated the use of public domain software in particular, which provided a perspective of possible reutilisation of the approach proposed in local decision problems.

3 Methods

3.1 The DSS Design and its implementation for the analysis of responses

The method applied for the evaluation of the responses to cope with flood risk is developed within the NetSyMoD framework designed for natural resources management in a participatory setting (refer to Fig. 1 in Chapter 6; Giupponi et al., 2008). NetSyMoD is based on the DPSIR causal framework (Driving forces, Pressures, State, Impacts, and Responses; EEA, 1999), which enables the organization of information, the structuring of issues, and the identification of solutions (i.e. Responses). The NetSyMoD approach is divided into six phases, two of these phases, DSS Design and Analysis of Options, were the object of the activities carried out at the two workshops discussed in this chapter. The DSS Design phase consists of system specification and development of software tools capable of managing the data required for informed and robust decisions. The Analysis of Options is performed with the mDSS software (Mulino DSS), a Decision Support System (DSS) tool providing capabilities for formalising, supporting and documenting the decision process and facilitating the adoption of Multi Criteria Decision Methods (MCDM) in a multi-actor context.

As an output of the implementation of the two steps mentioned above, substantial contributions to the design and evaluation of a set of alternative responses were obtained by means of group elicitation techniques and through the application of the DSS tool. The process for the identification of the IWRM strategies to be assessed, as described in Chapter 6, was based upon a series of workshops providing outputs that were organised in form of an Integrated Indicator Table (IIT). In the IIT (refer to Fig. 1) all the elements emerged from the interactions with stakeholders relevant for the identification of possible IWRM strategies and climate change adaptation were categorised as Responses (according to the DPSIR framework) and listed according to four broad categories:

1. ENG-LAND: Engineering Solutions and Land Management (response options would therefore include for example dam construction, river network maintenance, river training works, soil conservation practices, control of glacier lake outburst floods, forest management, renaturation, etc.);

2. GOV-INST: Investments in Governance and Institutional Strength (response options including accountability and transparency in government actions, enforcement of existing regulations, flood insurance, etc.);

3. KNOW-CAP: Knowledge Improvement and Capacity Building (response options including awareness raising activities, dissemination of scientific knowledge, strengthen traditional knowledge, training of public employees, environmental monitoring, etc.);

4. PLANNING: Solution based on planning instruments (response options would then include design and implementation of relief and rehabilitation plans, hazard zoning, disaster risk management, land-use planning etc.).

Two new workshops were organised, one for the UDRB in Salzburg, Austria (October 2008) and one for the UBRB in Kathmandu, Nepal (November 2008) to evaluate the relative effectiveness of the four Response categories. The workshops were divided into five phases.

1. First the goals of the workshop were defined, and then scenarios based on downscaled climate change model results were illustrated to introduce possible impacts of climate change at local level (see Dobler et al., 2011).

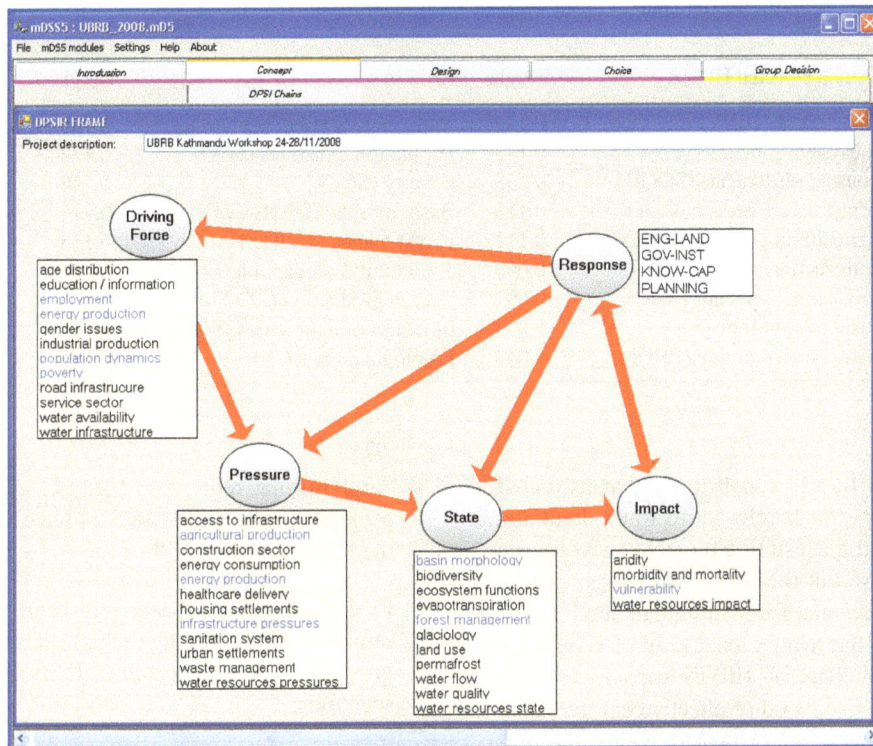

Figure 1. The conceptualisation of the information base stored in the IIT within the DPSIR framework (screenshot of the mDSS software).

2. The second component was a brainstorming session carried out to validate and specify the responses within the four categories that had been identified during previous workshops.

3. In the third phase participants selected the criteria for the evaluation of responses, attributing scores to the Sub-domains listed in the IIT.

4. In the fourth phase participants weighted the selected criteria.

5. In the fifth phase of the workshops the Analysis Matrix (AM) was created using criteria and responses. Participants compiled the Analysis Matrix to evaluate the potential effectiveness of each of the responses (columns) in coping with the issues expressed by the criteria (rows) applying a Likert scale ranging from 1 "very high effectiveness" to 5 "very low effectiveness".

All compiled AMs were imported into the mDSS software, for Multi-Criteria Analysis (MCA) and Group Decision Making (GDM), which enabled the evaluation of the relative effectiveness of alternative responses through MCA performed by decision rule ELECTRE III (Belton and Stewart, 2002). Following another possibility individual preferences were processed in the Group Decision Making component of mDSS using the Borda rule (de Borda, 1781).

3.2 Validation of response strategies

In order to validate the categorised possible response strategies identified by stakeholders (Sect. 3) against the relevant governance frameworks, an effort was made to compare these Responses to the governance and policy positions assessed in Chapter 4. The response strategies, at least in the short term, need to be seen in the context of these assessments because the governance and policy frameworks will have a strong bearing on the extent to which responses may be considered potentially successful or not (Hague Ministerial Declaration, 2000). The future scenarios into which this work has been incorporated are those proposed by the IPCC in 2000 (IPCC, 2000) in the Special Report Emission Scenarios (SRES) and their associated storylines (A1, A2, B1 and B2).

No projections as to the governance environment have been made in these storylines. In order therefore to evaluate the extent to which these preferred solutions would be practical over the period of the time slices identified in the project (up to 2080) and the time horizon envisaged by the IPCC scenarios, inferences were derived from the socio-economic and physical characteristics identified in the SRES storylines as regards the potential governance situation in 2100. This then allows an evaluation of those response strategies that seem most appropriate for the storylines based on the projected governance situations. This process was applied to the response strategies related to the Assamese context.

Table 1. Criteria selected by LAs from the Integrated Indicators Table.

Criteria selected UDRB	Weight	Criteria selected UBRB	Weight
Vulnerability (ENV)	0.144	Vulnerability (ENV)	0.145
Ecosystem functions (ENV)	0.143	Population dynamics (SOC)	0.132
Housing settlements (SOC)	0.138	Poverty (SOC)	0.125
Infrastructure pressures (SOC)	0.133	Basin morphology (ENV)	0.125
Agricultural production (ECON)	0.111	Forest management (ENV)	0.113
Construction sector (ECON)	0.099	Agricultural production (ECON)	0.103
Population dynamics (SOC)	0.097	Energy production (ECON)	0.101
Basin morphology (ENV)	0.091	Infrastructure pressures (SOC)	0.100
Energy consumption (ECON)	0.043	Employment (ECON)	0.056

The SRES storylines therefore had to be deconstructed to identify the particular strands relevant to water, land and disaster management and the resulting projected governance frameworks used to flesh-out the SRES storylines. These strands include the (i) potential for institutional and international co-operation; (ii) the relative balancing of economic, social and environmental concerns; (iii) the capacity for land use control; and (iv) the likelihood of effective enforcement. Each response strategy was then evaluated against the projected governance strengths and weaknesses derived from the SRES storylines, and against the legal and institutional reality in the relevant basin state.

3.3   Vulnerability scenarios

The methodology for the vulnerability scenarios comprises the following key-steps and has been carried out in the same way in the European and Asian case studies:

- Construction of a correlation analysis between the vulnerability score and the individual indicators and selection of five key variables that are highly correlated with the vulnerability score.

- With the key variables a multivariate regression analysis has been performed to identify the predictors of the level of vulnerability. Within the Salzach River basin case study two methodologies have been tested, a regression analysis identifying single predictors for the whole case study area (ordinary least squares method), and a geographically weighted regression which identifies for each location (in our case grid cells) individual, location-based predictors.

- A parallel step involves the correlation between past GDP and population data and past data of the key variables to identify their existing relationship.

- Taking the future GDP and population projections under the four SRES scenarios in consideration, values for the future key variables under the four scenarios have been calculated.

- Applying the regression formula identified in step 2, projected vulnerability indices have been calculated for the four scenarios for the time steps 2020 and 2050.

- In a final step the data has been normalised (scale range 0–100) according to the values of 2000 to identify growth and decline of vulnerability among the different scenarios.

- Visualisation and map production was the last step elaborated in this procedure.

4   Results and deliverables provided

4.1   Local actors evaluation of responses

Local actors (LAs) identified the three most important criteria for each of the three dimensions, economic, environmental and social, i.e. the three pillars of sustainable development, converging in both basins on the same five criteria out of nine, choosing from a set of 40 criteria listed in the IIT (15 social criteria, 17 environmental criteria, and 8 economic criteria) (Ceccato et al., 2010).

LAs then expressed the relative importance of every criterion, which will be used to rank the alternative IWRM responses. On average, in both river basins, the highest weight was given to the "Vulnerability" criterion (Environmental pillar) (see Table 1). Five out of nine criteria selected were common to both basins: Vulnerability, Population dynamics, Infrastructure pressure, Basin morphology, and Agricultural production. The elaboration of the average Analysis Matrix (AM) shown in Table 2 illustrates that no category of response prevails. All the average responses (listed in columns) are in a range between "very high effectiveness" and "medium effectiveness". We can, thus, say that all the responses are considered to be potentially effective to cope with flood risk.

Last but not least, the relative ranking of the alternative responses was carried out by performing Multi Criteria Analysis and Group Decision Making. The application of

Table 2. Analysis Matrices: average values of LAs' evaluations on the potential effectiveness of each response in coping with the issues expressed by the criteria (rows) by means of a Likert scale ranging from 1 "Very high effectiveness" to 5 "Very low effectiveness".

Analysis Matrix UDRB	PLANNING	KNOW-CAP	GOV-INST	ENG-LAND
Vulnerability (ENV)	2.33	2.67	2.50	2.67
Ecosystem functions (ENV)	2.86	2.43	2.29	3.43
Housing settlements (SOC)	2.00	2.43	2.57	2.71
Infrastructure pressures (SOC)	2.43	2.14	2.57	2.00
Agricultural production (ECON)	2.86	3.14	2.71	2.57
Construction sector (ECON)	2.14	3.29	2.57	2.43
Population dynamics (SOC)	2.86	3.00	2.29	3.29
Basin morphology (ENV)	2.71	2.57	3.43	3.29
Energy consumption (ECON)	2.86	2.43	2.57	2.86
Average	2.56	2.68	2.61	2.80

Analysis Matrix UBRB	PLANNING	KNOW-CAP	GOV-INST	ENG-LAND
Vulnerability (ENV)	1.71	2.43	2.24	1.95
Population dynamics (SOC)	1.76	2.52	2.33	3.19
Poverty (SOC)	2.43	2.62	2.00	3.33
Basin morphology (ENV)	2.38	2.67	3.10	2.43
Forest management (ENV)	1.86	2.10	2.10	1.95
Agricultural production (ECON)	2.15	2.50	2.48	2.29
Energy production (ECON)	2.19	3.00	2.43	2.10
Infrastructure pressures (SOC)	2.00	2.86	2.67	2.19
Employment (ECON)	2.43	2.57	2.43	3.52
Average	2.10	2.58	2.42	2.55

ELECTRE III (Fig. 2) shows that LAs of both river basins evaluated the PLANNING solution as the most promising one. Using the Group Decision Making (GDM) tool of mDSS, considering the Borda mark the PLANNING category is also the preferred solution (Ceccato et al., 2010). The comparison of these independent results confirmed that PLANNING instruments are the most promising responses in terms of effectiveness to cope with flood risk under climate change impacts. We recognise, therefore, that very similar results were recorded in the two river basins, confirming that, notwithstanding the differences in their environmental and socio-economic conditions, the areas present certain similarities not only regarding the problems to address, but also regarding the expectations of possible solutions.

From the governance perspective, the comparison of the responses against the legal and institutional frameworks in 2007 and over the scenario time periods revealed that the B1 storyline fitted best with the responses put forward by LAs in response to the local issues, with A2 being least appropriate (Table 3). Based on the number of strategies which were best suited for each storyline, B1 again came out as the winner, being most associated with ten strategies, but this time B2 clearly emerged as the worst, being best associated with only one. In Assam, however, the local governance context is currently strongest in relation to strategies that enforce an A1 scenario, suggesting there is a mismatch between what stakeholders believe are the policy and strategic approaches that should be taken in order to alleviate vulnerability on the one hand, and the approach taken, at least in the short to medium term, by government and regulatory authorities.

4.2 Implementation for the Salzach River basin case study

In the modelling of socio-economic vulnerability in the Salzach River catchment 52 indicators have been identified describing various domains of vulnerability (see Kienberger et al., 2009a). Through the application of spatial correlation those indicators have been selected which have a higher correlation value than 0.5. Out of the 14 remaining indicators those have been selected which show a compromise with a high correlation value (>0.7) and a significant number of correlating indicators. The following five key indicators have been identified:

– Number of houses with 1 or 2 households (×1)

– Number of industrial buildings (×2)

– Number of labours in agriculture (×3)

– Number of academics (×4)

– Number of male full-time employees (×5)

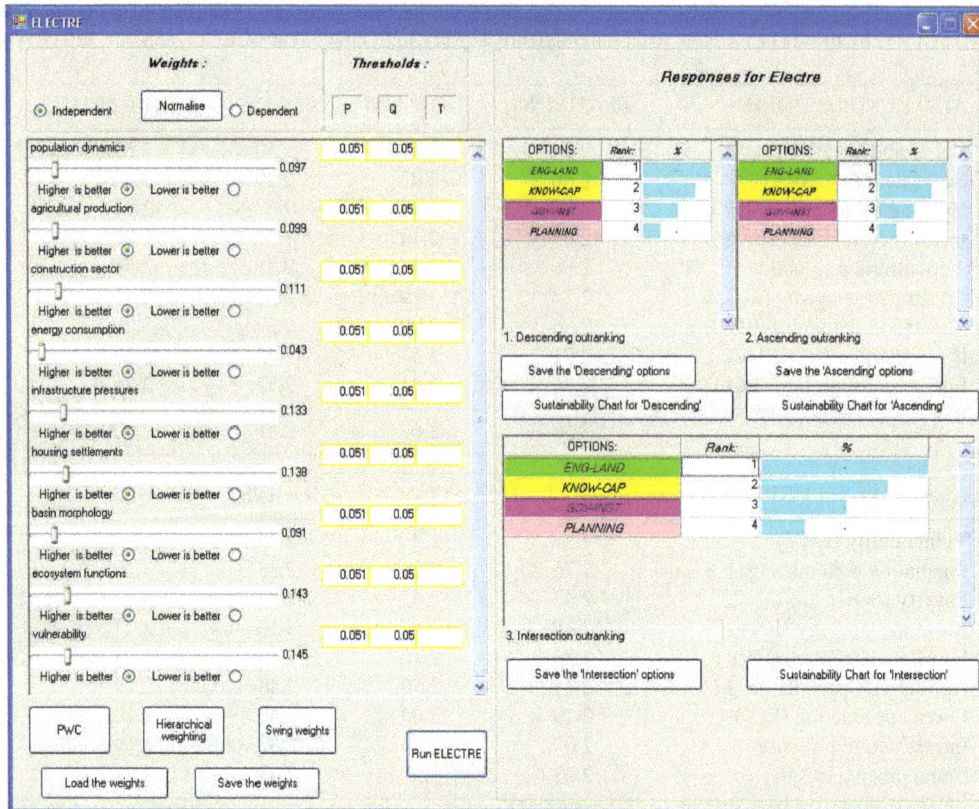

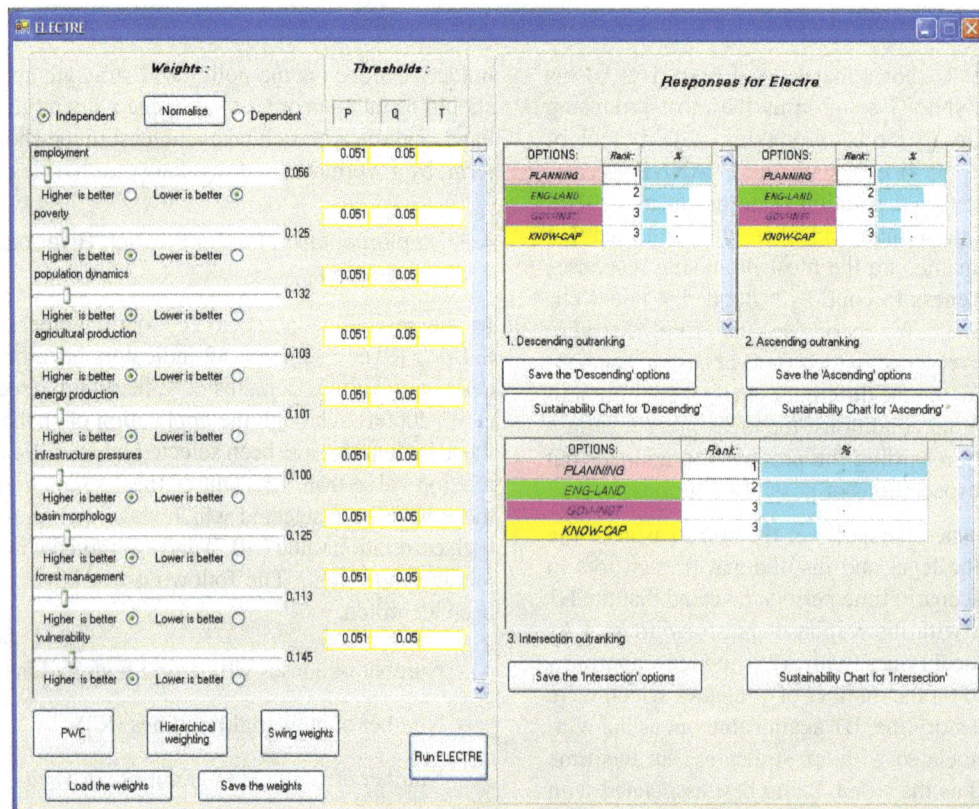

Figure 2. UDRB (top) and UBRB (bottom): ELECTRE III Analysis of alternative Responses. On the left side we can see the applied criteria weights and thresholds, while on the right side the ELECTRE III window appears with the final ranking (screenshot of the mDSS software).

Table 3. Evaluation of suitability of Assam response strategies against projected governance characteristics of SRES Scenarios.

Issue	Response strategy	A1	A2	B1	B2	Time Slice
Awareness of population on risks, conservation and WRM	Increase awareness of the population on risks, conservation and WRM	2	1	4	3	2001–2020
Integration of research in decision-making	Integration and coordination among different sectors of research and decision making	3	2	4	1	2001–2020
Community involvement in decision making	Improve community involvement and foster participatory processes for decision making	1	3	2	4	2001–2020
	Foster livelihood practices based on conservation, rehabilitation and sustainability	2	1	4	3	2020–2050
Early warning system	Early warning system	4	1	3	2	2001–2020
	Disaster risk management	4	1	3	2	2001–2020
	Hazard zonation	3	1	4	2	2001–2020
IWRM	Design and implement IWRM plans	3	1	4	2	2001–2020
Long term vision and measures vs. Short term engineering solutions	Multi-purpose dam construction	4	2	3	1	2020–2050
	Flood and erosion control	3	4	2	1	2020–2050
	Land use planning	2	1	4	3	2001–2020
	Environmental impact assessment for new dams	3	1	4	2	2001–2020
Relief and rehabilitation	Design and implement relief and rehabilitation plans	3	1	4	2	2001–2020
	Soil conservation efforts	1	4	2	3	2001–2020
	Renaturation	1	4	2	3	2020–2050
Policy making and implementation of laws	Accountability and transparency in government actions	3	2	3	2	2020–2050
	Implement and enforce existing laws and design new and more effective laws	3	2	3	2	2001–2020
Coordination among institutions	Resolve conflicts and strengthen coordination among institutions	3	2	4	1	2001–2020
Inter-state conflict, cross boundary issues	Inter-state coordination and conflict resolution	4	2	3	1	2020–2050
	Totals	**52**	**36**	**62**	**41**	

It is interesting to note that these indicators have gained high LA weights in the sub-domain ranking. So therefore the indictors do not only represent from the statistical point of view an appropriate selection but reflect some of the highest ranked indicators. However, it is important to note that the weights have not been considered in the correlation analysis.

Concluding the steps outlined above, in the final results it can be observed that the general pattern among the different scenarios shows a similar distribution, with slight changes in its peak values. First of all it is important to note, that some areas show a vulnerability value of zero. This is due to the fact that the selected five key indicators derive from the census data. This means that only vulnerability values higher than zero exist in those areas where population is present.

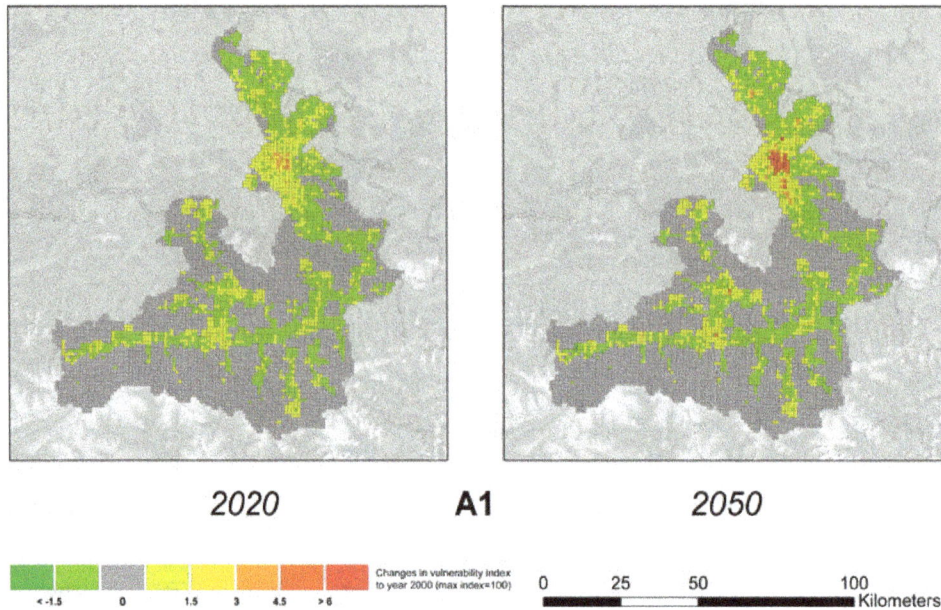

Figure 3. Change of vulnerability value within the A1 scenario to the reference year 2000.

This is an interesting fact following the general discussion on vulnerability, where it is argued that vulnerability only exists where humans are affected. In the vulnerability analysis carried out in Chapter 4 it was also assumed that vulnerability can exist in general everywhere and is also constituted by land use assets. However, for the purpose of specific vulnerability scenarios, which still carry some uncertainty itself, this is a valid approach to follow.

The highest vulnerability values can be observed around the city of Salzburg, which of course is the most densely populated area in the case study, where the indicators of houses with one to two households, male full-time employees, academics and industrial buildings show high values. Additionally to that rural areas with a high proportion with single houses and a higher number of labours in the agricultural sector have slightly higher values than surrounding areas. This concentrates in the case study around important central towns and villages. The highest vulnerability scores can be observed within the scenarios A1 and B1, whereas the absolutely highest vulnerability score (113.79) can be observed in the A1 scenario (see Fig. 3), followed by B1 with 111.65. Lower values show the 2-group scenarios A2 and B2 which have a more regional oriented focus than the globalised 1-group scenarios. The lowest vulnerability score is represented through the B2 scenario with a value of 107. The A2 scenario has a maximum value of 109.39 for 2050.

Examining the change rates among the different scenarios for 2020 and 2050 in reference to the baseline year 2000 a general increase in and around the city of Salzburg can be observed (which show the highest increase in vulnerability) and other areas with higher values of the key indicators. Those areas show all an increase in vulnerability, whereas the urban agglomeration has the highest values for 2050 within the scenarios A1 and B1. A decrease in vulnerability can be observed in strongly rural dominated areas. This is also due to the fact that regression analysis shows negative trends for the indicators of labours in the agricultural sector and interestingly in the number of male fulltime employees.

A similar picture as described above applies for the maximum change rates within the different scenarios for the time span between 2000 and 2050. Highest maximum increases show the A1 (+13.79%), B1 (+11.65%) followed by A2 (+9.39%) and B2 (+7%) scenario. A significant decrease in vulnerability can be observed in the A1 scenario with a maximum decrease value of −2.24%. Therefore it can be observed that the A1 scenario shows a larger dispersion of its value range than the other scenarios. Maximum decrease values are followed by B1 (−1.89%), A1 (−1.52%) and A2 (−1.14%). The A2 scenario shows both, low increase but also low decrease values. However, in general it can be observed that the mean value of change rates is ~0%. It shows (Fig. 4) that most of the raster cells do have a low decrease or increase and that the majority of units decrease. This also follows the observation in the change maps, where the increase is limited to highly urbanised areas, which only occupy a small area.

It can be summarised that urban and central villages in rural areas show a significant increase in vulnerability. However, from a spatial point of view most areas show a decrease in vulnerability, which are mostly less asset driven because of its rural characteristic. The methodology applied gives an overall estimation of vulnerability, but as those estimations inherit an unspecified high uncertainty they might only be applied to identify general future trends.

Table 4. Scenarios are based upon the work carried out by TERI (India) showing the projected GDP and Population outcomes for all India based upon the SRES scenarios (GDP ×1013 Rupees, Population in millions).

	A1		A2		B1		B2	
	GDP	Pop	GDP	Pop	GDP	Pop	GDP	Pop
1990	0.886	846	0.886	846	0.886	846	0.886	846
2020	8.924	1.291	5.094	1.102	5.866	1.228	3.833	1.012
2050	33.426	1.572	14.298	1.646	19.027	1.298	9.304	1.646

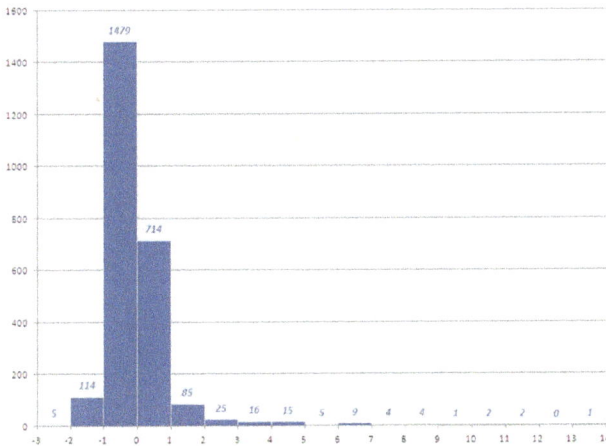

Figure 4. Histogram of change rates for the A1 scenario between 2000 and 2050.

4.3 Implementation for the Assam case study

The aim of the Assam component of the study is to investigate how different scenarios of socioeconomic development will mitigate the impact of climate change in the Assam test site. The basis of this analysis is adopted from socioeconomic scenarios developed for India by TERI (TERI, 2006), an independent not-for-profit research institution in India (Table 4). The scenarios are developed on the basis of the Intergovernmental Panel on Climate Change (IPCC) Special Report on Emissions Scenarios (SRES) scenarios. The conceptualization of the scenarios is based on two dimensions of policy directions and social values. The framework focuses on where policy direction is either inward-looking or globally integrated and where social values focus on economic growth or more localised social values and environmental consequences. The combinations results of socioeconomic scenarios can be used to investigate future impacts of climate change vulnerability which relates primary to governments focus and priorities. Scenarios A2 and B2 reflect a more inward-looking policy, while scenarios A1 and B1 reflect stronger integration with the global community for regulation and economic growth.

TERI used six factors (changes in population growth, GDP projections, food grain demand, demand for water, demand for electricity and demand for wood) to investigate how the four scenarios could impact on socioeconomic vulnerability to climate change for India for the 1990s and projected to the 2020s and 2050s (Table 4). The values are adopted for this study to investigate how government priorities could impact on socio-economic vulnerability in the Assam Study Area under the four scenarios.

In Chapter 4 estimates of socioeconomic vulnerability in 2001 (and also for specific domains of sensitivity and adaptive capacity) to climate hazards (e.g. floods, droughts, bank erosion) were derived for communities and Tehsils in the Assam Study Area. The scores were exponentially scaled such that they range between 0.001 and 100 (the higher the score the higher the level of vulnerability), with emphases on the tail of the distribution to identify the most vulnerable communities. Figure 5 shows vulnerability quintiles for the Assam study area in 2001 which is the last time a clear picture of vulnerability based upon the census and Landsat imagery is available (see Chapter 4).

In this follow-up study to the work presented in Chapter 4, the main aim is to investigate by how much the level of the estimated vulnerability for each Tehsil will increase or decrease depending on governments policy directions and social values under the four scenarios developed by TERI. To do this, we first identify all the individual variables that are highly correlated (> ±0.5) with the vulnerability score. In all 18 individual variables were identified to have a high correlation with the overall vulnerability score. Tehsils with high engagement in subsistence agriculture and poor housing materials are more likely to be vulnerable. Ownership of assets such as television, telephone, scooter, motor, cycle or moped is negatively correlated with vulnerability scores. It is interesting to note that Tehsils with high dependency of forest ecosystems e.g. using firewood for cooking are more likely to be vulnerable compared to those who use LPG for cooking. A multivariate (regression) analysis is then used to identify the predictors of level of the vulnerability. To satisfy the assumptions of normality and constant variance, the vulnerability scores were log transformed. It is important to note that there was a high level of collinearity between some of the

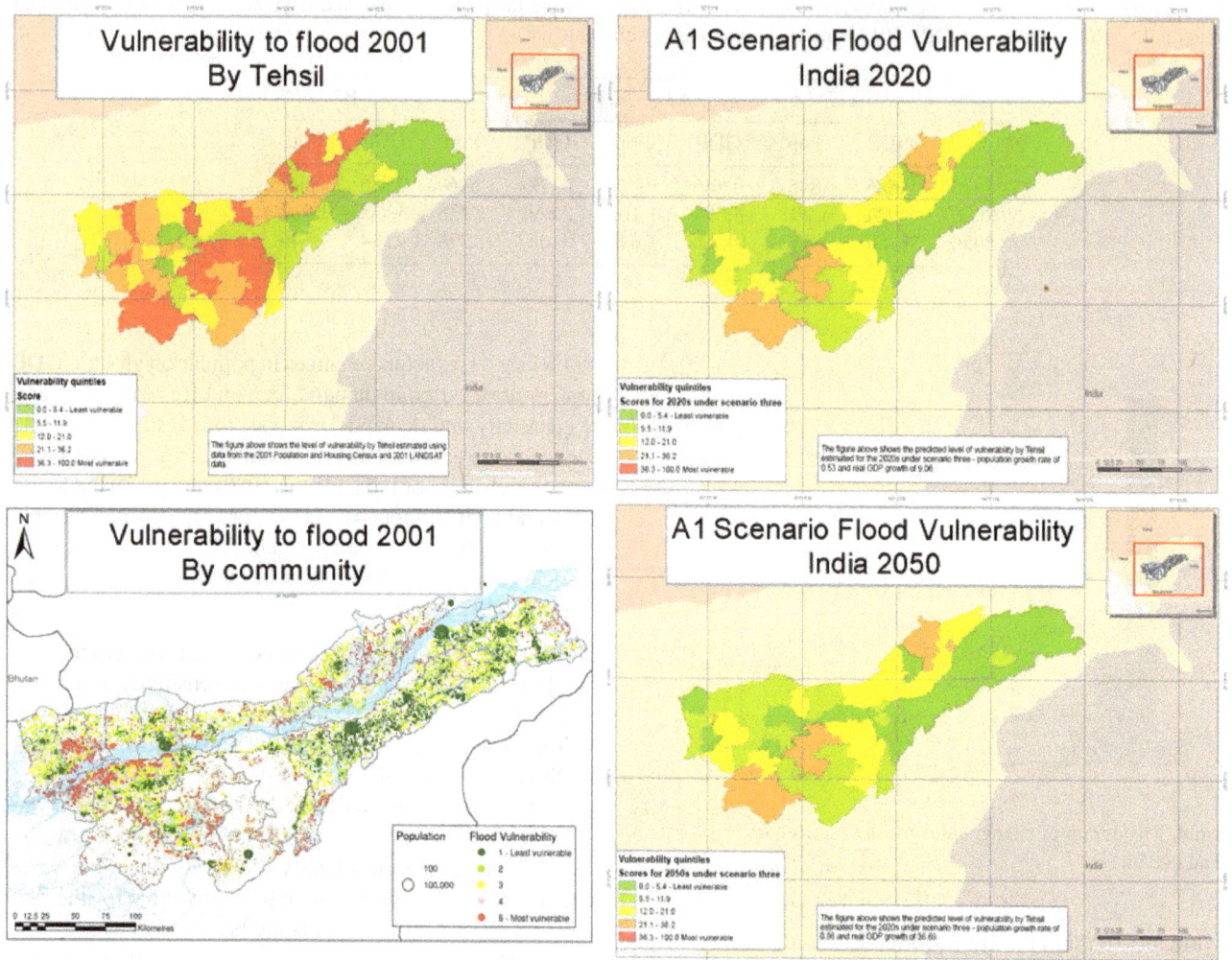

Figure 5. Estimates of vulnerability based upon know variables for 2001 at both a community and Tehsil level. The estimates of vulnerability to flood in 2020 and 2050 under SRES A1 (example) utilise TERI estimates of GDP and poulation and model the impact of these estimates on specific indicators of overall vulnerability.

variables. Where two or more variables were collinear, only the strongest predictor was included in the model. The estimated Adjusted R-square indicates that the five significant indicators explain 91.7% of the variability in vulnerability scores.

5   Contribution to sustainable IWRM

The results of BRAHMATWINN show that the implementation of NetSyMoD is useful for developing responses which are then evaluated as effective. The development of responses is, in fact, based on an iterative process which integrates knowledge coming from different disciplines and local actors. The two parallel participatory processes, on the one hand, allowed the understanding of the visions and preferences of LAs regarding the sustainable management of water resources. On the other hand, highlighted that the infor-

mation and tools proposed by the researchers was adequate to address local actors (e.g. decision makers and end-users) needs.

The methodology used enabled to frame the issues in a coherent manner and, thus, to focus the discussion. This, in a subsequent phase of the project, led to further refinements of the responses to cope with flood risk.

This result validates the motivations which triggered the BRAHMATWINN project design and led to develop a twinning river basin research approach, characterised by a strictly coordinated and combined series of participatory activities in the two twinning basins.

6   Conclusions and recommendations

The experimental application of the NetSyMoD approach to the twinned river basins provided the BRAHMATWINN project with an effective interface between the research activities and potential beneficiaries, in the case studies located in Asia and Europe.

The participative activities presented in this chapter made it possible to maintain an open communication interface with LAs, allowing the BRAHMATWINN researchers to learn from them and orient research activities. The phase of DSS design that was carried out by means of the mDSS software was followed and understood by local actors, who were able to influence the development of the following project phases.

From the perspective of assessing the potential effectiveness and feasibility of response strategies, the mismatch between what stakeholders believe are the policy and strategic approaches that should be taken in order to alleviate vulnerability on the one hand, and the approach taken, at least in the short to medium term, by government and regulatory authorities must raise questions as to the quality of the involvement of stakeholders in decision-making processes in Assam.

The vulnerability analysis shows that GDP and population growth impacts on household and community factors that predict socioeconomic vulnerability to climate hazards, such as the proportion of the population working in agriculture, proportion of roads that are metalled, proportion of households with a television, proportion of houses with burnt brick wall and proportion of households using firewood for cooking. The impact of GDP and population growth is highest in areas where levels of vulnerability are already high. The results depict that a slow growth in population with a concurrent rapid growth in GDP is important in reducing levels of vulnerability.

Acknowledgements. We would like to acknowledge the following BRAHMATWINN research partners for helping organize the workshops: Institute for Atmospheric and Environmental Sciences, Goethe University Frankfurt (Germany); Department for Geography, Ludwig-Maximilians University, Munich, Germany; ICIMOD, Kathmandu, Nepal; Royal University of Bhutan, Thimphu (Bhutan); Indian Institute of Technology Roorkee, Roorkee (India).

The interdisciplinary BRAHMATWINN EC-project carried out between 2006–2009 by European and Asian research teams in the UDRB and in the UBRB enhanced capacities and supported the implementation of sustainable Integrated Land and Water Resources Management (ILWRM).

References

Belton, V. and Stewart, T. J.: Multiple criteria decision analysis, Kluwer Academic Publishers, Boston, 2002.

Ceccato, L., Giannini, V., and Giupponi, C.: A participatory approach to assess the effectiveness of responses to cope with flood risk, FEEM Working Paper, 28, 2010.

de Borda, J.-C.: Mathematical derivation of an election system, Isis, 44, 42–51, 1781 (English translation by A. de Grazia, 1953).

de la Vega-Leinert, A., Schröter, D., Leemans, R., Fritsch, U., and Pluimers, J.: A stakeholder dialogue on European vulnerability, Reg. Environ. Change, 8, 3, 109–124, 2008.

Dobler, A., Yaoming, M., Sharma, N., Kienberger, S., and Ahrens, B.: Regional climate projections in two alpine river basins: Upper Danube and Upper Brahmaputra, Adv. Sci. Res., this special volume, 2011.

EEA: Environmental Indicators: typology and overview, edited by: European Environment Agency (EEA), Technical report n. 25, available at: http://reports.eea.eu.int/TEC25/en/tab_content_RLR (last access: March 2011), Copenhagen, 1999.

Giupponi, C., Sgobbi, A., Mysiak, J., Camera, R., and Fassio, A.: NetSyMoD – An Integrated Approach for Water Resources Management, in: Integrated Water Management, edited by: Meire, P., Coenen, M., Lombardo, C., Robba, M., and Sacile, R., Springer, Netherlands, 69–93, 2008.

Hutton, C. W., Kienberger, S., Amoako Johnson, F., Allan, A., Giannini, V., and Allen, R.: Vulnerability to Climate Change: People, Place and Exposure to Hazard, Adv. Sci. Res., this special volume, 2011.

IPCC: Emission Scenarios, edited by: Intergovernmental Panel on Climate Change (IPCC), Cambridge University Press, 2000.

Kienberger, S., Lang, S., and Zeil, P.: Spatial vulnerability units – expert-based spatial modelling of socio-economic vulnerability in the Salzach catchment, Austria, Nat. Hazards Earth Syst. Sci., 9, 767–778, doi:10.5194/nhess-9-767-2009, 2009a.

Kienberger, S., Amoako Johnson, F., Zeil, P., Hutton, C., Lang, S., and Clark, M.: Modelling socio-economic vulnerability to floods: Comparison of methods developed for European and Asian case studies. Sustainable Development – a Challenge for European Research, Brussels, 2009b.

Ministerial Declaration of The Hague on Water Security in the 21st Century, 22 March 2000.

Mysiak, J., Giupponi, C., and Rosato, P.: Towards the development of a decision support system for water resource management, Environ. Modell. Softw., 20, 2, 203–214, 2005.

Reed, M.: Stakeholder participation for environmental management: A literature review, Biol. Conserv., 141, 2417–2431, 2008.

Renn, O.: Participatory processes for designing environmental policies, Land Use Policy, 23, 1, 34–43, 2006.

Sgobbi, A. and Giupponi, C.: Models and decision support systems for participatory decision making in integrated water resource management, in: CIHEAM-IAMB, Water saving in Mediterranean agriculture and future research needs, Options Méditerranéennes: Série B. Etudes et Recherches 56, Bari, Italy, 259–271, 2007.

TERI: Socio-economic Scenarios for Climate Change Impacts in India, Key sheet 3, DEFRA publication (UK), 2006.

11

A Non-Linear Mixed Spectral Finite-Difference 3-D model for planetary boundary-layer flow over complex terrain

W. Weng and P. A. Taylor

Department of Earth and Space Science and Engineering, York University, 4700 Keele Street, Toronto, Ontario, M3J 1P3, Canada

Abstract. The Non-Linear Mixed Spectral Finite-Difference (NLMSFD) model for surface boundary-layer flow over complex terrain has been extended to planetary boundary-layer flow over topography. Comparisons are made between this new version and the surface layer model. The model is also applied to simulate an Askervein experimental case. The results are discussed and compared with the observed field data.

1 Introduction

The Mixed Spectral Finite-Difference (MSFD) model was originally developed by Beljaars et al. (1987). It is based on the idea that the topography produces a perturbation to a steady, neutrally stratified, non-evolving flow over horizontally homogeneous flat terrain. A number of efforts have been made to improve the model calculation of the turbulent boundary-layer flow over complex terrain. Ayotte et al. (1994) evaluated the model predictions with a number of different closure schemes which range from the simple first-order $\kappa - Z$ closure to the full second-order closure. Xu and Taylor (1992) and Xu et al. (1994) made the non-linear extension of the model by including all the neglected terms. In the non-linear version of the MSFD model (NLMSFD), the model equations were solved iteratively. Another model improvement is the extension to the stable boundary layer (MSFD-STAB, see Weng et al., 1997).

Although the MSFD and NLMSFD models have been improved since the late 80's, these models can only formally apply to the surface-layer flow due to the model assumption that upwind or zero-order profiles of mean and turbulent variables are simple logarithmic surface-layer profiles, e.g., wind speed is logarithmic, shear stresses are constant and the effect of Coriolis force is absent. Ayotte and Taylor (1995) made the first effort to extend the model to the planetary boundary-layer flow with the full second-order turbulence closure (MSFD-PBL) but it was still a linear model.

2 The model

2.1 The model equations

For the PBL boundary-layer flow, the model uses the Reynolds-averaged equations for steady-state, neutrally stratified incompressible flow. They are, in tensor notation including use of the summation convention,

$$U_j \frac{\partial U_i}{\partial x_j} = -\frac{1}{\rho}\frac{\partial p}{\partial x_i} + f\epsilon_{ij3}\left(U_j - U_{gj}\right) - \frac{\partial \langle u_i u_j \rangle}{\partial x_j}, \qquad (1)$$

$$\frac{\partial U_i}{\partial x_i} = 0, \qquad (2)$$

where U_i and u_i are the i-th component of the mean and turbulent flow respectively; f is the Coriolis parameter; ϵ_{ij3} is the alternating unit tensor; and angle bracket ($\langle \; \rangle$) denotes an ensemble mean. The pressure gradient force consists of a nonhydrostatic mesoscale pressure component, P, and a synoptic-scale component, $f\epsilon_{ij3}U_{gj}$, where U_{gj} is the j-th component of the geostrophic wind.

To close the system of the equations, a turbulent closure scheme is needed. Weng and Taylor (2003) have shown that the so-called simple $E - \ell$ turbulence closure scheme performs quite well in modelling the PBL flow in most atmospheric conditions compared with more sophisticated schemes. The $E - \ell$ turbulence closure is a $1\frac{1}{2}$-order scheme in which the prognostic equations for the turbulent kinetic energy (E) and a diagnostic equation for the turbulent length scale (ℓ) are used. The turbulent fluxes are locally related to mean vertical gradients and an eddy diffusivity, see Weng and Taylor (2003) for details.

2.2 Numerical scheme and boundary conditions

As with the previous versions of MSFD/NLMSFD model, the model equations are transformed from the original standard right-hand coordinate system (x,y,z) with z in the vertical direction to a new system (X,Y,Z), by using a terrain-following coordinate transform. In addition, to ensure sufficient resolution near the surface and to resolve strong gradients, a log-linear coordinate transform is further used for the vertical coordinate Z.

Fourier transformation is performed in the horizontal directions to the coordinate transformed governing equations. The model variables are decomposed into an unperturbed or zero-order part, independent of x and y, corresponding to equilibrium flow over uniform flat terrain and a perturbation part due to topographic forcing. Collecting the first-order perturbation terms and solving the resulting system of equations, forms the linear version of the model (MSFD-PBL). Treating the neglected high-order terms as the source terms and solving the resulted equations iteratively leads to the NLMSFD-PBL.

After all these transformations, equations are discretized. A staggered vertical grid is used, where U-grid points are located midway between neighboring W-grid points. Variables stored at U-grid points are U, V, and P; while W and turbulent quantities are at W-grid points. The lower and upper boundaries are at W-grid locations. The resulting set of difference equations are solved using a block **LU** factorization algorithm (Karpik, 1988).

The surface boundary conditions used are a non-slip condition for velocity, the vertical derivative of the perturbation pressure is zero and local equilibrium condition for the turbulent quantities. At the upper boundary, the effects of the topography vanish. We set the perturbations of mean variables and the vertical derivatives of turbulent quantities to zero. The model uses periodic boundary conditions in X and Y directions.

2.3 Upstream profiles

The upstream or undisturbed profiles for the current PBL model are the results of an integration of a 1-D unsteady-state form of the model equations to quasi-steady state. For the given conditions of the site location (f), the geostrophic wind speed (U_g, V_g) and the surface roughness length (z_0), we can obtain the necessary equilibrium profiles by running the 1-D PBL model of Weng and Taylor (2003). This model will be used for providing all required upstream profiles in our simulations.

3 Results and discussions

Model runs have been carried out for boundary-layer flow over an idealized isolated 3-D terrain and the Askervein Hill

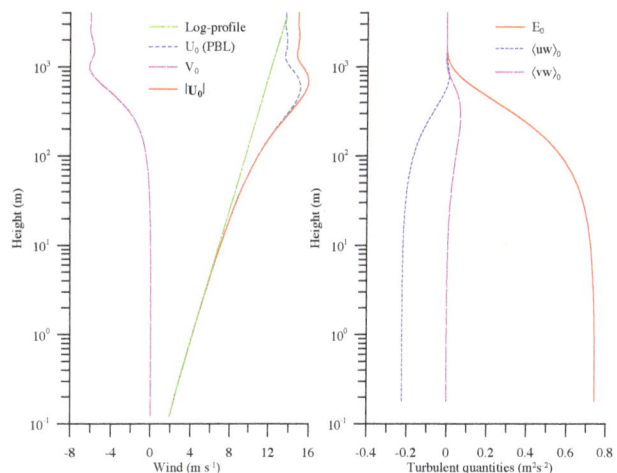

Figure 1. Initial input profiles of U_0, V_0, $|\mathbf{U}_0|$, E_0, $\langle uw \rangle_0$ and $\langle vw \rangle_0$, which is the result of 1-D PBL model runs of Weng and Taylor (2003) for the given condition of $z_0 = 0.03$ m, $f = 10^{-4}$ s^{-1}, $(U_g, V_g) = (13.77, -5.95)$ m s^{-1}, neutral thermal stratification. Logarithmic wind profile for the surface layer model is also included.

– the site of a detailed and much referenced field study of boundary-layer flow over low hills in the 1980s.

3.1 Flow over idealized terrain

For the idealized case, an "isolated", "cosine-squared" terrain surface is used, which is described by

$$z_s(x,y) = \begin{cases} h\cos^2(\pi r/\lambda), & \text{for } r < \lambda, \\ 0, & \text{for } r \geq \lambda, \end{cases} \tag{3}$$

where $r = \sqrt{x^2 + y^2}$ and the maximum slope is $\pi h/\lambda$. For our test case, the values of $h = 75$ m and $\lambda = 1500$ m are used and the maximum slope is 0.157. Our computational domain is set as $(X,Y) \in [-3000, 3000]$ m and 4000 m in the vertical and $129 \times 129 \times 101$ grid points are employed.

Figure 1 shows the initial background profiles of U_0, V_0, $|\mathbf{U}_0|$, E_0, $\langle uw \rangle_0$ and $\langle vw \rangle_0$. This is the result of the 1-D PBL model runs of Weng and Taylor (2003) for surface roughness $z_0 = 0.03$ m, Coriolis parameter $f = 10^{-4}$ s^{-1}, geostrophic wind is constant with height and set to $|\mathbf{U}_g| = 15$ m s^{-1} (the components are selected as $(U_g, V_g) = (13.77, -5.95)$ m s^{-1} so that the near surface (at $z = 10$ m) wind direction is approximately 0) and we assume neutral thermal stratification. This leads to the surface friction velocity, $u_* \approx 0.47$ m s^{-1}, which is used to calculate the logarithmic wind profile for the surface-layer model runs. As can be seen clearly from the figure, the PBL has a near logarithmic mean wind profile associated with a well developed constant stress layer to a depth of about 40 m (where TKE is about 90% of its surface value). Above this surface layer, the mean wind profiles show a smooth blending to geostrophic values at the upper boundary of the model. The PBL has a depth of about 900 m. The turbulent

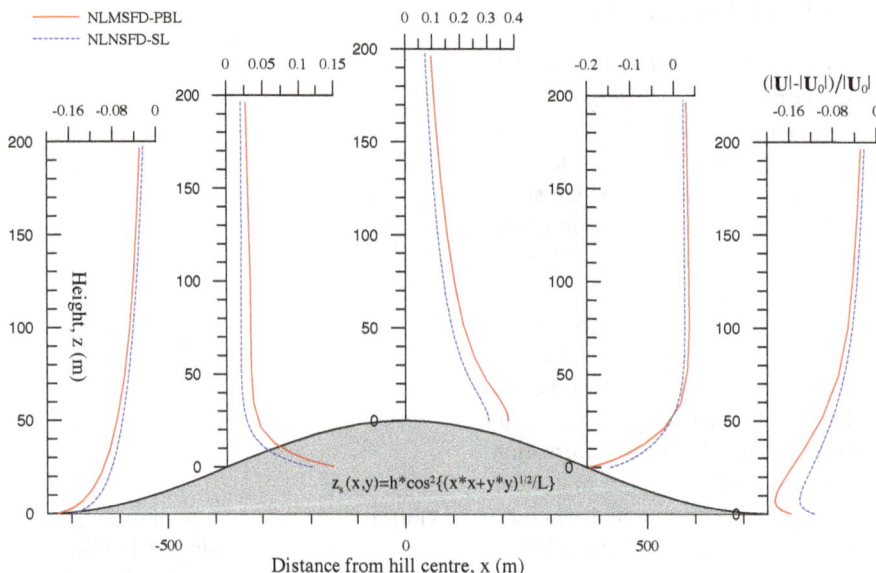

Figure 2. Comparison of vertical fractional speed-up profiles at five different locations along the central line ($y = 0$) of a circular cosine-squared hill from NLMSFD-PBL and NLMSFD-SL model runs.

quantities (TKE and shear stress) decrease appreciably over the depth of the boundary layer and diminish to near zero in the absence of shear at the top of the PBL. There are also supergostrophic winds around $z = 650$ m and the wind vector has Ekman spiral behaviour.

Figure 2 shows the comparison of vertical fractional speed-up (ΔS) profiles at five different locations along the central line ($y = 0$) from **NLMSFD-PBL** and **NLMSFD-SL** model runs. ΔS is defined as the difference between the local ($|\mathbf{U}(X,Y,Z)|$) and the upstream ($|\mathbf{U}_0(Z)|$) wind speeds divided by the upstream wind speed. It can be clearly seen that the PBL version of the model predicts larger speed-up around the hill top areas and larger wind reduction at both upwind and downwind hill foot areas than the surface-layer model. Even at this lower maximum slope of 0.157, the difference between the two models is apparent.

3.2 Flow over Askervein Hill

Askervein is a 116 m high (126 m above the sea level) hill on the west coast of South Uist, one of the islands of the Outer-Hebrides (Scotland). For our model computation, the terrain data was originally prepared by Walmsley, see Walmsley and Taylor (1996). There were a lot of very small terrain features which lead to a very noisy FFT (Fast Fourier Transform) field in the original, so-called Map B. For our model runs, a 9-point smoothing was used. There is a slightly decrease in amplitude in the resulting topography. The maximum height of the hill becomes 115.73 m and the maximum slope is about 0.39, see Fig. 3, where three tower lines and 10 m measurement locations are overlaid. Line-A and Line-AA go through the hill top (HT) and hill centre (CP) respec-

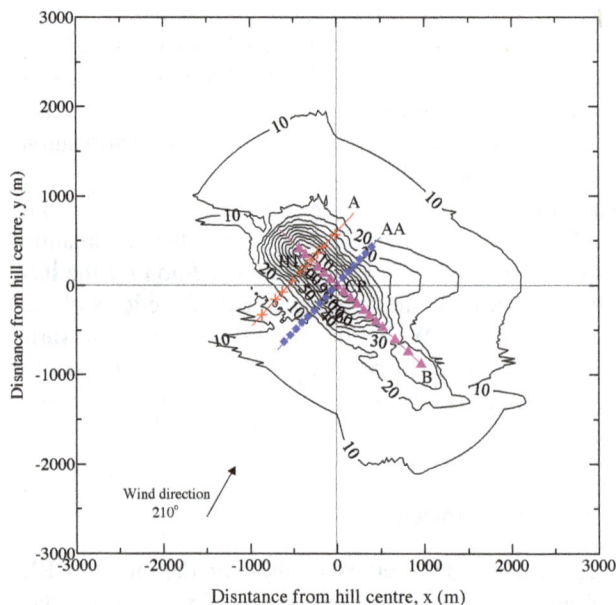

Figure 3. Computational domain of the Askervein hill and measurements tower along the A-line, AA-line and B-line.

tively and are perpendicular to the ridge, and Line-B goes through HT and CP along the hill ridge.

Our computational grid consists of 129×129 points uniformly distributed over an area of 6000×6000 m^2. This makes the resolution 46.875 m and the domain sufficiently large to avoid upstream effects due to the periodic boundary conditions. The model run is for the wind direction of 210°, $-13°$ from the 223° orientation of lines A and AA. The surface roughness is assumed uniform and taken as $z_0 = 0.03$ m.

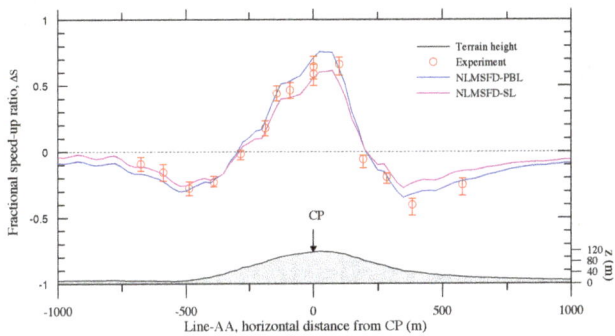

Figure 4. Fractional speed-up ratio, ΔS, for flow over Askervein hill at a height of 10 m above topography along line AA. Comparison of model results and experimental data.

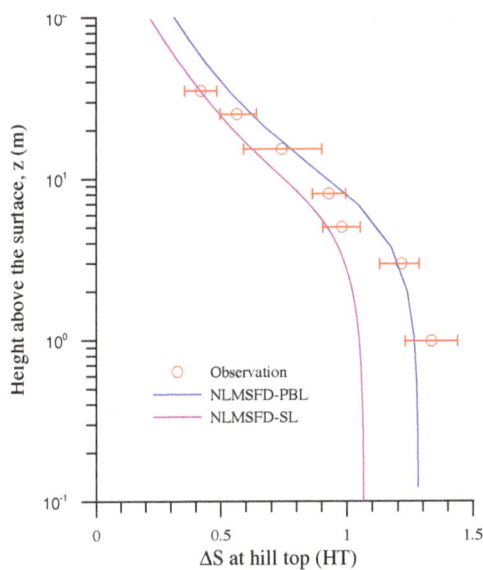

Figure 5. Vertical profile of fractional speed-up ratio, ΔS, at the hill top (point HT) of Askervein hill. Comparison of model results and experimental data.

To closely match the observational data, $|\mathbf{U_g}| = 19.25 \, \mathrm{m \, s^{-1}}$ is used. Again the 1-D PBL model of (Weng and Taylor, 2003) is used to compute the upstream profiles, which are very similar to those in Fig. 1.

Model results of fractional speed-up ratio at 10 m along the line AA together with observational data are shown in Fig. 4. The NLMSFD-PBL predicts slightly smaller values of ΔS over the upwind hill foot and lee side ares and slightly larger value over hill top (HT) area and agrees better with the measured data than NLMSFD-SL results. Figure 5 shows the comparison of vertical profiles of ΔS at HT. The PBL version of the model predicts a larger value of ΔS and again gives a better agreement with the observational data than the surface-layer version of the model.

4   Summary

A Non-Linear Mixed Spectral Finite-Difference model for neutral planetary boundary-layer flow over complex terrain has been developed. The model uses $E - \ell$ turbulence closure. Some of early limitations on the surface-layer version of model are removed and model can simulate flows that are more representative of the real atmosphere. This new model has good potential in wind energy applications.

Acknowledgements. This research has been funded by grants from Mathematics for Information Technology and Complex Systems (MITACS) and Zephyr North of Canada.

Edited by: E. Batchvarova
Reviewed by: A. Beljaars and another anonymous referee

References

Ayotte, K. W. and Taylor, P. A.: A mixed spectral finite-difference 3D model of neutral planetary boundary-layer flow over topography, J. Atmos. Sci., 50, 3523–3537, 1995.

Ayotte, K. W., Xu, D. ,and Taylor, P. A.: The impact of turbulence clousre schemes on predictions of the mixed spectral finite-difference model of flow over topography, Bound.-Lay. Meteorol., 68, 1–33, 1994.

Beljaars, A. C. M., Walmsley, J. L., and Taylor, P. A.: A mixed spectral finite-difference model of neutrally stratified boundary-layer flow over roughness changes and topography, Bound.-Lay. Meteorol., 38, 273–303, 1987.

Karpik, S. R.: An improved method for integrating the mixed spectral finite difference (MSFD) model equations, Bound.-Lay. Meteorol., 43, 273–286, 1988.

Walmsley, J. L. and Taylor, P. A.: Boundary-layer flow over topography: impacts of the Askervein study, Bound.-Lay. Meteorol., 78, 291–320, 1996.

Weng, W. and Taylor, P. A.: On modelling the one-dimensional atmospheric boundary layer, Bound.-Lay. Meteorol., 107, 371–400, 2003.

Weng, W., Chan, L., Taylor, P. A., and Xu, D.: Modelling stably stratified boundary-layer flow over low hills, Q. J. Roy. Meteorol. Soc., 123, 1841–1866, 1997.

Xu, D. and Taylor, P. A.: A non-linear extension of the mixed spectral finite-difference model for neutrally stratified turbulent flow over topography, Bound.-Lay. Meteorol., 59, 177–186, 1992.

Xu, D., Ayotte, K. W., and Taylor, P. A.: Development of the NLMSFD model of turbulent boundary-layer flow over topography, Bound.-Lay. Meteorol., 70, 341–367, 1994.

12

Estimating the photosynthetically active radiation under clear skies by means of a new approach

W. Wandji Nyamsi, B. Espinar, P. Blanc, and L. Wald

MINES ParisTech, PSL Research University, O. I. E. – Centre Observation, Impacts, Energy,
Sophia Antipolis CEDEX, France

Correspondence to: W. Wandji Nyamsi (william.wandji@mines-paristech.fr)

Abstract. The k-distribution method and the correlated-k approximation of Kato et al. (1999) is a computationally efficient approach originally designed for calculations of the broadband solar radiation by dividing the solar spectrum in 32 specific spectral bands from 240 to 4606 nm. This paper describes a technique for an accurate assessment of the photosynthetically active radiation (PAR) from 400 to 700 nm at ground level, under clear-sky conditions using twelve of these spectral bands. It is validated against detailed spectral calculations of the PAR made by the radiative transfer model libRadtran. For the direct and global PAR irradiance, the bias is $-0.4 \, \text{W m}^{-2}$ (-0.2%) and $-4 \, \text{W m}^{-2}$ (-1.3%) and the root mean square error is $1.8 \, \text{W m}^{-2}$ (0.7%) and $4.5 \, \text{W m}^{-2}$ (1.5%). For the direct and global Photosynthetic Photon Flux Density, the biases are of about $+10.3 \, \mu\text{mol m}^{-2} \, \text{s}^{-1}$ ($+0.8\%$) and $1.9 \, \mu\text{mol m}^{-2} \, \text{s}^{-1}$ (-0.1%) respectively, and the root mean square error is $11.4 \, \mu\text{mol m}^{-2} \, \text{s}^{-1}$ (0.9%) and $4.0 \, \mu\text{mol m}^{-2} \, \text{s}^{-1}$ (0.3%). The correlation coefficient is greater than 0.99. This technique provides much better results than two state-of-the-art empirical methods computing the daily mean of PAR from the daily mean of broadband irradiance.

1 Introduction

Photosynthetically active radiation, abbreviated in PAR, is the solar radiation in the range [400, 700] nm that can be used by organisms via the process of photosynthesis. PAR is defined as the incident power per unit surface for this spectral interval and may be expressed in W m^{-2}. PAR is also a measure of the photosynthetic photon flux density, abbreviated in PPFD and expressed in $\mu\text{mol m}^{-2} \, \text{s}^{-1}$, and is defined as the number of the incident photons per unit time per unit surface. Both units are linked by the widely used approximation $1 \, \text{W m}^{-2} \approx 4.57 \, \mu\text{mol m}^{-2} \, \text{s}^{-1}$ (McCree, 1972). PAR is a portion of the total, also known as broadband, solar irradiance. Whatever the spectral interval, the solar radiation available at ground level on a horizontal plane is called the global radiation. The global is the sum of the direct component that comes from the direction of the sun and the diffuse component that comes from the rest of the sky vault. Let note respectively G, P_G and Q_P, the global broadband irradiance, the global PAR irradiance, and the global PPFD at ground level.

In situ measurements of PAR are rare in space and time. This scarcity leads researchers and practitioners to calculate PAR from the global broadband solar irradiance by empirical means. For example, Udo and Aro (1999) proposed a ratio of 2.079 between the daily mean of G and the daily mean of Q_P:

$$Q_P = 2.079G \tag{1}$$

where the constant 2.079 is in $\mu\text{mol J}^{-1}$. Jacovides et al. (2004) suggested a ratio of 1.919. These authors acknowledge that the actual ratio depends on the sky conditions and atmospheric properties.

Other approaches to PAR assessment and more generally to assessment of the solar radiation in any spectral interval are atmospheric radiative transfer models (RTM). Besides the difficulty in knowing all inputs requested by RTMs, their main disadvantage is the computational load because many spectral calculations must be performed. Several methods have been proposed to reduce the number of calculations. Among them, are the k-distribution method and the correlated-k approximation proposed by Kato et al. (1999)

Table 1. KB covering and close to PAR spectral interval, selected sub-intervals $\delta\lambda_i$, slopes and intercepts of the affine functions between the clearness indices in KB and sub-intervals $\delta\lambda_i$ obtained from libRadtran simulations.

KB	Interval $\Delta\lambda$, nm	Sub-interval $\delta\lambda$, nm (#i)	Global		Direct normal	
			Slope a_i	Intercept b_i	Slope c_i	Intercept d_i
6	363–408	385–386 (#1)	0.9987	−0.0023	1.0030	−0.0032
7	408–452	430–431 (#2)	1.0026	−0.0004	0.9995	0.0013
8	452–518	484–485 (#3)	1.0034	0.0005	0.9979	0.0000
9	518–540	528–529 (#4)	0.9998	−0.0005	1.0008	−0.0013
10	540–550	545–546 (#5)	1.0001	0.0003	1.0003	−0.0003
11	550–567	558–559 (#6)	1.0004	0.0004	0.9997	0.0012
		569–570 (#7)	0.9960	−0.0119	1.0024	−0.0100
12	567–605	586–587 (#8)	1.0123	0.0064	0.9929	0.0267
		589–590 (#9)	0.9568	−0.0109	0.9804	−0.0434
		602–603 (#10)	1.0150	0.0167	1.0051	0.0212
13	605–625	615–616 (#11)	1.0004	0.0009	0.9977	0.0033
		625–626 (#12)	1.0104	−0.0174	1.0622	−0.0551
14	625–667	644–645 (#13)	1.0072	0.0029	0.9960	0.0154
		656–657 (#14)	0.9915	0.0068	0.9698	0.0205
15	667–684	675–676 (#15)	1.0006	0.0007	0.9978	0.0036
		685–686 (#16)	1.0473	0.0212	0.9681	0.1036
16	684–704	687–688 (#17)	0.9602	−0.0130	1.0041	−0.0531
		694–695 (#18)	0.9828	−0.0153	1.0323	−0.0642
17	704–743	715–716 (#19)	1.0262	0.0121	0.9771	0.0596

whose approach is implemented in several RTMs. This approach was originally designed as a very efficient way to speed up computations of G and its direct component by using 32 specific spectral intervals across the solar spectrum from 240 to 4606 nm. Hereafter, these spectral intervals are abbreviated in KB. This article deals with the assessment of the PAR – irradiance and PPFD – using the irradiance of each KB covering the PAR spectral range in clear sky conditions.

2 Problem statement

The global PAR irradiance P_G is mathematically defined as:

$$P_G = \int_{400}^{700} G_\lambda d\lambda \qquad (2)$$

where G_λ is the global spectral irradiance, λ the wavelength and the integration is made between 400 and 700 nm. The global PPFD Q_P is similarly defined as:

$$Q_P = \frac{1}{hc}\int_{400}^{700} G_\lambda \lambda d\lambda \qquad (3)$$

where h is the Planck's constant and c the velocity of light.

The direct normal irradiance is the irradiance received on a plane always facing the sun rays with a normal incidence. Let note respectively B, P_B, and Q_{PB}, the direct normal broadband irradiance, the direct normal PAR irradiance, and the direct normal PPFD:

$$P_B = \int_{400}^{700} B_\lambda d\lambda \qquad (4)$$

$$Q_{PB} = \frac{1}{hc}\int_{400}^{700} B_\lambda \lambda d\lambda \qquad (5)$$

where B_λ is the direct normal spectral irradiance.

The integral may be replaced by a Riemann sum using very narrow spectral intervals or bands, hereafter abbreviated NB. Here, we chose $\delta\lambda = 1$ nm, assuming that the optical properties of the atmosphere do not change over 1 nm. If λ_i denotes now the center wavelength of each NB of width $\delta\lambda$, it comes:

$$P_G = \sum_{i=1}^{300} G_{\delta\lambda i} \qquad (6)$$

$$Q_P = \frac{1}{hc}\sum_{i=1}^{300} G_{\delta\lambda i}\lambda_i . \qquad (7)$$

Similar equations hold for P_B and Q_{PB}.

The PAR spectral band [400, 700] nm is covered by 11 KB, from #6 [363, 408] nm to #16 [684, 704] nm (Table 1). Wandji Nyamsi et al. (2014) demonstrated that as a whole the approach of Kato et al. (1999) offer accurate estimates of the spectral irradiance in most of the 32 KB when compared to detailed spectral calculations in clear sky and cloudy conditions, and especially for the KB #6 to 16. It follows that the

PAR may be computed by a Riemann sum based on 11 KB instead of 300 NBs. KB #6 and #16 are partly outside the PAR range. One solution is a weighted sum based on the overlap between KB_j and the PAR interval. The weight w_j of KB_j may be defined as follows:

$$w_j = 1, \text{ if } j \text{ is not 6 or 16}$$
$$w_6 = (408 - 400)/(408 - 363) = 0.1778$$
$$w_{16} = (700 - 684)/(704 - 684) = 0.80$$

and

$$P_G = \sum_{j=6}^{16} G_{KB_j} w_j \qquad (8)$$

$$Q_P = \frac{1}{hc} \sum_{j=6}^{16} G_{KB_j} w_j \lambda_{KB_j} \qquad (9)$$

where G_{KB_j} is the global irradiance for KB_j and λ_{KB_j} the center wavelength of KB_j. Similar equations hold for P_B and Q_{PB}. However, Wandji Nyamsi et al. (2014) reported a relative root mean square error less than 2 % between detailed spectral calculations and the approach by Kato et al. (1999) for each KB, from #6 to #16. Though small this error may be decreased for the PAR by the technique proposed in this article.

3 Description of the technique

Actually, the bandwidth in several KB is larger than 30 nm and may be considered large for estimating PAR in an accurate manner. The concept underlying the proposed technique is to determine several narrower spectral bands NB whose transmissivities are correlated to those of the KB and then use these transmissivities in a linear interpolation process to compute the PAR. This technique is elaborated and validated by the means of the RTM libRadtran (Mayer and Kylling, 2005).

The clearness index KT_i and the direct clearness index KT_{Bi} – also called atmospheric transmissivity and direct atmospheric transmissivity – are defined as follows:

$$KT_i = \frac{G_{\delta\lambda i}}{E_{o_i} \cos(\theta_s)} \qquad (10)$$

$$KT_{Bi} = \frac{B_{\delta\lambda i}}{E_{o_i}} \qquad (11)$$

where θ_s is the solar zenithal angle; E_{o_i} is the irradiance at the top of atmosphere on a plane normal to the sun rays for the ith NB. Several solar spectra E_{o_i} have been published. That of Gueymard (2004) is available in libRadtran and has been used here. By introducing the clearness index in Eqs. (6) and (7), it comes:

Table 2. Ranges and distributions of values taken by the solar zenith angle, the ground albedo and the 7 variables describing the clear atmosphere.

Variable	Value
– Solar zenith angle θ_s	– Uniform between 0 and 89 (degree)
– Ground albedo ρg	– Uniform between 0 and 0.9
– Total column content of ozone	– Ozone content is: $300 \times \beta + 200$, in Dobson unit. Beta distribution, with A parameter = 2, and B parameter = 2, to compute β
– Total column content of water vapor	– Uniform between 0 and 70 (kg m^{-2})
– Elevation of the ground above mean sea level	– Equiprobable in the set: {0, 1, 2, 3} (km)
– Atmospheric profiles (Air Force Geophysics Laboratory standards)	– Equiprobable in the set: {"Midlatitude Summer", "Midlatitude Winter", "Subarctic Summer", "Subarctic Winter", "Tropical", "US Standard"}
– Aerosol optical depth at 550 nm	– Gamma distribution, with shape parameter = 2, and scale parameter = 0.13
– Angström coefficient	– Normal distribution, with mean = 1.3 and standard-deviation = 0.5
– Aerosol type	– Equiprobable in the set: {"urban", "rural", "maritime", "tropospheric", "desert", "continental", "Antarctic"}

$$P_G = \cos(\theta_s) \sum_{i=1}^{300} E_{o_i} KT_i \qquad (12)$$

$$Q_P = \frac{\cos(\theta_s)}{hc} \sum_{i=1}^{300} E_{o_i} KT_i \lambda_i. \qquad (13)$$

Similar equations hold for P_B and Q_{PB}. A set of 60 000 clear sky atmospheric states is built by the mean of Monte-Carlo technique that will be input to libRadtran. Table 2 reports the nine input variables selected with seven of them describing the clear sky atmosphere: θ_s, ground albedo, total column content of water vapor and ozone, the vertical profile of temperature, pressure, density, and volume mixing ratio for gases as a function of altitude, the aerosol optical depth at 550 nm, Angström coefficient, and aerosol type, and the elevation of the ground above sea level. The random selection of inputs takes into account the modelled marginal distribution established from observation proposed by Lefevre et al. (2013) and Oumbe et al. (2011). More precisely, the uniform distribution is chosen as a model for marginal probability for all parameters except aerosol optical thickness, Angstrom coefficient, and total column content of ozone. The chi-square law for aerosol optical thickness, the normal law for the Angstrom coefficient, and the beta law for total column content of ozone have been selected. The selection of these parametric probability density functions and their corresponding parameters have been empirically determined from the analyses of the observations made in the AERONET network for aerosol properties and from meteorological satellite-based ozone products (Lefevre et al., 2013).

Several plots were made superimposing KT_{KB} and KT_i obtained every nm for the interval [363, 743] nm. A visual inspection of the differences between KT_{KB} and KT_i helps in establishing a set of selected NB, taking into account that the number of these sub-intervals should be as small as possible but still retaining a high accuracy when using a linear interpolation between the sub-intervals to compute the PAR, as explained later. Table 1 reports the 12 KB and the 19 sub-intervals NB. The KB #17 is necessary to obtain NBs enclosing the PAR interval. All KBs contain one NB, except KB #12, 14 and 16, where 4, 3 and 3 NBs were found respectively. These bands exhibit strong variations of KT_i that cannot be accounted for with a single NB.

In each selected NB, an affine function is determined by least-square fitting technique:

$$KT_i = a_i KT_{KB_j} + b_i \qquad (14)$$

$$KT_{Bi} = c_i KT_{BKB_j} + d_i. \qquad (15)$$

Table 1 also reports the slope and intercept for the global and direct clearness indices for each selected NB. This set of affine functions is established once for all. For any atmospheric state, given the twelve values of KT_{KB_j} and KT_{BKB_j}, the nineteen KT_i and KT_{Bi} are computed for each corresponding NB using the affine functions. Then, KT and KT_B are computed for each nm between 400 and 700 nm using a linear interpolation with KT_i and KT_{Bi} as nodes. Finally, Eqs. (12)–(13) provide P_G and Q_P. A similar process yields the direct normal PAR: P_B and Q_{PB}.

4 Numerical validation

A comparison of the results of the proposed technique against the results from the detailed spectral calculations made by libRadtran considered as a reference is performed to assess the performances of the proposed technique for P_G, Q_P, P_B and Q_{PB}. Then, these performances are compared to those obtained for Q_P by the methods proposed by Jacovides et al. (2004), Udo and Aro (1999) and the weighted sum (Eq. 9). Another sample of 15 000 atmospheric states has been constructed and used for validation.

4.1 Performance of the proposed technique

Deviations: estimates minus reference, are computed for each state of the validation sample for P_G, Q_P, P_B and Q_{PB}. They are synthesized by the bias, the root mean square error (RMSE) and the correlation coefficient. The relative bias and RMSE are computed relative to the mean value of the reference. Figure 1 exhibits these statistical parameters for the global and direct normal PAR irradiance and PPFD. For the direct component, the bias for PAR irradiance, respectively PPFD, is $-0.4\,\mathrm{W\,m^{-2}}$, i.e. $-0.2\,\%$ in relative value, and $+10.3\,\mu\mathrm{mol\,m^{-2}\,s^{-1}}$, i.e. $+0.8\,\%$ in relative value. The RMSE is respectively 1.8 W m^{-2} (0.7 %) and

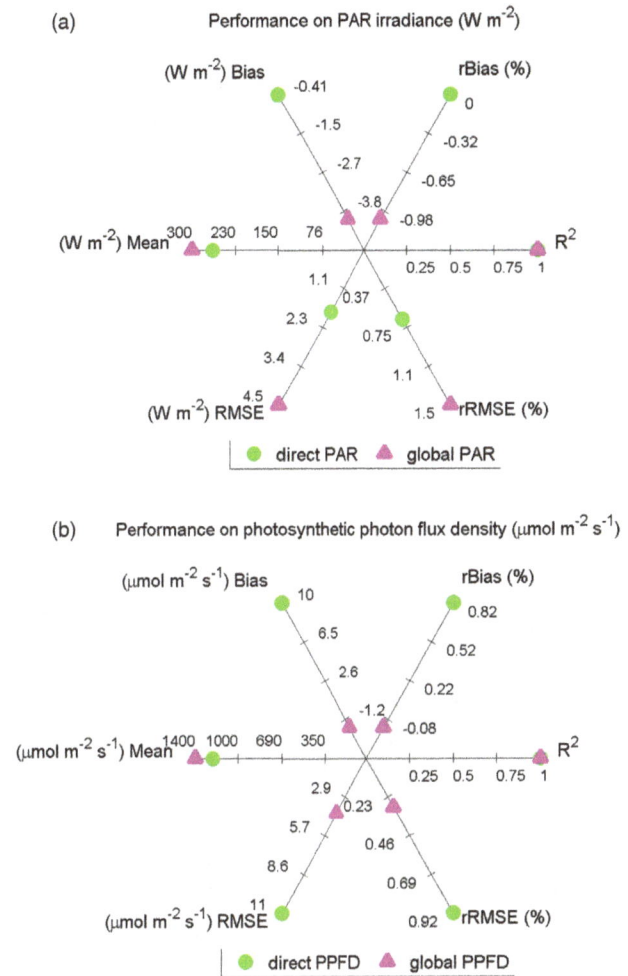

Figure 1. Synthesis of the performance of the proposed technique.

$11.4\,\mu\mathrm{mol\,m^{-2}\,s^{-1}}$ (0.9 %). For the global, the bias for PAR irradiance, respectively PPFD, is $-4.0\,\mathrm{W\,m^{-2}}$, i.e. $-1.3\,\%$ in relative value, and $1.9\,\mu\mathrm{mol\,m^{-2}\,s^{-1}}$, i.e. $-0.1\,\%$ in relative value. The RMSE is respectively 4.5 W m^{-2} (1.5 %) and $4.0\,\mu\mathrm{mol\,m^{-2}\,s^{-1}}$ (0.3 %). The coefficient of determination R^2 is greater than 0.99. These figures prove the good level of performance of the proposed technique.

4.2 Comparison with other methods

Figure 2 exhibits the statistical indicators for Q_P obtained by the methods of Jacovides et al. (2004), Udo and Aro (1999), weighted sum and the proposed technique. The method of Jacovides et al. underestimates the PAR by $-7.4\,\%$; the relative RMSE is 9.6 %. The method of Udo and Aro shows better results with a relative bias of 0.3 % and a relative RMSE of 4.7 %. The weighted sum exhibits very low relative bias: $-0.2\,\%$, and relative RMSE: 0.3 %. The proposed technique shows also a very good agreement with a relative bias of

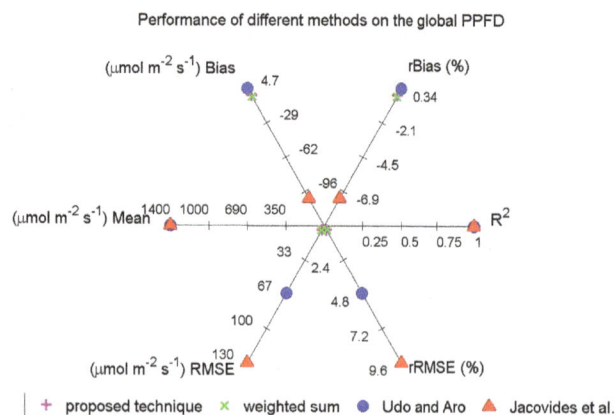

Figure 2. Performance of different methods: Jacovides et al. (2004); Udo and Aro (1999), weighted sum and the proposed technique.

-0.1% and relative RMSE of 0.3 % and offers the same results than the weighted sum.

There are two causes for this similarity. The first cause is that all KB contain one $\delta\lambda_i$, except KB #12, 14 and 16. For these bands, variations of KT_i with λ_i are very small. If it were the case for all KB, from #6 to #17, the proposed technique and the weighted sum would agree and provide similar results. But small discrepancies happen between both techniques due to the bands KB #12, 14 and 16, which are subdivided by respectively 4, 3 and 3 $\delta\lambda_i$. However, the contribution of these bands to the PAR outside the atmosphere is only 25 % approximately. As a consequence, these small discrepancies have a small influence on the final result. Both the weighted sum and the proposed technique exhibit better performances than the empirical method because both take into account the actual atmospheric effects by the means of the Kato et al. approach.

5 Conclusions

The k-distribution method and the correlated-k approximation of Kato et al. (1999) is a computationally efficient approach originally designed for calculations of the broadband solar radiation at ground level by dividing the solar spectrum in 32 specific spectral bands from 240 to 4606 nm. This paper describes a technique for an accurate assessment of the PAR under clear-sky conditions using the irradiance estimated in twelve of these spectral bands. The validation against numerical simulation exhibits very good performances. For the direct and global PAR irradiance, the bias is $-0.4\,W\,m^{-2}$ (-0.2%) and $-4\,W\,m^{-2}$ (-1.3%) and the RMSE is $1.8\,W\,m^{-2}$ (0.7 %) and $4.5\,W\,m^{-2}$ (1.5 %). For the direct and global PPFD, the bias of is $+10.3\,\mu mol\,m^{-2}\,s^{-1}$ ($+0.8\%$) and $1.9\,\mu mol\,m^{-2}\,s^{-1}$ (-0.1%) and the RMSE is $11.4\,\mu mol\,m^{-2}\,s^{-1}$ (0.9 %) and $4.0\,\mu mol\,m^{-2}\,s^{-1}$ (0.3 %). The correlation coefficient is greater than 0.99. It is also

shown that the proposed technique provides better results than two state-of-the-art empirical methods estimating the global PPFD from the global irradiance (Jacovides et al., 2004; Udo and Aro, 1999).

The proposed technique is very useful for the operational estimation of PAR when computational load and great accuracy in PAR are major issues. In addition, this technique may be extended to be able to accurately estimate other spectral quantities taking into account spectral absorption of photosynthetic pigments found in plants and algae such as chlorophyll, carotenoids. The authors are aware of the heuristic way used for selecting the specific NBs for each KB in their work. Other ways are possible. For example, one may think of using the variance between KT_{KB} and all KT_i within a given KB to determine automatically the need for more than one sub-interval. The greater the number of sub-intervals, the higher the accuracy in PAR computation at the expense of a greater number of affine functions.

Acknowledgements. The research leading to these results has received funding from the ADEME, research grant No. 1105C0028.

Edited by: S.-E. Gryning
Reviewed by: two anonymous referees

References

Gueymard, C.: The sun's total and the spectral irradiance for solar energy applications and solar radiations models, Solar Energy, 76, 423–452, 2004.

Jacovides, C. P., Timvios, F. S., Papaioannou, G., Asimakopoulos, D. N., and Theofilou, C. M.: Ratio of PAR to broadband solar radiation measured in Cyprus, Agr. Forest. Meteorol., 121, 135–140, 2004.

Kato, S., Ackerman, T., Mather, J., and Clothiaux, E.: The k-distribution method and correlated-k approximation for shortwave radiative transfer model, J. Quant. Spectrosc. Ra., 62, 109–121, 1999.

Lefèvre, M., Oumbe, A., Blanc, P., Espinar, B., Gschwind, B., Qu, Z., Wald, L., Schroedter-Homscheidt, M., Hoyer-Klick, C., Arola, A., Benedetti, A., Kaiser, J. W., and Morcrette, J.-J.: McClear: a new model estimating downwelling solar radiation at ground level in clear-sky conditions, Atmos. Meas. Tech., 6, 2403–2418, doi:10.5194/amt-6-2403-2013, 2013.

Mayer, B. and Kylling, A.: Technical note: The libRadtran software package for radiative transfer calculations – description and examples of use, Atmos. Chem. Phys., 5, 1855–1877, doi:10.5194/acp-5-1855-2005, 2005.

McCree, K. J.: Test of current definitions of photosynthetically active radiation against leaf photosynthesis data, Agric. Meteorol., 10, 443–453, 1972.

Oumbe, A., Blanc, P., Gschwind, B., Lefevre, M., Qu, Z., Schroedter-Homscheidt, M., and Wald, L.: Solar irradiance in clear atmosphere: study of parameterisations of change with altitude, Adv. Sci. Res., 6, 199–203, doi:10.5194/ASR-6-199-2011, 2011.

Udo, S. O. and Aro, T. O.: Global PAR related to global solar radiation for central Nigeria, Agr. Forest. Meteorol, 97, 21–31, 1999.

Wandji Nyamsi, W., Espinar, B., Blanc, P., and Wald, L.: How close to detailed spectral calculations is the k-distribution method and correlated-k approximation of Kato et al. (1999) in each spectral interval?, Meteorol. Z., 23, 547–556, doi:10.1127/metz/2014/0607, 2014.

13

Development and implementation of an Integrated Water Resources Management System (IWRMS)

W.-A. Flügel[1] and C. Busch[2]

[1]Institute for Geography, University of Jena, Germany
[2]Codematix GmbH, Jena, Germany

Abstract. One of the innovative objectives in the EC project BRAHMATWINN was the development of a stakeholder oriented Integrated Water Resources Management System (IWRMS). The toolset integrates the findings of the project and presents it in a user friendly way for decision support in sustainable integrated water resources management (IWRM) in river basins. IWRMS is a framework, which integrates different types of basin information and which supports the development of IWRM options for climate change mitigation. It is based on the River Basin Information System (RBIS) data models and delivers a graphical user interface for stakeholders. A special interface was developed for the integration of the enhanced DANUBIA model input and the NetSyMod model with its Mulino decision support system (mulino mDss) component. The web based IWRMS contains and combines different types of data and methods to provide river basin data and information for decision support. IWRMS is based on a three tier software framework which uses (i) html/javascript at the client tier, (ii) PHP programming language to realize the application tier, and (iii) a postgresql/postgis database tier to manage and storage all data, except the DANUBIA modelling raw data, which are file based and registered in the database tier. All three tiers can reside on one or different computers and are adapted to the local hardware infrastructure. IWRMS as well as RBIS are based on Open Source Software (OSS) components and flexible and time saving access to that database is guaranteed by web-based interfaces for data visualization and retrieval. The IWRMS is accessible via the BRAHMATWINN homepage: http://www.brahmatwinn.uni-jena.de and a user manual for the RBIS is available for download as well.

1 Introduction and objectives

Integrated river basin studies as done in the BRAHMATWINN project require a common data and information platform that was provided by the River Basin Information System (RBIS) developed by at the FSU-Jena (Flügel, 2007, 2009; Kralisch et al., 2009). The IWRMS builds on this well tested toolset and enhanced the latter by integrating data and methods from the deliverables of the BRAHMATWINN project, like the time series of station data, the results of the DANUBIA modelling system (Mauser and Bach, 2009) and the analysis done by means of the NetSyMod-mulino decision support system (mDss) (Giupponi et al., 2008).

The *overall objective* of the IWRMS development was to offer users and decision makers the DANUBIA modelling results (Chapter 7) together with the "what-if?" scenarios elaborated by the mDss (Chapter 8). In addition analysis tools should enable users to support decision making for adaptive IWRM option design accounting for impacts of climate change on a basin scale. The overall objective was realised by elaborating on the following *scientific-technological objectives*:

1. Integration of all BRAHMATWINN research work results, data time series, GIS maps and documents from in the Upper Danube River Basin (UDRB) and the Upper Brahmaputra River Basin (UBRB) respectively into the IWRMS in such a way that they can be used by means of a web-based graphical user interface (GUI).

2. Making the comprehensive system assessment, modelling studies, spatial analysis of water balance components and the integrated socio-economic analysis and scenario evaluation available to local actors, decision makers and planners.

3. Provide additional GUI tools for analysing the information supplied by (i) and (ii) in the twinning basins for IWRM related decision support.

The challenges related to these objectives were met by building the IWRMS development on the sophisticated RBIS data model structure and designing the system as an easy to use application that extends the RBIS modular framework and deploys all basin specific data accounting for the local computer network structure.

2   Role within the integrated project

The RBIS as well as the IWRMS have a central role within the BRAHMATWINN project. RBIS provides the common data and information base accessible via the web to all project partners and authorized stakeholders, and the IWRMS in turn is making use of the RBIS functionalities for decision support. In addition IWRMS provides access to the modelling results of the DANUBIA model applied in the twinning UDRB and UBRB and the "what-if?" scenario evaluation done by means of the mDss. IWRMS furthermore offers different modules that elaborate on this information and provide knowledge based decision support.

The IWRMS thereby offers two important services to users and decision makers, as it *firstly* makes the knowledge elaborated by the project consortium (see Chapter 1 till Chapter 8) available to decision makers, and *secondly* provides them with sophisticated tools to make use of the knowledge for decision support related to the design of adaptive IWRM strategies to cope with impacts of climate change.

3   Scientific methods applied

3.1   Definition of IWRMS requirements

IWRMS requirements were defined in close cooperation with users and stakeholders on several workshops in Germany and Asia as follows:

i. New datasets had to be described by metadata according to ISO19115 standards.

ii. Description of locality had to be implemented applying the Open Geospatial Consortium (OGC) prescriptions.

iii. Different data types and their interrelations, i.e. how applications access data has been grouped as follows:

   – GIS data layers (raster and vector) i.e. for geology, land use, river network, or vegetation related to model input data.

   – Measured station time series data, i.e. discharge or climate provided by partners and stakeholders for the twinning UDRB and UBRB.

   – Raster data provided by the climate modellers (Chapter 2) and the DANUBIA model (Chapter 7) comprising:

   – regionalized data of Global Circulation Models (GCM) for different scenarios and modelled climate projections with a grid cell size of 50×50 km;

   – output data and parameterisations of the different DANUBIA runs, i.e. 15 modelled hydrometeorological parameter and input meteorological data downscaled to a grid cell of 1×1 km (Marke, 2008).

– *Results from the mDss studies*, i.e. analysis matrix, evaluation matrix, weights and options for indicators, sensitivity analysis, and decision making.

– *"What-if?" scenarios* based on the Special Report on Emissions Scenarios (SRES) of the IPCC (IPCC, 2000).

– *Indicators* selected in different Delphi exercises (Chapter 6).

– *Any type of documents*, i.e. reports, graphs, pictures, or spread sheets in standard formats.

– *Web based access* to original (e.g. xls) and derived (e.g. graphs) data.

These user requirements have been realised by respective modules and interfaces than can be added for use on demand. All data are stored in a set of databases and no distinctions are made between alphanumerical, location based and binary data items. The underlying database uses spatial based extensions (PostGIS) according to the Open Geospatial Consortium (OGC) standards.

3.2   Design of graphical user interfaces (GUIs)

The overall objective of the IWRMS was realized by developing new modules and GUIs to handle the data and information exchange between the DANUBIA hydrological model, the RBIS and the mDss. The innovative methodology developed for the data exchange between applications via the IWRMS interfaces defined descriptions of

– *how* applications want to access input and output data in the database, and

– *what* kind of API methods will be used by applications to access the data.

In the first part XML-based data dictionaries have been designed, which in the second part the API will make use of to read and write data in a failure-proof manner. Integration of the different kinds of data and indicators for the design of "what-if-scenarios" is provided for the evaluation of prognostic system management options during the decision making processes and when developing adaptive IWRM strategies.

The design of the different GUIs for the IWRM user support was discussed with project partners and stakeholders

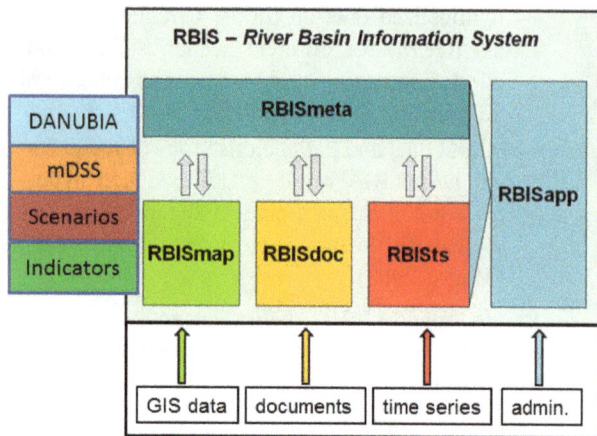

Figure 1. IWRMS data management architecture and data flow.

during IWRMS workshops. It was decided to keep them identical to those of the RBIS in terms of the cascading web map service (WMS) based definitions, and the user management. Hence the GUIs of further IWRMS data, functions and programs like the DANUBIA modelling results, the indicators or the mDss have been designed accordingly. They make use of a similar web based approach and apply a set of adaptable and RBIS compatible extension modules programmed in PHP.

4 Results achieved and deliverables provided

4.1 Enhanced RBIS structure

The schematically IWRMS model structure is shown in Fig. 1. It shows that the central part of the system is the enhanced RBIS with its four components managing and administrating GIS data, documents, time series and the administration modules. The IWRMS modular components are represented by the interface to the DANUBIA model and the mDss, a module for SRES scenario design and the definition and calculation of indicators to be used for IWRM decision support.

These modules can be selected by the user on demand and thereby reduce the complexity of the system accounting for the level of staff training available at the user side.

Each module of the IWRMS delivers a set of different functions, which can be combined to a new workflow structure. The user rights are administered by the system administrator of the respective stakeholder organization and depending on given user permissions the menu structure is adapted accordingly. This functionality is adjusting the system's complexity to the respective work level of each IWRMS application. A similar fine grained user permission hierarchy rights (hidden, read, download, write, owner) can be assigned to the management of time series data, documents, indicators, GIS layers or GIS maps.

4.2 Component interaction

The IWRMS integrates the results of deliverables produced in the BRAHMATWINN project by means of access definitions for semi-automated data interfaces. The latter provides information required by the user, i.e. metadata from file headers, aggregated time series data or coordinate reference systems with minimized controlling user interaction. The IWRMS interface interaction architecture and data fluxes between the IWRMS model components mDss and DANUBIA are shown in Fig. 2 and can be described as follows:

- The mDss needs a tight interface coupling, because data will be exchanged between both systems, especially if indicators from IWRMS are applied in the mDss.

- Opposite the DANUBIA model requires only a lose interface coupling. As the model applies raster input data with grid cell values for each parameter and for each time stamp the latter are linked and registered to the IWRMS only.

- DANUBIA input and output time series data can be extracted by the IWRMS for any given location by selecting the respective raster cell.

4.3 Enhanced RBIS data management architecture

Data management and data flow between the different IWRMS components is shown in Fig. 3. They reflect the data models needed to integrate the results generated in the BRAHMATWINN project and can be described as follows:

1. The GIS database management server is a core element of the system providing the different services offered.

2. At the left side modules provide functionalities to manage metadata, time series, raster and vector data.

3. At the right side the component modules provide Desktop GIS, networking (internet/intranet) and indicator management functionalities.

4. Extensive data fluxes between these components are managed by the central server core system which could be located remote or at the user site.

4.4 Registering new data

The implementation of the IWRMS modules is based on defined formats and data fluxes and accounts for the data volume and the user requirements towards a web based application. To read a parameter time series output by the DANUBIA model requires the processing of separate files for each time step and is done by means of asynchronous processes as shown in Fig. 4.

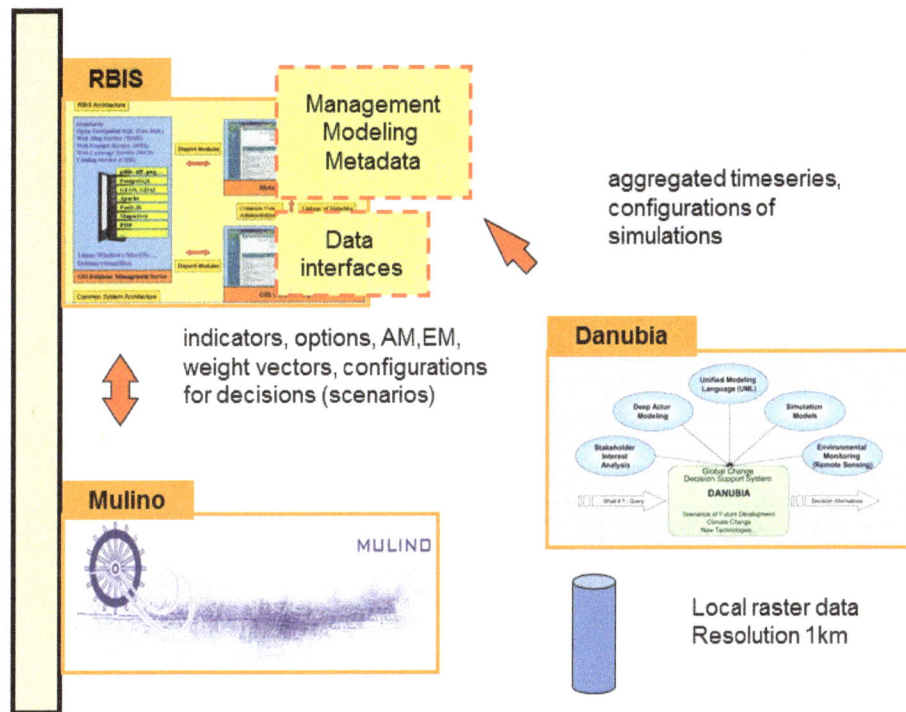

Figure 2. Implemented interaction structure between external applications.

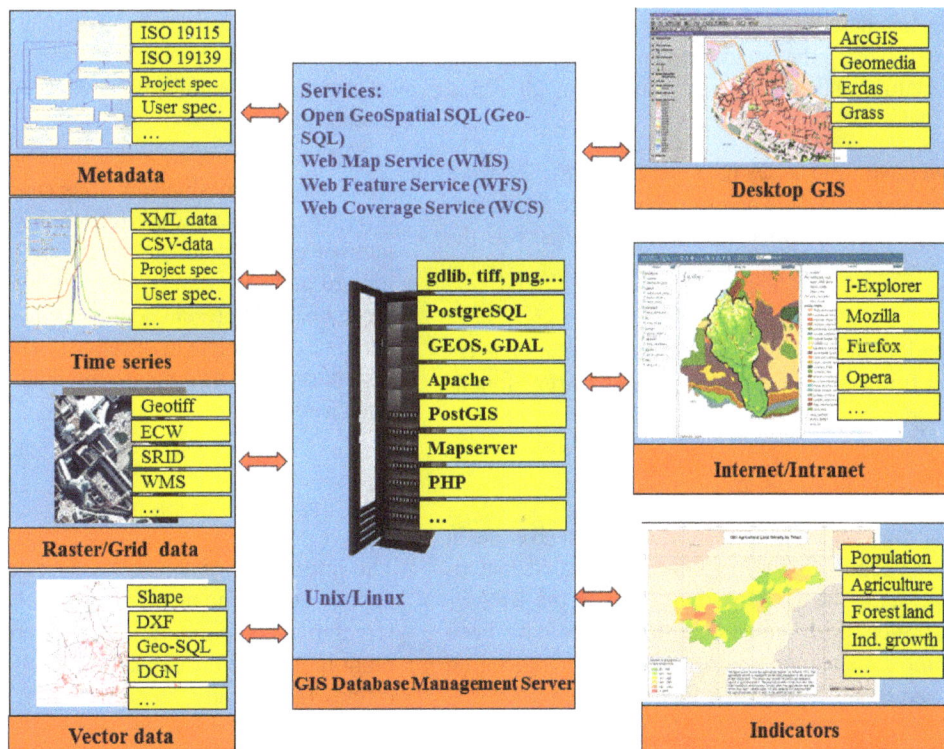

Figure 3. IWRMS data management architecture and data flow.

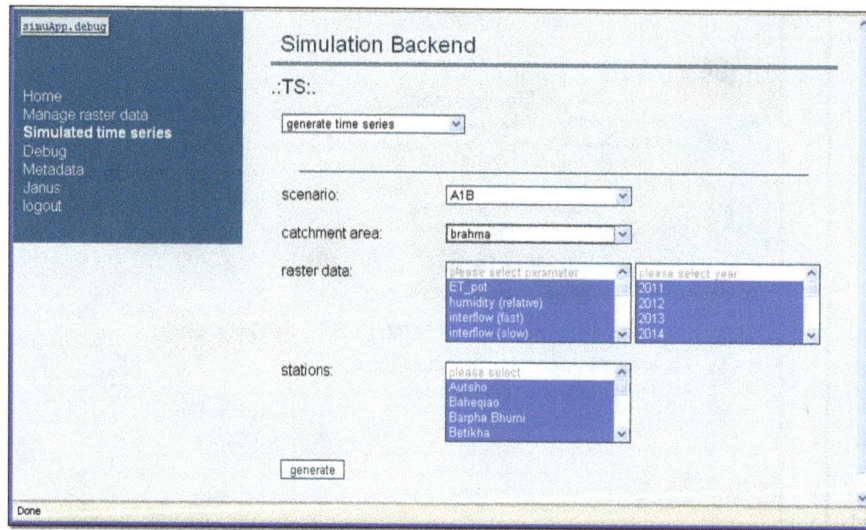

Figure 4. Registering new data to the IWRMS.

The user controlled work flow done by means of GUIs can be described as follows:

1. After registering new data, the first process generates the RBIS and ISO19115 conform metadata associated to the location and its time series.

2. The user in a next step is generating the simulated time series by selecting the parameters specifying time window, location, and name, i.e. precipitation, 2013 and station name.

3. The generated time series data are then extracted from the respective raster file and transposed into a parameter time series that either can be input to RBIS as the IWRMS database or can be downloaded for external use.

4. After loading the extracted time series into the IWRMS they can further be processed by means of functionalities provided by the RBIS.

4.5   Generating time series for virtual stations

Virtual stations are represented by grid cells of a river basin model domain. They either can store downscaled meteorological time series or model output data. In the BRAHMATWINN project downscaled data time series have been produced by the climate modelling on a 50×50 km resolution (Chapter 2) and by the DANUBIA model on a 1×1 km resolution respectively (Marke, 2008). The DANUBIA model is also using this resolution when writing the model output from historical or projected time periods, thereby producing many terra bytes of data volume.

The IWRMS function developed to extract selected cell data from this huge data volume is completely integrated in the *RBISmap* application component of RBIS. As shown in Fig. 6 the selection of the raster cell is done on *RBISmap* by means of a GIS data layer that can address each grid cell of the model domain. After the grid cell has been selected the service applies parts of the workflow described in Sect. 4.4.

This IWRMS service provides a processing method which combines the data from the remote central server with the local catchment server and thereby offers to the user the full information potential produced by the different DANUBIA model runs.

Metadata are added to the generated virtual station and all RBIS data processing functions can be used on the "virtual time series". To avoid an overloading of the central server the number of simultaneously running extraction processes are limited and selection requests exceeding the threshold limit are queued. At present it takes approximately four minutes to generate a complete time series (1970 to 2080, one hour time resolution) with a complete parameter set (20 parameters) for one virtual station.

4.6   Integration of the mDss system

The integration of the mDss package was implemented by applying the same modular approach. IWRMS manages different mDss core program versions, the basin specific configuration files and the associated results of the different Delphi rounds (Nevo and Chan, 2007). IWRMS can either be deployed to represent the results, together with all metadata as relevant information or to run the decision process again locally. Users can change general regulations and can adapt the mDss configuration to its local needs. Different runs can later be added together with new metadata to IWRMS depending on the permissions given to the user.

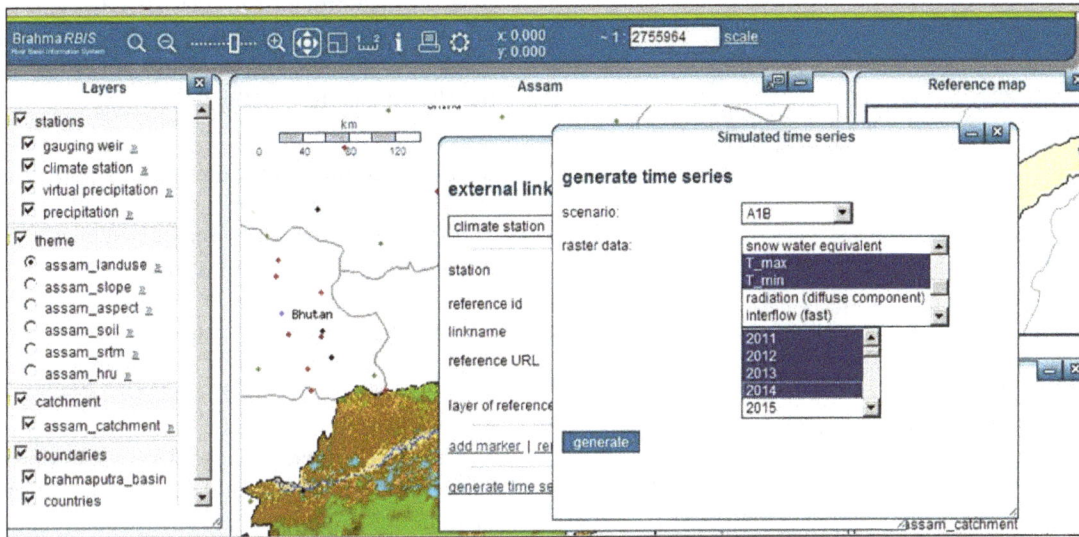

Figure 5. Extracting downscaled data time series from grid cells of a model domain.

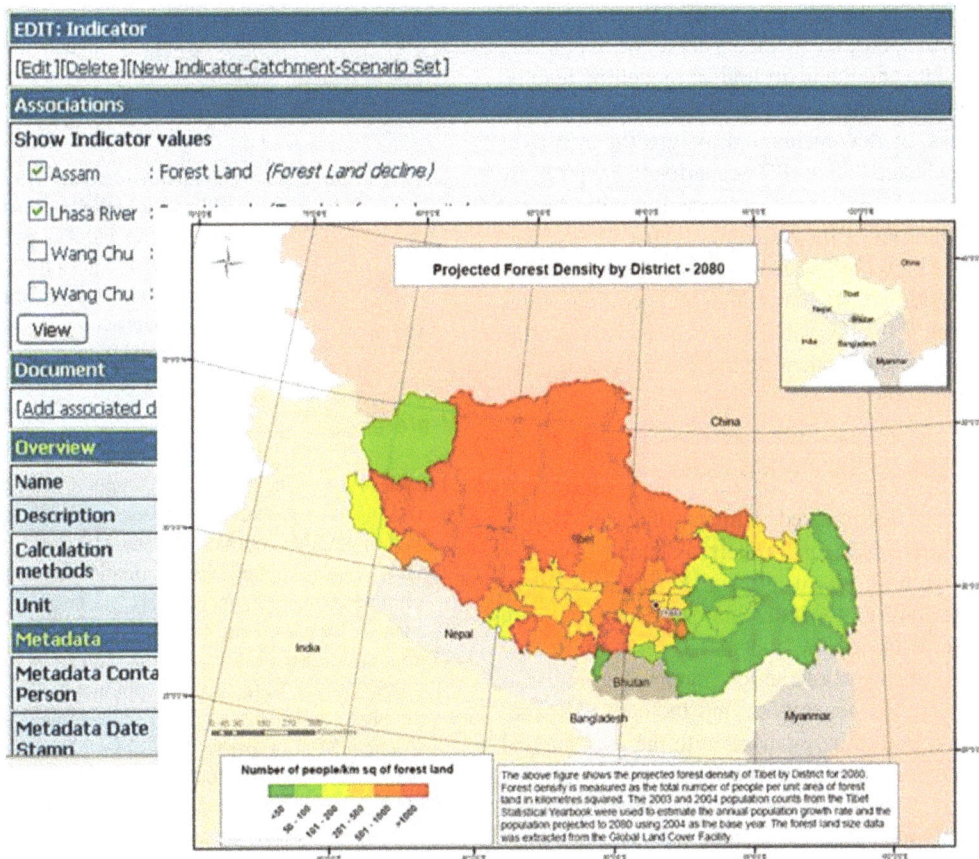

Figure 6. IWRMS components RBISscen and RBISind for indicator management.

Figure 7. Start GUI for the IWRMS demo-tour of the BRAH-MATWINN project.

4.7 Indicators and IPCC based "what-if?" scenarios

The use of indicators to analyse scenario based climate and hydrological model projections was realized by developing and implementing the modules *RBISscen* and *RBISind*. To meet expressed user needs, the results and dependencies of different indicators can be saved in different user selected formats.

RBISscen delivers an extended database structure and PHP application to integrate the different SRES scenarios (IPCC, 2000) together with additional metadata about the river basin, socio-economic data and the hydrological modelling results.

RBISind integrates indicators of different types as values, normalized values, or documents and assigns them to river basins and user selected "what-if?" scenarios (Chapter 8) by associating measured 'real world' information with the parameters and definitions of the respective "what-if?" scenarios. As shown in Fig. 6 the "what-if?" scenarios were linked to IWRMS metadata, i.e. stations and associated time series or binary documents, thus providing the required knowledge basis for developing and evaluating IWRM options for climate change adaptation.

4.8 Documentation of the IWRMS toolset

The system has been documented by means of a user friendly manual and comprehensive online help assistance for each GUI and item. A flash based interactive demo tour tutorial (Fig. 7) describes all functions and steps to work with the system, how to insert new metadata, upload GIS layers, create maps, manage time series, model data registration, virtual time series generation, and indicator management. The guided tour is harmonised with the user manual and can be executed either by a web browser or as a standalone application. The flash tutorial is available via the website http://www.brahmatwinn.uni-jena.de.

5 Contributions to sustainable IWRM

The development of the IWRMS presented herein can be seen as an innovative milestone for knowledge based decision support in IWRM. If used together with the many terra

bytes of data provided by the BRAHMATWINN project it offers a wide range of decision support for water and land managers, planners and all other kind of decision makers in the UDRB and the UBRB, respectively.

6 Conclusions and recommendations

The IWRMS, although presented herein as a "final product" of the BRAHMATWINN project is subject for enhancement and improvement that can be incorporated if applied in other IWRM related research projects. This process has already been started since the end of the project in December 2009.

Lessons learned in the BRAHMATWINN project are that a sophisticated training is a prerequisite to explore the potential of the system. Stakeholder organisations should associate permanent staff with the application of the system and this staff must be trained properly to form the core of further in-house training. If this is provided the system will offer its full potential also for other river basin IWRM challenges that are elaborated by the stakeholder organisation.

Acknowledgements. The cooperation and support from BRAH-MATWINN partners and stakeholders was highly appreciated and is acknowledged hereby. Acknowledgement is also given to the EC which funded the IWRMS development in the BRAHMATWINN project under the contract number 036952.

The interdisciplinary BRAHMATWINN EC-project carried out between 2006–2009 by European and Asian research teams in the UDRB and in the UBRB enhanced capacities and supported the implementation of sustainable Integrated Land and Water Resources Management (ILWRM).

References

Flügel, W.-A.: The Adaptive Integrated Data Information System (AIDIS) for global water research, Water Resources Management (WARM) Journal, 21, 199–210, 2007.

Flügel, W.-A.: Applied Geoinformatics for sustainable IWRM and climate change impact analysis, Technology, Resource Management & Development, 6, 57–85, 2009.

Giupponi, C., Sgobbi, A., Mysiak, J., Camera, R., and Fassio, A.: NetSyMoD – An Integrated Approach for Water Resources Management, in: Integrated Water Management, edited by: Meire, P., Coenen, M., Lombardo, C., Robba, M., and Sacile, R., Springer, Netherlands, 69–93, 2008.

IPCC, Intergovernmental Panel on Climate Change: Emissions Scenarios. A Special Report of IPCC Working Group III, 27 pp., 2000.

Kralisch, S., Zander, F., and Krause, P.: Coupling the RBIS Environmental Information System and the JAMS Modelling Framework, in: Proc. 18th World IMACS/and MODSIM09 International Congress on Modelling and Simulation, edited by: Anderssen, R., Braddock, R., and Newham, L., Cairns, Australia, 902–908, 2009.

Marke, T.: Development and Application of a Model Interface to couple Regional Climate Models with Land Surface Models for Climate Change Risk Assessment in the Upper Danube Watershed, Dissertation der FakultätfürGeowissenschaften, DigitaleHochschulschriften der LMU München, 188, available at: http://edoc.ub.uni-muenchen.de/9162/, München, 2008.

Mauser, W. and Bach, H.: PROMET – Large scale distributed hydrological modelling to study the impact of climate change on the water flows of mountain watersheds, J. Hydrol., 376, 362–377, 2009.

Nevo, D. and Chan, Y. E.: A Delphi study of knowledge management systems: Scope and requirements, Inform. Manage., 44, 583–597, 2007.

14

Data validation procedures in agricultural meteorology – a prerequisite for their use

J. Estévez[1], P. Gavilán[2], and A. P. García-Marín[1]

[1]University of Córdoba, Projects Engineering, Córdoba, Spain
[2]IFAPA Center "Alameda del Obispo", Junta de Andalucía, Córdoba, Spain

Abstract. Quality meteorological data sources are critical to scientists, engineers, climate assessments and to make climate related decisions. Accurate quantification of reference evapotranspiration (ET_0) in irrigated agriculture is crucial for optimizing crop production, planning and managing irrigation, and for using water resources efficiently. Validation of data insures that the information needed is been properly generated, identifies incorrect values and detects problems that require immediate maintenance attention. The Agroclimatic Information Network of Andalusia at present provides daily estimations of ET_0 using meteorological information collected by nearly of one hundred automatic weather stations. It is currently used for technicians and farmers to generate irrigation schedules. Data validation is essential in this context and then, diverse quality control procedures have been applied for each station. Daily average of several meteorological variables were analysed (air temperature, relative humidity and rainfall). The main objective of this study was to develop a quality control system for daily meteorological data which could be applied on any platform and using open source code. Each procedure will either accept the datum as being true or reject the datum and label it as an outlier. The number of outliers for each variable is related to a dynamic range used on each test. Finally, geographical distribution of the outliers was analysed. The study underscores the fact that it is necessary to use different ranges for each station, variable and test to keep the rate of error uniform across the region.

1   Introduction

Meteorological information is one of the most important tools used by agriculture producers in decision making (Weiss and Robb, 1986). Some of the applications for these climate data include: crop water-use estimates, irrigation scheduling, integrated pest management, crop and soil moisture modeling, design and management of irrigation and drainage system and frost and freeze warnings and forecasts (Meyer and Hubbard, 1992).

Andalusia is located in the south of the Iberian Peninsula. This region is situated between the meridians 1° and 7° W and the parallels 37° and 39° N, with an extension around 9 Mha. The climate is semiarid, typically Mediterranean, with very hot and dry summers. In Andalusia 900 000 ha are irrigated (around 20 % of the cultivated area) under very different conditions (Gavilán et al., 2006).

The Agroclimatic Information Network of Andalusia (RIAA in Spanish) was deployed to provide coverage to most of the irrigated areas of the region and to improve irrigation water management (De Haro et al., 2003). Its exploitation and maintenance are carried out by the IFAPA (Agricultural Research Institute of Regional Government of Andalusia). This network provides at present daily estimations of reference evapotranspiration (ET_0) using meteorological information collected by nearly one hundred automatic weather stations (Gavilán et al., 2008). This information is easily accessible due to it is published in the Web: http://www.juntadeandalucia.es/agriculturaypesca/ifapa/ria/.

Meteorological data validation is very important for hydrological designs and agricultural decision makings, concretely to estimate irrigation schedules. The quality control system discussed herein was applied to 85 stations, summarized in Table 1. The rest of the stations have been recently installed and their data series were too short. Quality control system consists of procedures or tests against which data are tested, setting data flags to provide guidance to end users. These flags give information about which tests have been applied satisfactorily or not to meteorological data.

Table 1. Summary of automated weather stations used in the study.

Stations (Province)	Elevation (m)	Latitude (°)	Longitude (°)
Basurta-Jerez (CÁDIZ)	60	36.75	−6.01
Jerez Frontera (CÁDIZ)	32	36.64	−6.01
Villamartín (CÁDIZ)	171	36.84	−5.62
Conil Frontera (CÁDIZ)	26	36.33	−6.13
Vejer Frontera (CÁDIZ)	24	36.28	−5.83
Jimena Frontera (CÁDIZ)	53	36.41	−5.38
Puerto Sta. María (CÁDIZ)	20	36.61	−6.15
La Mojonera (ALMERÍA)	142	36.78	−2.70
Almería (ALMERÍA)	22	36.83	−2.40
Tabernas (ALMERÍA)	435	37.09	−2.30
Fiñana (ALMERÍA)	971	37.15	−2.83
V. Fátima-Cuevas (ALMERÍA)	185	37.39	−1.76
Huércal-Overa (ALMERÍA)	317	37.41	−1.88
Cuevas Almanz. (ALMERÍA)	20	37.25	−1.79
Adra (ALMERÍA)	42	36.74	−2.99
Níjar (ALMERÍA)	182	36.95	−2.15
Tíjola (ALMERÍA)	796	37.37	−2.45
Bélmez (CÓRDOBA)	523	38.25	−5.20
Adamuz (CÓRDOBA)	90	37.99	−4.44
Palma del Río (CÓRDOBA)	134	37.67	−5.24
Hornachuelos (CÓRDOBA)	157	37.72	−5.15
El Carpio (CÓRDOBA)	165	37.91	−4.50
Córdoba (CÓRDOBA)	117	37.86	−4.80
Santaella (CÓRDOBA)	207	37.52	−4.88
Baena (CÓRDOBA)	334	37.69	−4.30
Baza (GRANADA)	814	37.56	−2.76
Puebla D.Fadriq. (GRANADA)	1110	37.87	−2.38
Loja (GRANADA)	487	37.17	−4.13
Pinos Puente (GRANADA)	594	37.26	−3.77
Iznalloz (GRANADA)	935	37.41	−3.55
Jerez Marques. (GRANADA)	1212	37.19	−3.14
Cádiar (GRANADA)	950	36.92	−3.18
Zafarraya (GRANADA)	905	36.99	−4.15
Almuñécar (GRANADA)	49	36.74	−3.67
Padul (GRANADA)	781	37.02	−3.59
Tojalillo-Gibraleón (HUELVA)	52	37.31	−7.02
Lepe (HUELVA)	74	37.24	−7.24
Gibraleón (HUELVA)	169	37.41	−7.05
Moguer (HUELVA)	87	37.14	−6.79
Niebla (HUELVA)	52	37.34	−6.73
Aroche (HUELVA)	299	37.95	−6.94
Puebla Guzmán (HUELVA)	288	37.55	−7.24
El Campillo (HUELVA)	406	37.66	−6.59
Palma Condado (HUELVA)	192	37.36	−6.54
Almonte (HUELVA)	18	37.15	−6.47
Moguer-Cebollar (HUELVA)	63	37.24	−6.80
Huesa (JAÉN)	793	37.74	−3.06
Pozo Alcón (JAÉN)	893	37.67	−2.92
S.José Propios (JAÉN)	509	37.85	−3.22
Sabiote (JAÉN)	822	38.08	−3.23
Torreblascopedro (JAÉN)	291	37.98	−3.68
Alcaudete (JAÉN)	645	37.57	−4.07
Mancha Real (JAÉN)	436	37.91	−3.59
Úbeda (JAÉN)	358	37.94	−3.29
Linares (JAÉN)	443	38.06	−3.64
Marmolejo (JAÉN)	208	38.05	−4.12
Chiclana Segura (JAÉN)	510	38.30	−2.95
Higuera Arjona (JAÉN)	267	37.95	−4.00

Table 1. Continued.

Stations (Province)	Elevation (m)	Latitude (°)	Longitude (°)
Santo Tomé (JAÉN)	571	38.03	−3.08
Jaén (JAÉN)	299	37.89	−3.77
Palacios-Villafran. (SEVILLA)	21	37.18	−5.93
Cabezas S. Juan (SEVILLA)	25	37.01	−5.88
Lebrija 2 (SEVILLA)	40	36.90	−6.00
Aznalcázar (SEVILLA)	4	37.15	−6.27
Puebla del Río II (SEVILLA)	41	37.08	−6.04
Écija (SEVILLA)	125	37.59	−5.07
La Luisiana (SEVILLA)	188	37.52	−5.22
Osuna (SEVILLA)	214	37.25	−5.13
La Rinconada (SEVILLA)	37	37.45	−5.92
Sanlúcar la Mayor (SEVILLA)	88	37.42	−6.25
Villan.Río-Minas (SEVILLA)	38	37.61	−5.68
Lora del Río (SEVILLA)	68	37.66	−5.53
Los Molares (SEVILLA)	90	37.17	−5.67
Guillena (SEVILLA)	191	37.51	−6.06
Puebla Cazalla (SEVILLA)	229	37.21	−5.34
Carmona-Tomejil (SEVILLA)	79	37.40	−5.58
Málaga (MÁLAGA)	68	36.75	−4.53
Vélez-Málaga (MÁLAGA)	49	36.79	−4.13
Antequera (MÁLAGA)	457	37.05	−4.55
Estepona (MÁLAGA)	199	36.44	−5.20
Archidona (MÁLAGA)	516	37.07	−4.42
Sierra Yeguas (MÁLAGA)	464	37.13	−4.83
Churriana (MÁLAGA)	32	36.67	−4.50
Pizarra (MÁLAGA)	84	36.76	−4.71
Cártama (MÁLAGA)	95	36.71	−4.67

2 Materials and methods

2.1 Source of data

The dataset used in the present study was obtained from the daily database of the RIAA and it was from 2004 to 2009. Each station is controlled by a CR10X datalogger (Campbell Scientific) and is equipped with sensors to measure air temperature and relative humidity (HMP45C probe, Vaisala), solar radiation (pyranometer SP1110 Skye), wind speed and direction (wind monitor RM Young 05103) and rainfall (tipping bucket rain gauge ARG 100). Air temperature and relative humidity are measured at 1.5 m and wind speed at 2 m above soil surface. Data from stations are transferred to the data-collecting seat (Main Center) by using GSM modems. This information is saved in a database. The Main Center is responsible for quality control procedures that comprise the routine maintenance program of the network, including sensor calibration and data validation.

Accuracy of ET_0 calculations depends on the quality and the integrity of meteorological data used (Allen, 1996), being necessary data quality control application. Different procedures for quality assurance have been described by Meek and Hatfield (1994), Allen (1996), Shafer et al. (2000) and Feng et al. (2004). These tests are based on some rules proposed

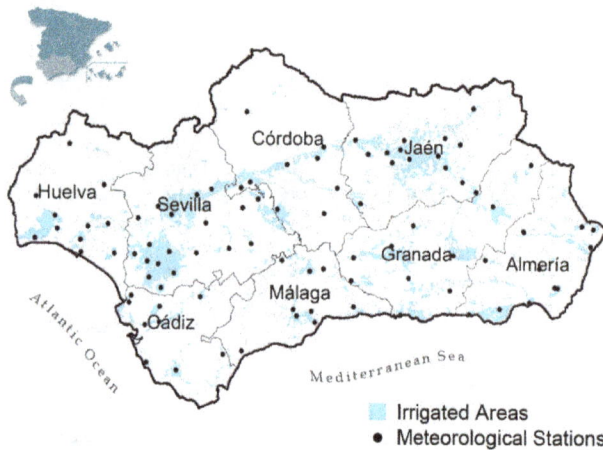

Figure 1. Agroclimatic Information Network of Andalusia (85 meteorological stations).

by O'Brien and Keefer (1985). However, the tests applied in this study are based on statistical decisions and they were conducted for 84 stations (Fig. 1), using data only from a single site. Three procedures were tuned to the prevailing climate: seasonal thresholds, seasonal rate of change and seasonal persistence (Hubbard et al., 2005). These tests are related to station climatology at the monthly level, using dynamic limits for each variable. The tests were applied to the following variables: maximum, minimum and mean air temperature (Tx, Tn, Tm), maximum, minimum and mean relative humidity (RHx, RHn, RHm), and precipitation (Preci).

2.2 Theory

The THRESHOLD test is a quality control approach that checks whether the variable x falls in a specific range for the month in question. The equation is

$$\overline{x} - f\sigma_x \leq x \leq \overline{x} + f\sigma_x \tag{1}$$

where $\overline{x}$ is the daily mean (e.g., mean of maximum daily temperature for December) and σ_x is the standard deviation of the daily values for the month in question. This relationship indicates that with larger values of f, the number of potential outliers decreases.

The STEP CHANGE test compares the change between successive observations. This test checks if the difference value of the variable falls inside the climatologically expected lower and upper limits on daily rate of change for the month in question. The step change test for variable x is given in Eq. (2):

$$\overline{d}_i - f\sigma_{d_i} \leq d_i \leq \overline{d}_i + f\sigma_{d_i} \tag{2}$$

where $d_i = x_i - x_{i-1}$, i is the day and σ_{d_i} is the standard deviation of d_i.

The PERSISTENCE test checks the variability of the measurements. When the variability is too high or too low, the

data should be flagged for further checking. If the sensor fails it will often report a constant value and the standard deviation (σ) will become smaller. When the sensor is out for an entire period, σ will be zero. If the instrument works intermittently and produces reasonable values interspersed with zero values, thereby greatly increasing the variability for the period. This test compares the standard deviation for the time period being tested to the limits expected as follows:

$$\overline{\sigma}_j - f\sigma_{\sigma_j} \leq \sigma_j \leq \overline{\sigma}_j + f\sigma_{\sigma_j} \tag{3}$$

where σ_j is the standard deviation from daily values for each month (j) and year and σ_{σ_j} is the standard deviation of σ_j for the month in question.

When the datum is valid and is rejected by the tests, a Type I error is committed. If the datum is not valid but it is accepted by the quality control procedures, a Type II error is committed. The results discussed in this paper only show the potential outliers of Type I error.

This system was developed in open source code, using GNU GPL (General Public License) support and it can be installed on any platform: Linux, Windows, Unix, Mac OS, Solaris, etc. PostgreSQL, PostGIS and PLpgSQL are the selected free technologies under the quality procedures were developed.

PosgreSQL is an object-relational database management system (ORDBMS) based on POSTGRES version 4.2, developed at the University of California at the Berkeley Computer Science Department (Stonebraker and Kemnitz, 1991). It supports a large part of the SQL standard and offers many modern features: complex queries, foreign keys, triggers, views, functions, procedures languages, etc. PostGIS is an extension to PostgreSQL which allows GIS (Geographic Information Systems) objects to be stored in the database. It includes support for a range important GIS functionality, including full OpenGIS support, advanced topological constructs (coverages, surfaces, networks), desktop user interface tools for viewing and editing GIS data, and web-based access tools. Finally, PLpgSQL is a powerful procedure language used to specify a sequence of steps that are followed to procedure an intended programmatic result. The use of SQL within PLpgSQL increases the power, flexibility, and performance of the quality tests. The most important aspect of using this language is its portability. Its functions are compatible with all the platforms that can operate de PostgreSQL database system.

These three tests were applied to data from selected stations, following Eqs. (1), (2) and (3).

3 Results and discussion

The next figures show the number of potential Type I errors that would occur when using the specified tests with various f factors. The fraction data flagged is represented on a log scale and related to the all the network tested (85 stations).

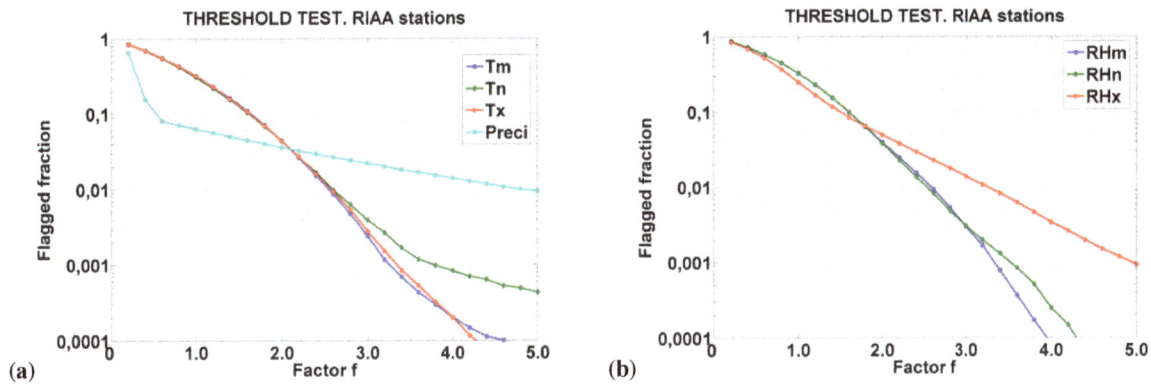

Figure 2. (a) Threshold Test – Maximum (Tx), minimum (Tn) and mean temperature (Tm) and Precipitation (Preci). (b) Threshold Test – Maximum (RHx), minimum (RHn) and mean relative humidity (RHm).

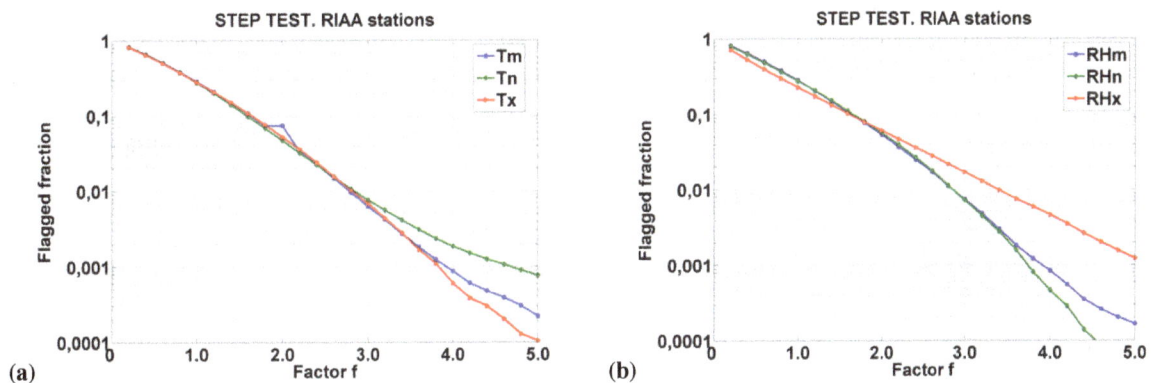

Figure 3. (a) Step Test – Maximum (Tx), minimum (Tn) and mean temperature (Tm). (b) Step Test – Maximum (RHx), minimum (RHn) and mean relative humidity (RHm).

The general shape of the relationship between f and the fraction of data flagged is shown in Figs. 2, 3 and 4. The results obtained in this work are similar to the results of Hubbard et al. (2005). The results for the threshold analysis indicate that approximately 2 % of the data would be flagged for maximum, minimum and mean temperature if an f value of 2.3 is used. For precipitation, 2 % of the data were flagged in this test for an f value of 3.1. These results are shown in Fig. 2a. The results on Fig. 2b show the same fraction data flagged for minimum and mean relative humidity when f value of 2.2 is used. In this figure and for maximum relative humidity, this percentage of data would be flagged with an f value of 2.7. Similar figures are shown for the step change test (Fig. 3a and b) and the persistence test (Fig. 4a and b). The results for the persistence analysis indicate that approximately 1 % of the data would be flagged for all the variables if an f value less than 2.0 is used. This is consequence of the need for longer series of data to calculate the variability from daily values for each month and year. For precipitation, the step test was not applied because of the discontinuous nature of rainfall. These results are related to the three tests applied to 85 automatic weather stations of the RIAA. It is impor-

tant to remark that the fraction flagged for each f value was different for each station. These results show that it will be possible to select dynamic f values for each station and temporal scale and to fix a specific rate of Type I errors across the region.

The spatial distribution of the fraction data flagged for an f value of 3 in threshold and step tests was estimated using GIS techniques for all the variables. This analysis is very useful to visually study the distribution of outliers across the region. The results for threshold test using ordinary krigging interpolation for maximum temperature are shown in Fig. 5. This map shows that the fraction data flagged is higher in coastal weather stations than in inland locations. This is caused by the different climate regime between them. The maximum temperatures are lower in locations near the coast than in inland locations where the air masses are not influenced by a nearby and large water body (Mediterranean Sea or Atlantic Ocean).

The quality control system can dynamically generate this type of maps using any GIS software at any time.

Sometimes, for scientific or other purposes we cannot reject too much data. It can be very useful to fix a rate of

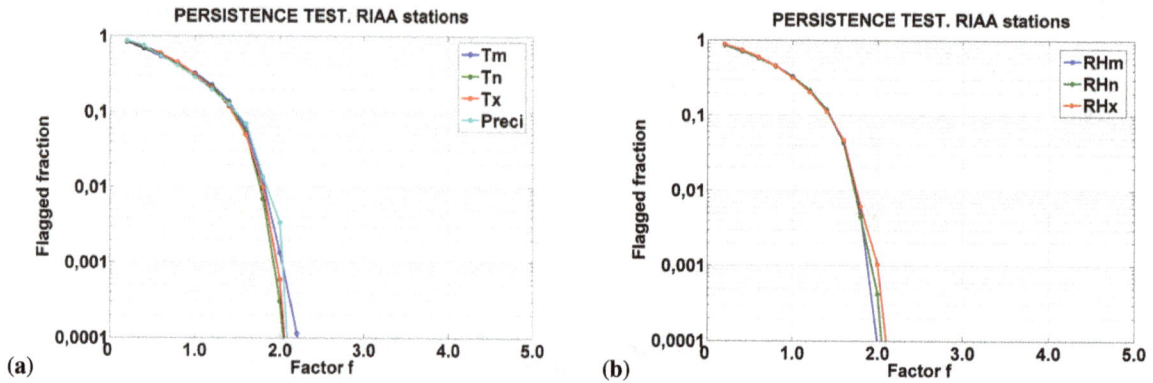

Figure 4. **(a)** Persistence Test – Maximum (Tx), minimum (Tn) and mean temperature (Tm) and Precipitation (Preci). **(b)** Persistence Test – Maximum (RHx), minimum (RHn) and mean relative humidity (RHm).

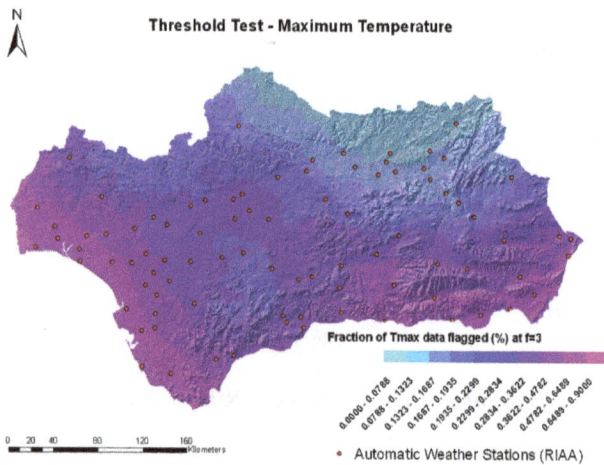

Figure 5. Fraction of maximum temperature data flagged at $f = 3$ for threshold test.

4 Summary and conclusions

In this study, the validation tests applied to daily climatic data from 85 automatic weather stations varied modestly with climate type and significantly with the variable tested. It is essential to test the capability of validation procedures because of quality control is a major prerequisite for using meteorological information. Several tests based on statistical decisions have been applied to meteorological data from the Agroclimatic Information network of Andalusia (RIAA). The validated variables were maximum, minimum and mean air temperature (Tx, Tn, Tm), maximum, minimum and mean relative humidity (RHx, RHn, RHm) and precipitation (Preci). Although daily precipitation is known to follow a gamma distribution, it was included in these tests to give a reference point. Results obtained from running the quality control procedures showed a high variability when different f values are used. It is essential to test the capability of these tests to produce flags if data are out of range or are internally or temporally inconsistent.

The use of open source code and General Public License technologies (GNU GPL) to develop the procedures allows any meteorological network to implement a similar system with zero cost. All the functions and algorithms can be read and rewritten or adapted for future users.

The possibility of dynamically mapping the percentage of errors for any variable is a powerful tool to visually study the spatial distribution of the fraction data flagged. These results show that it necessary to select dynamic f values for each station and test to preselect a fixed rate of error detection across the Andalusia region.

This quality control system can easily be used with any conventional GIS software. The treatment of the meteorological data like geographical variables using GIS techniques can be very useful for maintenance routines and sensors calibration.

Future works of the authors should include spatial consistency procedures and to introduce seeded random errors to examine the Type II errors detection.

potential outliers for not considering them in our model or study. For fixing a specific rate of fraction flagged in this example of maximum temperature (Tx), we should use different f values for each station. As it can be seen in Fig. 5, using $f = 3$, the fraction of Tx data flagged ranged from nearly 0 (station located at northeast of Jaén) to 0.6–0.9 approximately (coastal stations) across Andalusia region.

These automated validation procedures should be accompanied by other tasks such as: field visits for maintenance routines, sensors calibration and manual inspection (Feng et al., 2004; Shafer et al., 2000). This manual inspection is crucial and necessary for ensuring an appropriate flagging process, providing human judgment to it, catching subtle errors that automated techniques may miss (Shafer et al., 2000).

Edited by: B. Lalic
Reviewed by: V. Vucetic and two other anonymous referees

References

Allen, R. G.: Assessing integrity of weather data for reference evapotranspiration estimation, J. Irrig. Drain. Eng., 122(2), 97–106, 1996.

De Haro, J. M., Gavilán, P., and Fernández, R.: The Agroclimatic Information Network of Andalusia, Proceeding of the Third International Conference on Experiences with Automatic Weather Stations, Torremolinos, Spain, 19–21 February, 1–12, 2003.

Feng, S., Hu, Q., and Qian, Q.: Quality control of daily meteorological data in China, 1951-2000: a new dataset, Int. J. Climatol., 24, 853–870, 2004.

Gavilán, P., Lorite, I. J., Tornero, S., and Berengena, J.: Regional calibration of Hargreaves equation for estimating reference ET in a semiarid environment, Agric. Water Manag., 81, 257–281, 2006.

Gavilán, P., Estévez J., and Berengena, J.: Comparison of standardized reference evapotranspiration equations in southern Spain, J. Irrig. Drain. Eng. ASCE, 134(1), 1–12, 2008.

Hubbard, K. G., Goddard, S., Sorensen, W. D., Wells, N., and Osugi, T. T.: Performance of quality assurance procedures for an applied climate information system, J. Atmos. Oceanic Technol., 22, 105–112, 2005.

Meek, D. W. and Hatfield, J. L.: Data quality checking for single station meteorological databases, Agric. For. Meteor., 69, 85–109, 1994.

Meyer, S. J. and Hubbard, K. G.: Nonfederal automated weather stations and networks in the United States and Canada: a preliminary survey, B. Am. Meteorol. Soc., 73(4), 449–457, 1992.

O'Brien, K. J. and Keefer, T. N.: Real-time data verification, Proc. ASCE Special Conf., Buffalo, NY, American Society of Civil Engineers, 764–770, 1985.

PostGIS: http://postgis.refractions.net (last access: 5 December 2009), 2009.

PostgreSQL: http://www.postgresql.org (last access: 5 December 2009), 2009.

Shafer, M. A., Fiebrich, C. A., Arndt, D. S., Fredrickson, S. E., and Hughes, T. W.: Quality assurance procedures in the Oklahoma Mesonet, J. Atmos. Oceanic Technol., 17, 474–494, 2000.

Stonebraker, M. and Kemnitz, G.: The Postgres next-generation database-management system, Communicat. ACM., 34, 78–92, 1991.

Weiss, A. and Robb, J. G.: Results and interpretations from a survey on agriculturally related weather information, B. Am. Meteorol. Soc., 67(1), 10–15, 1986.

15

Prototype of a drought monitoring and forecasting system for the Tuscany region

R. Magno[1,2]**, L. Angeli**[1]**, M. Chiesi**[2]**, and M. Pasqui**[2]

[1]LaMMA Consortium, Sesto Fiorentino, Florence, Italy
[2]Institute of Biometeorology, National Research Council, Florence, Italy

Correspondence to: R. Magno (magno@lamma.rete.toscana.it)

Abstract. A system for drought monitoring and medium–long time forecasting in the Tuscany region (central Italy) is briefly introduced, which is based on ground and satellite data (1 km spatial resolution and 16-day temporal resolution). It is also shown how information about current conditions and future evolution of a drought event is periodically delivered on the LaMMA Consortium website, in collaboration with the Institute of Biometeorology (IBIMET-CNR).

1 Introduction

With respect to other extreme climatic events, drought is a creeping and complex phenomenon (Gillette, 1950; Tannehill, 1947), characterized by a slow and often long-lasting evolution: its onset and withdrawal are generally difficult to define (Iglesias et al., 2009); its intensity and extension are spatially and temporally extremely variable; and the impacts produced on the environment can arise later and persist even after its end (Vicente-Serrano et al., 2012). As demonstrated by other authors (e.g., Jayaraman et al., 1997; Brown et al., 2008; Jain et al., 2010), a comprehensive framework including a climate-based, satellite-derived monitoring and a seasonal weather forecast is the most reliable way to identify drought occurrence and trends and to deliver timely information for impact reduction. In this study a proactive, integrated drought monitoring and seasonal forecasting tool is briefly illustrated; it was implemented for the Tuscany region (Italy/central Mediterranean) by the LaMMA Consortium and the Institute of Biometeorology (IBIMET-CNR), and aims at filling the temporal gap between the development of a dry period and the response of final users in managing drought-related emergencies, by delivering maps and information in quasi-real-time.

2 The operational chain

The operational chain implemented to calculate drought indices and deliver final products for drought monitoring and forecasting in Tuscany is based on semi-automatic procedures (see Fig. 1).

2.1 The monitoring component

The monitoring component is developed by integrating the state-of-the-art science and technologies and by selecting a set of coupled rainfall-based and satellite-derived indices that follow several criteria: (1) types of drought, (2) availability and consistency of data, (3) geographical characteristics, (4) time and spatial variability, (5) main final users. For our operational framework two rainfall-based indices were identified, being considered more representative than others (Morid et al., 2006): the Standardized Precipitation Index (SPI) (McKee et al., 1993) and the Effective Drought Index (EDI) (Byun et al., 1999). The SPI, a robust and reliable index (Heim, 2002; Keyantash and Dracup, 2002), provides multiple timescale drought occurrence and detects its variation and duration; the EDI, which is calculated on a daily basis, is thus more sensitive to each single rainfall event and tracks the influence of precipitation on the recovery from an accumulated deficit. Additionally EDI is effective to spatially recognize the onset of a drought episode (Morid et al., 2006). The satellite-derived indices are focused on the vegetation

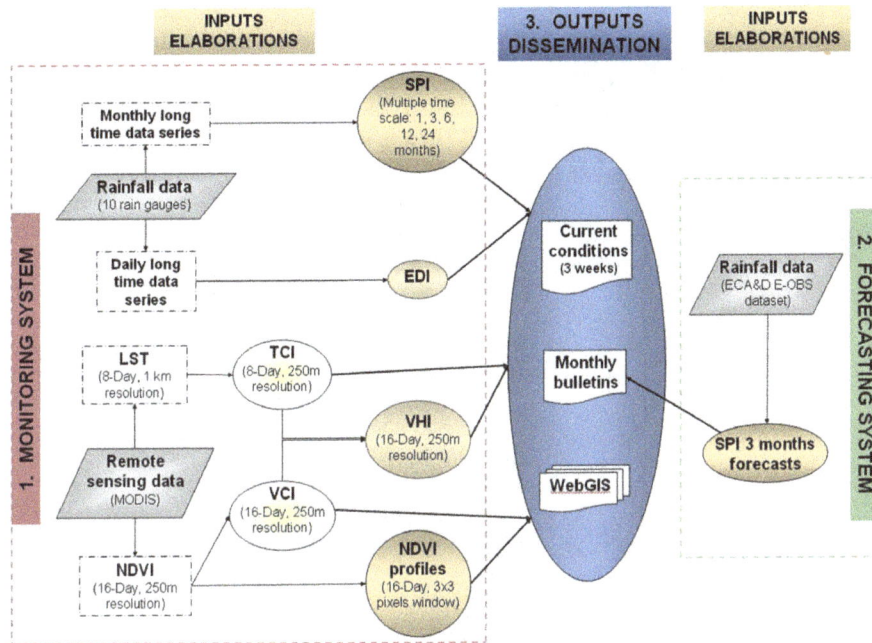

Figure 1. Schematic representation of the drought monitoring and forecasting system components.

performances related to temperature and moisture stress and are based both on the Normalized Difference Vegetation Index (NDVI) and the Land Surface Temperature (LST), elaborated from Terra MODIS (Moderate Resolution Imaging Spectroradiometer) images to obtain the comprehensive Vegetation Health Index (VHI) (1 km spatial resolution) (Kogan, 1995) (see Fig. 2). These indices represent an indirect drought responsive way to analyze the phenomenon and are widely applied due to their capability to describe in more detail the spatial characteristics of temporal drought dynamics (Wan et al., 2004). They constitute the core of our monitoring system during the growing season (spring and summer), when problems related to the cloud cover, typical of the northern part of the Mediterranean region, are generally reduced with respect other periods.

2.2 The seasonal forecast component

An empirical multi-regressive approach, based on observed climate indices, is adopted to produce seasonal forecasts for the 3-month SPI. This component provides an estimate of the spatial and temporal distribution of climate anomalies up to 3 months into the future. These seasonal forecasts follow a physically based empirical approach based on a multi-regressive method (Pasqui et al., 2009): it estimates the multi-linear relations of a data set of observed oceanic and atmospheric predictors, on a monthly and 3-monthly basis, with SPI values with respect to the 1981–2010 training period. Oceanic and atmospheric predictors, along with their relative leading time are selected according to the best correlation with the target SPI values over the target area

(see Pasqui et al. (2009) for the complete list of predictors). The multi-regressive coefficients, computed over the training period, are then used to forecast the expected SPI values according to the last observed values of selected predictors. The 3-month SPI dates are computed from the daily E-OBS gridded precipitation data set (period from 1950 to 2013) from the ECA&D (European Climate Assessment & Dataset) project, providing the reference framework for seasonal drought evolution outlooks.

3 Product dissemination

Index analysis and final products must be easily delivered to ensure the end users useful, effective and timely information for their final needs: the internet is the best way to disseminate drought monitoring and forecasting alerts. To this aim, information is delivered at different time steps:

– During the growing season, vegetation conditions of the previous 16 days are updated on a specific page of the Consortium website, with a lag of about 1 week. This is done analyzing the NDVI profiles and the VHI index (http://www.lamma.rete.toscana.it/siccita-situazione-corrente).

– Online monthly bulletins (http://issuu.com/consorziolamma), on the other hand, provide a more detailed description of drought evolution throughout a joint analysis of satellite- and climate-based drought indices during the previous 30 days, with focus on forest types and main tree crops. A forecast of the SPI index for the next months is also provided.

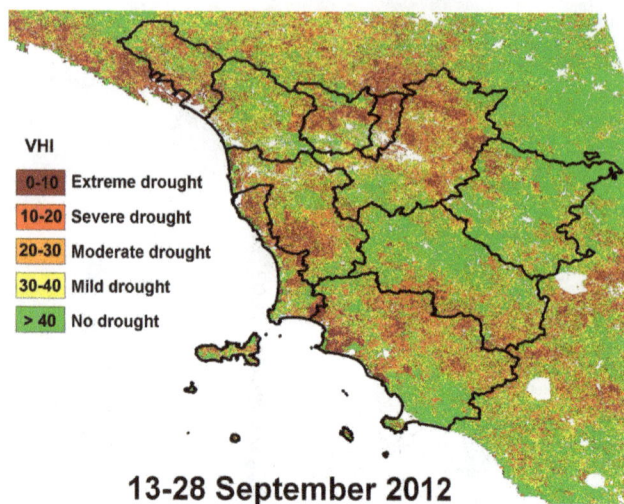

13-28 September 2012

Figure 2. Vegetation Health Index (VHI) referred to the bi-weekly period 13–28 September 2012.

- A WebGIS application (http://www.lamma.rete. toscana.it/webgis-siccita) based on open source solutions has been customized in order to integrate different data sets and share maps of drought indices with decision-makers and other stakeholders (Rocchi et al., 2010).

Statistics of the online monthly bulletin web accesses after 18 months of operational drought monitoring and forecast indicate that, on average, there are about 1200 specific contacts, with peaks during the end of spring and summer. Even in cold seasons when drought events occur (like in 2012), a large number of contacts were registered (see Fig. 2).

4   Conclusions and future perspectives

The described monitoring and forecasting systems are active all over the year, following and assessing the temporal and spatial evolution of possible drought events and integrating heterogeneous data: climate-based and vegetation indices. Bulletins emission follows a monthly to bi-weekly basis scheme providing information on the vegetation response to possible drought conditions. Due to the importance of the terrestrial water budget and in order to provide a better tool for natural and agricultural resources management, especially during drought events, a simplified "water balance model" has been also implemented and is under verification; it provides an estimate of actual evapotranspiration (ET_A) with a high spatiotemporal resolution, based on ground and remote sensing data (Chiesi et al., 2013). The model combines estimates of potential evapotranspiration (ET_0) and of fractional vegetation cover derived from NDVI, in order to simulate both transpiration and evaporation processes. Current results and further ongoing validations indicate a promising valuable operational use on Tuscany areas where vegetation cover is

fragmented and agriculture is often represented by annual rotating crops.

Acknowledgements. The authors thank their colleagues Fabio Maselli and Luca Fibbi for assisting during the development of the simplified water balance model and Bernardo Gozzini for continuously supporting new ideas. This research was partly supported by the project "LInking Long Term Observatories with Crop Systems Modeling For a better understanding of Climate Change Impact and Adaptation StRategies for Italian Cropping Systems" (IC-FAR; www.icfar.it), and it also contributes to the knowledge hub "Modelling European Agriculture with Climate Change for Food Security" (MACSUR; www.macsur.eu) within the Joint Programming Initiative on Agriculture, Food Security and Climate Change (FACCE-JPI; www.faccejpi.com).

Edited by: C. Buontempo
Reviewed by: two anonymous referees

References

Brown, J. F., Wardlow, B. D., Tadesse, T., Hayes, M. J., and Reed, B. C.: The Vegetation Drought Response Index (VegDRI): A New Integrated Approach for Monitoring Drought Stress in Vegetation, GIScience & Remote Sensing, 45, 16–46, 2008.

Byun, H. R. and Wilhite, D. A.: Objective Quantification of Drought Severity and Duration, J. Climate, 12, 2747–2756, 1999.

Chiesi, M., Rapi, B., Battista, P., Fibbi, L., Gozzini, B., Magno, R., Raschi, A., and Maselli, F.: Combination of ground and satellite data for the operational estimation of daily evapotranspiration, European J. Remote Sens., 46, 675–688, 2013.

Gillette, H. P.: A creeping drought under way, Water and Sewage Works, March, 104–105, 1950.

Heim Jr., R. R.: A review of twentieth-century drought indices used in the United States, B. Am. Meteorol. Soc., 83, 1149–1165, 2002.

Iglesias, A., Garrote, L., Cancelliere, A., Cubillo, F., and Wilhite, D. A.: Coping with drought risk in agriculture and water supply systems. Drought Management and Policy Development in the Mediterranean Series: Advances in Natural and Technological Hazards Research, 26, XVIII, 322 pp., 2009.

Jain, S. K., Keshri, R., Goswami, A., and Sarkar, A.: Application of meteorological and vegetation indices for evaluation of drought impact: a case study for Rajasthan, India. Natural Hazards, 54, 643–656, 2010.

Jayaraman, V., Chandrasekhar, M. G., and Rao, U. R.: Managing the natural disasters from space technology inputs, Acta Astronautica, 40, 291–325, 1997.

Keyantash, J. and Dracup, J. A.: The quantification of drought: an evaluation of drought indices, B. Am. Meteorol. Soc., 83, 1167–1180, 2002.

Kogan, F. N.: Application of vegetation index and brightness temperature for drought detection, Adv. Space Res., 15, 91–100, 1995.

McKee, T. B., Doesken, N. J., and Kliest, J.: The relationship of drought frequency and duration to time scales, Proceedings of the 8th Conference of Applied Climatology, Anaheim, CA, 179–184, 1993.

Morid, S., Smakhtin, V., and Moghaddasi, M.: Comparison of seven meteorological indices for drought monitoring in Iran, Int. J. Climatol., 26, 971–985, 2006.

Pasqui, M., Primicerio, J., Benedetti, R., Crisci, A., Genesio, L., and Maracchi, G.: Seasonal forecasting precipitation in the Mediterranean basin. Managing water in a changing world, International Conference hosted by the Commission for Water Sustainability International Geographical Union (IGU), Torino, Italy. 2009.

Rocchi, L., De Filippis, T., and Magno, R.: An open source general-purpose framework for implementing webGIS applications. FOSS4G International Conference for Open Source Geospatial Software. Barcelona (Spain), 2010.

Tannehill, I. R.: Drought: Its Causes and Effects, Princeton University Press, Princeton, New Jersey, 1947.

Vicente-Serrano, S. M., Beguería, S., Lorenzo-Lacruz, J., Camarero, J. J., López-Moreno, J. I., Azorin-Molina, C., Revuelto, J., Morán-Tejeda, E., and Sanchez-Lorenzo, A.: Performance of drought indices for ecological, agricultural and hydrological applications, Earth Interactions, 16, 1–27, 2012.

Wan, Z., Wang, P., and Li, X.: Using MODIS land surface temperature and normalized difference vegetation index for monitoring drought in the southern Great Plains, USA, Int. J. Remote Sens., 25, 61–72, 2004.

16

Analysis of present IWRM in the Upper Brahmaputra and the Upper Danube River Basins

W.-A. Flügel and A. Bartosch

Department of Geoinformatics, Friedrich Schiller University Jena, Germany

Abstract. Integrated Water Resources Management (IWRM) is a process which strives towards the sustainable management of water resources in river basins. The approach integrates insights and knowledge from various scientific disciplines comprising natural, socio-economic, and engineering sciences. These three pillars of sustainability are important components of this approach integrating the environmental, economic and social dimension. In the ideal IWRM case planning is based on the river basin scale and therefore is comparatively discussed herein for the two twinning BRAHMATWINN river basins, i.e. the Upper Danube River Basin (UDRB) in Europe and the Upper Brahmaputra River Basin (UBRB) in South Asia. In this chapter major challenges for the implementation of the IWRM process towards a sustainable management of water resources in the two UDRB and UBRB twinning river basins of the BRAHMATWINN project are analysed. The study revealed that in the UDRB the IWRM approach is already part of water management planning and the implementation of the EU Water Framework Directive (WFD) is a good example in this regard. Contrary in the UBRB the implementation of IWRM is just at the beginning phase, only recently is being discussed in the riparian states but has not been implemented in any way so far on the basin scale.

1 Introduction and objectives

IWRM is understood as a continuous process of coordinating sustainable land and water resources management with the aims (1) to maximize the socio-economic development and social welfare without (2) compromising the sustainability of vital ecosystems (GWP, 2000). Thus IWRM internationally is considered as the appropriate way to implement sustainable management of river basin water resources and to adapt respective strategies to impacts of climate change (Flügel, 2010). Applied in river basins IWRM has to provide the administrative and technological means to *firstly* manage the sustainable use of available surface and subsurface water resources, *secondly* to guarantee their sustainable recharge dynamics both in terms of water quantity and quality, and *thirdly* to protect water users and the society against destructive flood and drought hazards.

Climate change is a major challenge for the implementation of IWRM in both twinning basins and describes the overall objective of the studies carried out in the BRAHMATWINN project. In view of these challenges the results presented herein provide a comprehensive assessment, analysis and evaluation of present IWRM practices and their adaptive potential in respect to mitigate likely impacts from climate change. Focus is given on urban and industrial water demands, irrigation and hydropower, water distribution policies, water pollution and water quality issues.

2 Role within the integrated project

Within the context of the BRAHMATWINN project an analysis of applied IWRM strategies and practices as well as an assessment of the related institutional framework is delivering the required information and knowledge base to identify indicators with relevance for IWRM (Chapter 6), to design IWRM "what-if?" scenarios with respect to climate change (Chapter 8) and to develop appropriate IWRM adaptation strategies (Chapter 10). The study strongly elaborates on input from the assessment of the natural environment (NE) and its human dimension (HD) described in Chapter 3 and 4, respectively. Adaptation to climate change and mitigation of climate change impacts are crucial challenges in both the UDRB and the UBRB. The study is providing a analysis of existing IWRM practices and potential IWRM application as a prerequisite for the development of IWRM strategies options presented in Chapter 10. By combining the IWRM analysis results with the development of Water Resources Response Units (WRRU) a conceptual methodology is provided to analyse the spatial distribution of runoff generation as a prerequisite for improved flood management.

3 Scientific methods applied

The analysis of present IWRM practices was based on expert knowledge from local project partners and the evaluation of numerous reports published in the UDRB and the UBRB respectively. Information was also derived from a comprehensive literature review including most recent publications. Focus was given on existing water governance systems in the regions, implemented water management projects, strategies to achieve stated Millennium Development Goals (MDG) (UN, 2005) and the development of river basin plans and management initiatives to generate awareness of climate change impacts. An analysis and comparison of IWRM practices in the twinning river basins of the UBRB and the UDRB was carried out based on information available from the different stakeholders and project partners. The concept of Water Resources Response Units (WRRU) is introduced and described with respect to the delineation of Hydrological Response Units (HRU) in Chapter 3. Their distribution in the UBRB with respect to regional differences of runoff generation is briefly discussed.

4 Results achieved and deliverables provided

The twinning UDRB and UBRB both are representative in a global perspective for trans-boundary basins having alpine mountain headwater catchments and supplying their forelands with water resources to sustain food production, socio-economic development and the environment. In both basins millions of people depend on fresh water of high quality and sufficient quantity, and each basin experienced flood and drought hazards claiming human lives and destroying settlements and infrastructures. Investigations on climate change in both basins indicate that the risk of natural hazards might rise in the near future, and the need for adaptive water management strategies is increasingly be appreciated as a required IWRM measure.

Related to the glacier retreat in their alpine headwater catchments the UDRB and UBRB twinning basins have many consequent processes in common, which can jointly be analysed in a comparative strategic IWRM analysis:

- Runoff regimes range from glacial-nival to pluvial with snow and glacier melt driving flood hydrographs during spring till early summer and establishing base flow during summer.

- Glacier lake outburst floods (GLOFs) or floods from storm rainfall as well as summer droughts, are typical hydrological threats to the livelihoods of people and water management infrastructures.

- Thawing permafrost is increasing the risk of slope instability triggering landslides and rockfall processes and increasing the sediment load of rivers.

- Water quality is deteriorated by urban and industrial point sources as well as by non-point subsurface seepage from agricultural lands.

- Hydropower potential is high and competes with demands from other water users and the environment.

- Present climate change impacts are likely to exaggerate during the forthcoming years triggering hydrological changes that will impact present water management.

- IWRM related trans-boundary conflicts and water related disputes exist and are in part linked to national water management regulations and policies.

Besides these common challenges both basins also differ distinctively in other issues that require the regional adaptation of IWRM modelling tools provided by the BRAHMATWINN partners:

- Present climate is of an oceanic temperate type in the UDRB and monsoonal with complement pre-monsoon storm rainfall in the UBRB.

- The monitoring network is dense in the UDRB, while it is sparse in the UBRB.

- A complex, and sometimes conflicting, legal framework exists within the UDRB at the national level (national water legislation), the basin level (1994 Danube Convention), and the regional level (EU and UN ECE); whereas in the UBRB legal frameworks on all levels are not well developed.

- Trans-boundary conflicts in the UDRB are being dealt with through the regional and basin-wide legal framework, but in the geo-political sensitive UBRB disputes are still on-going and there is still a long way to go before a similar setup for conflict solution will be established.

- Socio-economic development is based on agriculture, industry, forestry, hydropower generation and tourism in the UDRB, and mainly on agriculture, forestry, hydropower generation, and mining in the UBRB.

- Impacts of climate change to the common natural environment are likely to be different regarding the human dimension and socio-economic environment in the UDRB and the UBRB, respectively.

4.1 Urban and industrial water demands and water quality issues

Urban, industrial, and agricultural water users and the environment generate the majority of water demand within the two twinning basins but have different weighting in each basin. Thus a comprehensive assessment of major water users and their water quality requirements is a focal element for the development and implementation of IWRM strategies and plans.

The population in the UDRB reaches about 11.5 Million people, while the UBRB is home for about 118 Million people, all depending on access to water resources in sufficient quantity and quality to meet their needs.

Present water supply in the UDRB can be described as well managed and secure. The per capita water consumption is about $136 \, l \, day^{-1}$ in the German part of the UDRB and is showing a decreasing trend during the last decades. Rising water price and more efficient technologies in industrial multiuse water cycles are major reasons to be mentioned in this regard. About 95% of the potable water supply is produced from groundwater and spring water extractions. Industrial branches with intensive water demands like the semiconductor, aerospace, chemical, armaments and automobile industry have been established in the basin with concentrations in highly industrialized centres around the cities of Augsburg, Ingolstadt, Regensburg and Munich. Water in these industries is mainly used for cooling purposes and for energy generation. In the year 2001 the extracted water for industry added up to about $750 \, Mio \, m^3$. Two third of this demand was satisfied by surface water extractions (Bayerisches Landesamt für Wasserwirtschaft, 2005). A high degree of water use efficiency has been achieved by applying multiple-shift and circuit usage technologies. Because of the long term water resources expertise and sophisticated management strategies the UDRB actually has no serious IWRM problems in respect to water quantity and water supply. With the construction of the Main-Danube channel water is even transferred to the Rhine River basin to sustain water availability over there. However, as the Upper Danube River headwaters are glacier fed and the latter are likely to melt away to a large extend in the forthcoming decades impacts on the runoff regime of the river might change the present water availability and water balances impacting future IWRM.

The water supply-demand situation in the UBRB is quite different in all of the riparian states. While Tibet is characterized by a sparse population and respectively low water demands, Bhutan shows high potential of hydropower production for the Indian market and increasing demand for irrigation supply. Assam has the highest population and a high population density in the flood prone regions of the Brahmaputra flood plain. Water demand is high comprising public water supply and irrigation agriculture. Present water supply is facing various challenges and insufficient water treatment is causing pollution of surface and subsurface water resources. India's sanitation record is extremely poor and according to the United Nations Human Development Report (UNDP, 2006) only 33% of India's population has access to improved sanitation facilities.

4.2   Water distribution policies in respect to floods and droughts

IWRM in a trans-boundary river system has to account for different policy and governance aspects relevant in the transboundary context to cope with natural hazards, i.e. landslides, GLOFs, floods and droughts. Fresh water demand from user groups in the forelands have to be satisfied in sufficient quantity and quality. Both twinning basins are characterized by seasonal runoff but in the UDRB dams and reservoirs buffer the seasonal variation of water availability and establish sufficient base flow during the year. In the UBRB such infrastructures are missing and consequently during low flow water demands for the public, industry, and the environment cannot always be met. In flood periods, however, inundation and bank erosion cause serious problems and poses risks for human livelihoods, socio-economic development and existing infrastructures.

IWRM in the UDRB is mainly based on bilateral agreements and is embedded into the regional co-operation of the Danube countries within the framework of the IHP UNESCO. The process was further driven by the implementation of the European Water Framework Directive (WFD) in 2000 and the consequent establishment of the International Commission for the Protection of the Danube River (ICPDR). The ICPDR acts as an umbrella body that is linked to the UNDP and receives scientific advice from six associated expert groups, e.g. on floods, river management or ecology (ICPDR, 2005, 2007). Water resources management in the UDRB is complex and distributes across regional, provincial, national and international scales involving bordering countries, GOs, NGOs and stakeholder groups representing urban and rural development, industries, and the environment.

The overall aim of IWRM in the UDRB is to manage flood protection, hydropower generation and water allocation by means of runoff regulation in all major tributaries and in the reservoirs located in the Alps according to the snow and glacier melt dynamics. However, in spite of the many infrastructures built in the last 100 years in the UDRB floods are still a significant risk. They mostly occur in summer time when snow- and glacier melt coincides with extreme precipitation events. With the completion of the Main-Danube-Canal in 1992 the upper course of the Danube River improve the river's water balance considerably linking the UDRB draining to the Black Sea in the East with the Rhine River Basin draining towards the Northern Sea. Controversial discussed regarding its ecological impact till today the canal transfers water from the Danube to the Main thereby improving (i) the water quality during low flow, (ii) compensates for evaporation losses caused by the operation of the thermal power stations, and (iii) cuts down the number of floods in the valley of the middle Altmühl River in summer. Another significant IWRM measure is the implementation of an effective flood warning system jointly managed by the Bavarian and the Baden-Wuerttemberg Environment

Agencies in Germany. Risk management, protection forecasting systems and public information were driven in the tributary Inn River by the cooperation between the Bavarian "Hochwassernachrichtendienst" in Germany and the Austrian "Hydrographischer Dienst" (ICPDR, 2007).

The Himalayan region of the UBRB is quite sensitive to climate change and the projected warming in Tibet is increasing faster than anywhere else in the world (IPCC, 2007). Consequently snow and glacier covered regions are at risk to melt away and will not fulfil their present water storing function in the near future which at present buffers the seasonal rainfall variation and secures the water supply in the downstream forelands. Despite the urgent need to account for climate change impacts when developing water management activities, the governments in the riparian states of the UBRB have just recently attended to these challenges.

In *Bhutan* the National Environment Commission is a high level autonomous agency responsible for environmental policy. It is also responsible for monitoring the impact of socioeconomic development on the environment and puts in place necessary controls and regulations for water resources and environment protection. Several hydropower projects have been implemented in the past decades to exploit the high hydropower potential of the Himalayan Mountains in Bhutan and to achieve the political objective of "electricity for all" by 2020. Related to this development several projects have been implemented in this regard to protect water resources. For example, the Wang River in Bhutan has been exploited for hydroelectricity to improve the sustainable management of the water resources within the basin and the Wang Watershed Management Project (WWMP) has been initiated in 2002.

Within *India* there are numerous government agencies involved in managing the various inter-related IWRM aspects of the Brahmaputra system including the pressing issues of protecting against flood inundation and associated bank erosion. With 40% of its land surface susceptible to flood damage, the Brahmaputra valley represents one of the most hazard-prone regions in India, having a total flood prone area of 3.2 million hectares (Das, 2005). However, still today there is no effective mechanism in place to coordinate interlinked IWRM activities undertaken by these agencies. In many cases IWRM became a casualty in letter and spirit as can be seen for example by reservoirs built for hydropower by the North Eastern Electric Power Corporation Limited (NEEPCO) without integrating obvious demands for flood control and irrigation water supply. Flood protection management by means of rainfall and snow melt based predictive runoff modelling and the integration of natural wetland flood retention in the NE-Indian flood plains is not implemented so far and needs special attention when striving towards IWRM and integrated flood management.

Trans-boundary IWRM in the UBRB also relates to the sensitive geo-political situation with on-going territorial disputes between the riparian states. Present water management therefore is mainly based on state and national regulations and co-operation between riparian countries is still in the initial phase, demanding further improvement. Nevertheless, agreements at the trans-boundary level already exist, for example between India and Bhutan related to joint hydropower production.

Because of the unresolved water resources issues meteorological and discharge data are mostly classified and were not made accessible for joint IWRM initiatives to neighbouring countries. Encouraging initiatives for improvements are the agreement between the Government of India and the People's Republic of China for the sharing of hydrological information of the Brahmaputra River during the monsoon flood season. Based on this agreement hydrological information, i.e. water level, discharge and rainfall from the three Tibetan stations Nugesha, Yangcun and Nuxia located at the Yarlung Tsangpo is provided to India from 1 June to 15 October every year (National Portal of India, 2010). However, other conflicts remain unresolved such as the construction of huge hydropower dams in China and India that are located in the earth-quake prone Himalaya region of the UBRB or water transfer plans related to the National River Linking Project (NRLP) of India (Jain et al., 2008).

4.3 Water consumers and polluters

In headwater and upstream river ranges water quality in general is quite good, but deteriorates downstream as they receive waste water inflows from point and non-point sources. Especially densely populated areas and such with high industrial activity contribute significantly to the pollution of surface and subsurface water resources.

In the UDRB water quality is basically in good condition. Waste water treatment is common practice and effluences into the river are strictly regularized and controlled. As a consequence countries downstream of the UDRB receive good water quality by the Danube River. The ICPDR is responsible for the implementation of the EU water framework directive and as a superior authority attempts to establish good water quality and sufficient water availability for downstream water users. Water quality risks are caused by diffuse pollution sources from the agricultural sector in form of fertilizers and pesticides as well as from hazards from suddenly spilling poisoning waste waters into the river as it has happened twice in Hungary during the last decade.

In the UBRB waste water discharge into the river is a common practice and is not regularly monitored by the authorities. The river consequently suffers from a declining water quality and especially the north-eastern Indian states have to cope with polluted surface- and groundwater resources. Missing waste water treatment and insufficient waste disposal are major IWRM challenges with respect to water quality in these regions as they are in India in general. Most rural settlements and even large cities as Guwahati are not sufficiently connected to waste water treatment systems and untreated sewage flows into rivers polluting surface and groundwater bodies. Waste water is also used for irrigation posing considerable risks to both irrigators and consumers. Uncontrolled solid waste disposal is even exaggerating the situation and contributes to the environment pollution that is obvious in many riparian states of the UBRB.

4.4   Irrigation agriculture, fertilization and crop pattern

In the UBRB water use for irrigation agriculture accounts for about 70–80% of the total water consumption. Agriculture is an important employer providing jobs to some 80% of the Bhutanese population although only 12.5% of the arable land (7% of Bhutan's state area) is under irrigation. Traditional irrigation is dominant with well-known inefficiencies, i.e. high water losses by evapotranspiration. Water management needs to focus on improving water supply to the irrigated areas, especially in times of water shortage. Because of shortfalls in irrigated rice production today some 50% of the rice consumed in Bhutan has to be imported from India. Hence increasing of self-sufficiency has high priority and the Government of Bhutan, and the Ministry of Agriculture (MoA) policies have recently stressed the need to widen the crop pattern that are grown under irrigation, and to introduce new forms of modern micro irrigation techniques for more efficient farm management.

The governance system consists of formal and informal institutions and organizations complemented by public and local farmer associations that generate the frame for an effective agricultural water management. Irrigation management in most parts of the UBRB is a matter of local governments and farmers associations. In *Bhutan* for example the latter control most of the irrigation schemes through Water User's Associations (WUAs), and only two larger schemes in the south are managed by the local government. Water use for irrigation is bound by the National Irrigation Policy of Bhutan, and essential components of this policy include the establishment and integrated participation of water user groups and the operation and management of the irrigation schemes by farmer associations with only a supportive role played by the Government of Bhutan.

In *India/Assam* there is an urgent need to improve the agricultural production from its current level to meet the demands from the ever growing population. Existent irrigation schemes in Assam had been designed as river diversion run-of-the-river-schemes, hence irrigation water is not always readily available at the time of need. There is not a single storage type irrigation system which could buffer flood surplus to support low flow by provide the required storage capacity to significantly increase cropping intensity and crop productivity to higher levels. The Ministry of Water Resources is responsible for the monitoring and technical guidance of the irrigation schemes and the Department of Irrigation in Assam is responsible to improve the situation based on a step wise irrigation development program.

4.5   Groundwater resources, quality and exploitation

Groundwater use in Assam, India, is suffering from constraints due to bad water quality resulting either from natural bedrock sources, from solid water disposals or agricultural fertilizer, herbicide and pesticide applications. If groundwater is suitable it is often overexploited and in Assam both processes are frequently combined. Demand driven development of ground water resources by different user groups is dominant and little or no management strategies based on proper understanding of local ground water regimes is applied leading to progressing depletion of the resources and degradation of groundwater quality.

A literature review reveals that groundwater resources in Assam are highly ferruginous and often have high fluoride and arsenic contents. As a consequence of uncontrolled use of such groundwater fluorosis is of public concern and has been widely reported. The situation of the water quality in the region and continuous consumption of contaminated groundwater is alarming and poses serious health hazard to the local population. This unfavourable situation is aggravated by input from agriculture due to inappropriate application of fertilizers, pesticides and herbicides which can seep into the groundwater aquifer and move towards the groundwater wells.

The Central Ground Water Board carries out regional hydro-geological studies which provide information on ground water occurrence in different terrains and are essential for sustainable future planning of ground water development and management. One headquarter is located in Guwahati and is responsible for the groundwater management in the North Eastern Region (NER) of India. Because of the hydro-geological situation common to all states in the NER it is a must for such state authorities to jointly establish a common IWRM platform to address all issues related to surface and subsurface water resources management by applying an integrated and holistic systems approach supported by respective assessment and management decision methodologies.

4.6   IWRM enhancement of the RU regionalization approach

For improving flood management the Water Resources Response Units (WRRU) approach was developed, which basically identifies areas with a high runoff contribution that generate floods. Basis for the delineation of WRRU are the Hydrological Response Units (HRUs) introduced in Chapter 3. HRUs are defined as distributed spatial model entities which have a similar hydrological process dynamics and therefore generate similar outputs as response to a given rainfall input (Flügel, 1995). The hydrological system analysis is providing the definition of process based criteria to delineate HRUs by means of GIS analysis which represent distributed landscape entities of unique hydrological system response. The

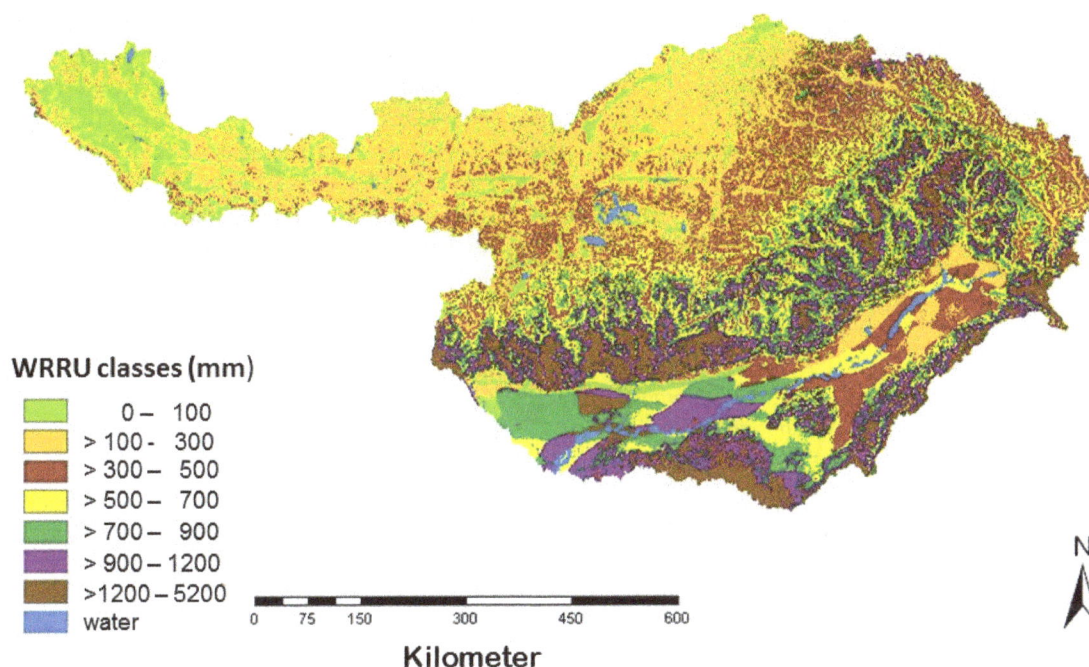

Figure 1. Water Resources Response Units (WRRU) distributed within the UBRB.

application of the RU conceptual landscape model offers substantial progress for a hydrological and process oriented water balance analysis in river basins.

The mean annual water balance components obtained from the hydrological modelling done by means of the DANUBIA hydrological model (Chapter 7) were used as inputs for the delineation of the WRRUs. By classifying HRU with respect to discharge generation by surface runoff and interflow on the one side and groundwater recharge on the other WRRU are generated. In result each WRRU class is merging dominant water balance components from their different HRU components with the respective land use and land cover (LULC), topography, and soil attributes of the HRU. Depending on the in-depth GIS analysis the elaborated results will achieve different degree of detail and quantification.

The WRRU have been delineated based on the annual average of the DANUBIA modelling period 1960 till 2000 and are shown in Fig. 1.

The distribution of WRRU indicating the distributed runoff generation reveals the following results with respect to flood generation:

1. The barrier function of the Himalaya mountain ridge for the monsoon rainfall is obvious and influences the runoff generation in luv and lee of the mountain ridge.

2. The western part of the Yarlung Tsangpo (name of the Brahmaputra River in Tibet) with a semi-arid cold temperate climate has no or only little annual runoff contri-

bution mostly not exceeding 200 mm as annual sum or $6.34 \, \mathrm{l \, s^{-1} \, km^{-2}}$.

3. In the middle part moderate runoff generation is dominant ranging between 300 and 500 mm ($9.51 \, \mathrm{l \, s^{-1} \, km^{-2}}$ till $15.85 \, \mathrm{l \, s^{-1} \, km^{-2}}$) in the middle mountain range and in insular places of the alpine ridge even reaching up to 900 mm or $28.54 \, \mathrm{l \, s^{-1} \, km^{-2}}$.

4. The main runoff generation, however, is occurring in the luv side of the Himalayan in the Nort-East Region (NER) states of India and by heavy monsoon rain in the floodplains of the rivers like the Brahmaputra ranging between 700 mm till 5200 mm ($22.20 \, \mathrm{l \, s^{-1} \, km^{-2}}$ till $164.89 \, \mathrm{l \, s^{-1} \, km^{-2}}$).

5 Contributions to sustainable IWRM

The detailed analysis of present IWRM practices and strategies reveals their potential to cope with impacts from climate change. Adaptation to climate change is a high priority challenge in the UBRB, which is characterized by high climate variations impacting human life, infrastructures and the environment. Assam for example has to cope with serious IWRM problems like floods, droughts and bank erosion but still lacks the appreciation to employ the required knowledge and allocate respective budgets to bring the appropriate measures in place. The analysis of present IWRM practices in the twinning basins UDRB and UBRB is a prerequisite to understand likely impacts from climate change on water management and allows the development of appropriate adaptation measures for mitigation. The identification of areas with high runoff generation by using the WRRU approach provides the means for effective implementation of protective or mitigating measures with respect to flood generation and forecast. The study on IWRM practices constitutes the basis for developed IWRM options, described in Chapter 10.

6　Conclusions and recommendations

The entire UBRB is characterized by increasing pressure on the available water and land resources due to competing water demands from different user groups, e.g. domestic consumption, agriculture, industry, tourism and recreation, hydropower, navigation and settlements. Rapid population growth in Assam, Bhutan and Tibet has been identified as a major challenge with respect to sustainable IWRM and preservation of the environment. Continuous population increase, the finite extent to which further land can be converted into agricultural use and limited water availability are serious constraints with respect to the supply of water for irrigation and are of major concern. Such constraints are already obvious in the foothills and valleys of Bhutan and Tibet. The increasing need for water and the decrease in water availability are high priority challenges for IWRM in this basin and different attempts have already be initiated to better the situation.

The Bhutan Water Policy for example puts high priority to the supply of drinking water, and the fundamental right of every individual to access good quality water. This will drive Bhutan's commitment to meet the MDG targets and underlines the importance of efficient water use and proper wastewater management. It furthermore highlights the importance of water resources protection by means of the "polluter pays" principle which is also considered in IWRM planning done in China/Tibet and India/Assam.

Although river basin commissions for the largest river basins have been implemented in China no such commission has been established so far for the Yarlung Tsangpo as the Brahmaputra is named in Tibet. Management of water resources and water supply in this part of the UBRB has to improve considerable. Water pollution is a minor concern for IWRM in Tibet but indicators show that this will become soon a matter of future concern as decreasing water quality is already apparent in various places due to uncontrolled disposal of solid wastes and insufficient treatment of waste water discharging from various sources.

With regards to the problem of flood management and respective land management options, a high priority problem in Assam, the concept of WRRU has been developed. Enhancing the RU concept toward WRRU permits to relate flood vulnerabilities to spatially distributed runoff generation and groundwater recharge dynamics on the macro-scale of the UBRB. Calculating the individual runoff generation of each WRRU class will provide water and land managers with the

means to link the natural landscape environment, e.g. LULC, soil, geology and climate with the water yield produced from each geo-referenced WRRU entity. The WRRU concept thereby provides the Geoinformatics means for a coordinated land and water management in IWRM and the assessment of climate change impact on river basin water balances.

IWRM in the UBRB at present is also restricted due to state border disputes, i.e. between China and India. The activities carried out at upstream areas are both challenges, i.e. soil erosion, landslides, and deforestation as well as IWRM opportunities, i.e. provision of reliable amounts of good quality water to downstream user communities and irrigation schemes.

Acknowledgements. The cooperation and support from all BRAHMATWINN partners and stakeholders was highly appreciated and is acknowledged hereby. Acknowledgement is also given to the EC which funded the IWRMS development in the BRAHMATWINN project under the contract number 036952.

The interdisciplinary BRAHMATWINN EC-project carried out between 2006–2009 by European and Asian research teams in the UDRB and in the UBRB enhanced capacities and supported the implementation of sustainable Integrated Land and Water Resources Management (ILWRM).

References

Bayerisches Landesamt für Wasserwirtschaft: Bericht zur Bestandsaufnahme gemäß Art. 5, Anhang II, sowie Art. 6, Anhang IV, der WRRL für das Deutsche Donaugebiet, München, 2005.

Das, P. J.: Integrated Water Resources Management (IWRM): A Northeast Indian Perspective, 2005.

Flügel, W.-A.: Climate impact analysis for IWRM in Man-made landscapes: Applications for Geoinformatics in Africa and Europe, Initiativen zum Umweltschutz, Bd. 79, 101–134, 2010.

Flügel, W.-A.: Delineating Hydrological Response Units (HRU's) by GIS analysis for regional hydrological modelling using PRMS/MMS in the drainage basin of the River Bröl, Germany, Hydrol. Process., 9, 423–436, 1995.

GWP TAC: Background Paper No. 4, Integrated Water Resources Management, Global Water Partnership, Stockholm, Sweden, 2000.

International Commission for the Protection of the Danube River (ICPDR): Development of the Danube River Basin District Management Plan – Strategy for coordination in a large international river basin, 2005.

International Commission for the Protection of the Danube River (ICPDR): Germany facts and figures, http://www.icpdr.org (last access: 27 March 2011), 2007.

IPCC: Climate Change 2007: Synthesis Report, Contribution of Working Groups I, II and III to the Fourth Assessment Report of the Intergovernmental Panel on Climate Change (Core Writing Team, Pachauri, R. K. and Reisinger, A. (Eds.)), IPCC, 2007.

Jain, S. K., Kumar, V., and Panigrahy, N.: Some issues on interlinking of rivers in India, Curr. Sci. India, 95, 6, 728–735, 2008.

National Portal of India: International Cooperation: http://india.gov.in/sectors/water_resources/international_corp.php (last access: 28 March 2011), 2010.

UN: The Millennium Development Goals Report, New York, 2005.

UNDP: Human Development Report 2006, New York, 2006.

17

Soil moisture initialization effects in the Indian monsoon system

S. Asharaf, A. Dobler, and B. Ahrens

Institute for Atmospheric and Environmental Sciences, Goethe University, Frankfurt, Germany

Abstract. Towards the goal to understand the role of land-surface processes over the Indian sub-continent, a series of soil-moisture sensitivity simulations have been performed using a non-hydrostatic regional climate model COSMO-CLM. The experiments were driven by the lateral boundary conditions provided by the ERA-Interim (ECMWF) reanalysis. The simulation results show that the pre-monsoonal soil moisture has a significant influence on the monsoonal precipitation. Both, positive and negative soil-moisture precipitation (S-P) feedback processes are of importance. The negative S-P feedback process is especially influential in the western and the northern parts of India.

1 Introduction

The Earth's surface plays a key role in weather and climate because of the large energy and water exchange with the overlying atmosphere (Zhang et al., 2004). The water reaching the land surface is distributed into soil water storage, runoff, or recycled into the atmosphere by evapotranspiration. Exchange of radiation, sensible heat, latent heat, and momentum has direct impacts on the wind vector, precipitation, and surface soil (Sellers, 1991). Hence, a proper understanding of the land atmosphere interaction is necessary.

The present work focuses on the investigation of the soil moisture-precipitation (S-P) feedback process over the Indian region. The structure of this paper is as follows. Section 2 presents the model setup, and experimental design. The results and discussion are addressed in Sect. 3 and followed by conclusions (Sect. 4).

2 Model and experimental design

The simulations are performed with the non-hydrostatic limited-area climate model COSMO-CLM (Dobler and Ahrens, 2008). As model input, the initial and lateral boundary conditions are taken from the ERA-Interim reanalysis. In the present study, the model horizontal resolution is set to 0.25° with 32 vertical layers. The simulation domain encompasses the entire Indian region (Fig. 1). More de-tails about the model are given at the community website (http://www.clm-community.eu/).

In order to assess the influence of soil moisture, we have performed several simulations: a reference simulation (CTL) for the period 1989 to 2007 and perturbed simulations. For each year on 2 April, there is a DRY run with an initialized soil 50 % drier and a WET run which is 50 % wetter than CTL (with perturbations limited by the field capacity and the wilting point of the soil type). These perturbations are small in absolute terms, as the pre-monsoonal soils are dry. However, the pre-monsoonal perturbation allows for the investigation of the impacts on the simulated monsoon. The perturbed experiments are driven by the same lateral boundary condition as CTL. Therefore, these set of experiments explain the soil water initialization impacts to the model simulations at the regional scale, as well as at the Indian summer monsoon scale. This setup is motivated by Schär et al. (1999) and Pielke Sr. et al. (1999).

3 Results and discussions

The simulations are carried out over the Indian sub-continent for the period 1989 to 2007, where a spin-up time of one year is neglected in the subsequent analysis to mitigate the initial value errors. The present analysis is focused on 1990 to 2007 for the seasonal average (June to September, JJAS) of monthly accumulated monsoonal precipitation.

Table 1. Mean values of (JJAS, 1990 to 2007) moisture influx (IN), precipitation efficiency (χ), and recycling ratio (β). Changes in precipitation (ΔP) between the perturbed and the control simulations as defined in Eq. (1) are also shown.

Region	χ			β			IN (mm/month)			$\Delta P_{WET-CTL}$ (mm/month)		$\Delta P_{DRY-CTL}$ (mm/month)	
	WET	CTL	DRY	WET	CTL	DRY	WET	CTL	DRY	Direct	Indir.	Direct	Indir.
E	0.51	0.49	0.49	0.14	0.13	0.12	440	437	427	5.0	7.0	−7.0	3.0
W	0.22	0.18	0.17	0.14	0.09	0.08	184	198	206	−1.3	7.5	1.0	−3.0
N	0.53	0.51	0.51	0.29	0.23	0.21	144	156	158	0.8	3.9	−1.3	0.5
CE	0.30	0.29	0.29	0.10	0.09	0.09	481	483	479	0.5	3.1	−1.6	−1.5

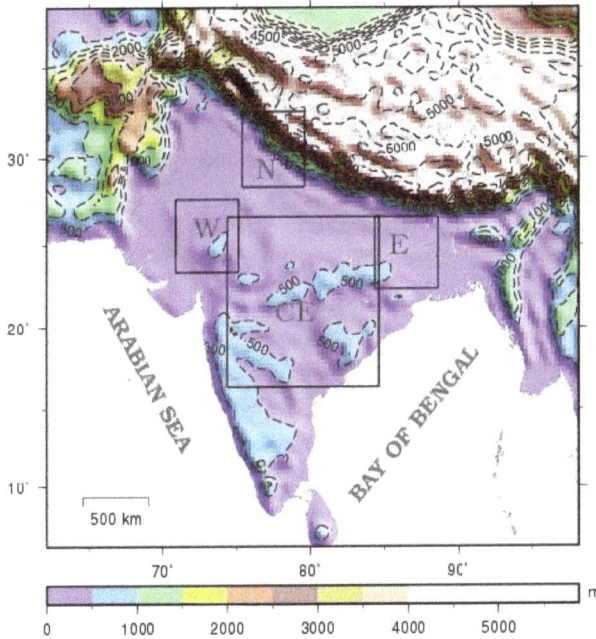

Figure 1. Simulation domain and analysis domains N, W, CE and 〈E. The shading and contour lines show the model orography.

The sensitivity of summer monsoon rainfall with response to the change in initial soil moisture is shown in Fig. 2. The precipitation changes between the WET and CTL experiments are about +10 % to +20 % in the eastern region, while they are larger than +40 % over the northwest part of India. These spatial variations in precipitation are related to the available water in the soils (Kim and Wang, 2007) and to the regional circulations as discussed in the following paragraph. Apart from this, some places, especially the northern Himalayan foothills and the Thar desert (northwest part of India) indicate a negative S-P feedback process, which is consistent with the Bowen ratio (not shown).

The wet (dry) soil moisture perturbations cause an increment (decrement) in the surface pressure by cooling (warming) of the Earth surface through the partitioning of the surface heat fluxes. As a result, there is a decrement (increment)

in the geopotential height in the middle of the troposphere (Fig. 3). These local changes especially over the northwestern region further modify the large scale circulation as shown by vertically integrated moisture flux vectors. Decrement in the geopotential height at 500 hPa (Fig. 3) can be depicted by the increasing surface pressure over the northwestern region in the WET experiment. This change (increment in the surface pressure) caused a decrement in transportation of water vapor from the Arabian Sea to the northwest region in the wet soil condition.

To study the land-atmosphere interaction quantitatively, we have calculated the recycling ratio $\beta = ET/(ET + IN)$ and precipitation efficiency $\chi = P/(ET + IN)$ following Schär et al. (1999), where ET is evapotranspiration, IN influx and P precipitation. For the analysis we have divided the Indian domain into four sub-regions: East (E), West (W), Central (CE), and North (N), as depicted in the Fig. 1. Figure 4 shows the considerable inter-annual variation in the recycling ratio and precipitation efficiency of the CTL and the soil moisture sensitivity simulations (WET and DRY) for the analysis domains E and W. Here, the bulk characteristics β and χ are predominantly increasing and decreasing in a regular way. In general, the recycling ratios are in the range of 0.1 to 0.4 with largest values for the northern region (cf. Table 1).

The precipitation efficiency χ is of moderate sensitivity to the soil moisture initialization in most of the years with a small positive feedback. However, some years (e.g., 1992 in E, 1993 in W) experience a negative feedback to soil moisture initialization perturbation. This result is opposite to the results in Schär et al. (1999), where χ always increases with an increment in the initial soil moisture.

The precipitation changes in the sensitivity experiments (WET and DRY) are estimated by the following equation (see Schär et al., 1999):

$$\Delta P = \chi'(\Delta ET + \Delta IN) + \Delta \chi (ET + IN) \qquad (1)$$

Where ' denotes the perturbed simulation case and the Δ-terms indicate differences between the perturbed and the control simulations. The first term on the right hand side of the Eq. (1) reflects the precipitation change through direct (recycling) process and the second term depicts the indirect

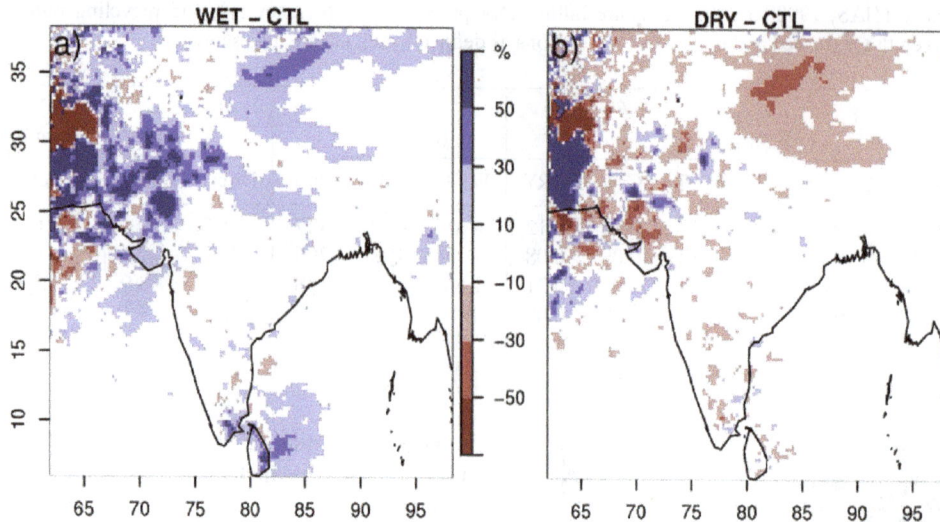

Figure 2. Average summer monsoon (JJAS, 1990 to 2007) relative precipitation difference (in percent) in the WET (left) and DRY (right) experiments relative to the CTL experiment. The red color represents less precipitation and the blue color represents more precipitation than the CTL precipitation.

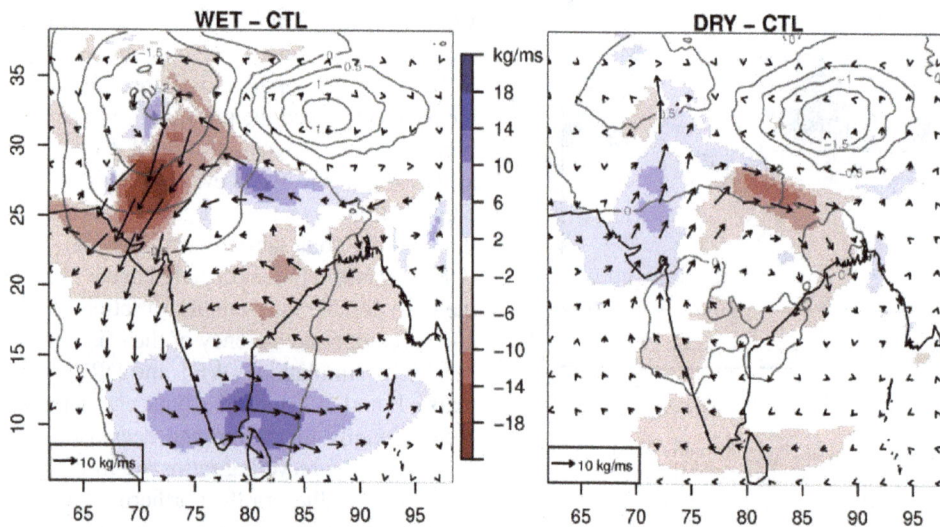

Figure 3. Average summer monsoon (JJAS, 1990 to 2007) vertically integrated moisture flux differences (shaded and vector) and geopotential height differences at 500 hPa (contour, isoline distance 0.5 m) in the WET (left) and DRY (right) experiments relative to the CTL experiment.

(feedback) contribution. The results for the different analysis domains (Fig. 1) are made comparable by normalization following Zangvil et al. (2010).

Figure 5 represents the time series of the aforementioned (direct and indirect) processes for the sub-regions E and W. The results show that the changes in the monsoonal precipitation are clearly dominated by the indirect process in most of the years. The direct recycling process, however, dominates in some of the years, especially in the eastern analysis region E. In this region, the years, e.g., 1992, 1999, 2000, contain higher magnitudes of the direct process than the in-

direct process. This high magnitude of the direct process in dry soil conditions can be linked with the high influx differences. Here, the influx differences with respect to the CTL are around -10 mm month^{-1} and $+3$ mm month^{-1} in dry and wet soil conditions respectively. Often wet soil enhances precipitation and vice versa (positive feedback). But a negative feedback process is also present in all sub-domains, which is more pronounced in the western region. In this case, the drier soil leads to the higher sensible heat flux and an expansion of the boundary layer, which may in turn lead to increased rainfall (Collini et al., 2008). On the other hand, moist conditions

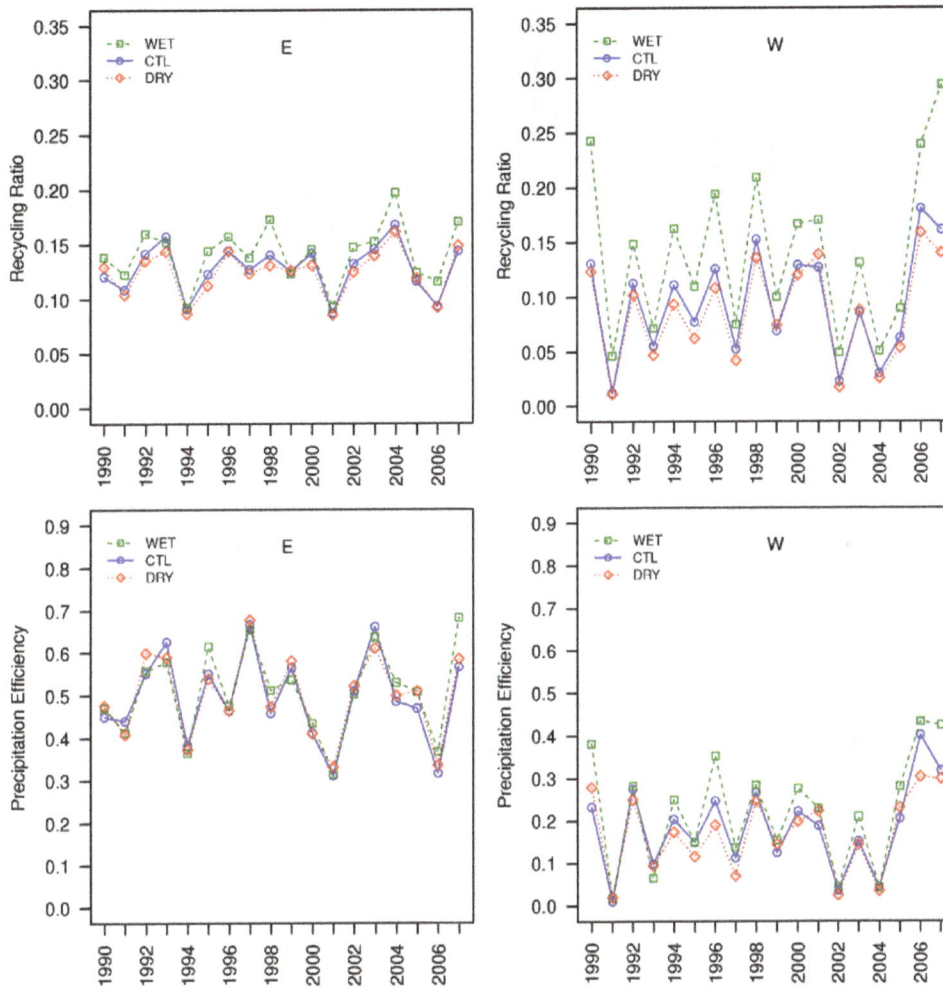

Figure 4. Inter-annual variation of recycling ratio and precipitation efficiency averaged over the monsoon season (JJAS) in the analysis domains East (E) and West (W).

stabilize the atmosphere, inhibiting the vertical movement of air parcels, and as a result decrease in rainfall is observed (Cook et al., 2006).

4 Conclusions

Precipitation recycling and feedback processes were investigated through perturbation simulations with the COSMO-CLM over India. The results suggest that soil moisture has significant impact on the precipitation formation. In the simulations in which the initial soil moisture is increased by a factor of two (WET), a decrease in the surface Bowen ratio, and thereby a surface cooling was observed. This cooling may be responsible for weakening the strength of moisture transportation from Arabian Sea to the northwest region (Fig. 3). Furthermore, the moisture budget has been examined for different sub-regions to interpret the leading factor as precipitation originating from local evapotranspiration (direct process) or being advected from external sources (indi-

rect). As a first result, the Indian summer monsoon contains both processes, where the dominant one varies spatially and temporally. The direct process is more pronounced in the dry soil conditions for the eastern analysis region E, whereas in the wet experiment, the precipitation is controlled by the indirect process. Additionally, the decreasing moisture influx with the increasing soil moisture is related to a negative S-P feedback process, which is more frequent in the western and the northern Indian regions than in the other two regions.

Acknowledgements. The authors acknowledge funding from the Hessian Initiative for the Development of Scientific and Economic Excellence (LOEWE) through the Biodiversity and Climate Research Centre (BiK-F), Frankfurt am Main. The COSMO-CLM community for providing the model and the German Climate Computing Center (DKRZ) for supporting part of the calculation. The authors also thank two anonymous referees, as well as Herbert Formayer (Topical Editor) for their constructive comments and suggestions to the manuscript.

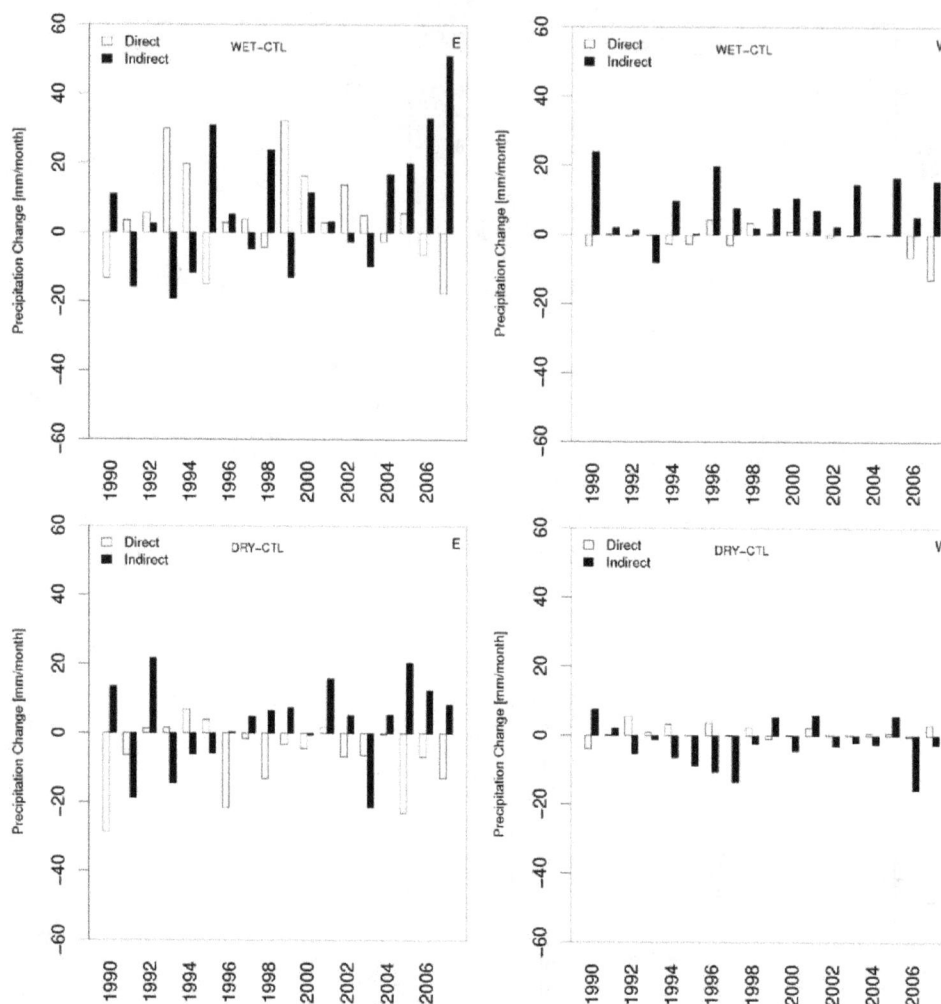

Figure 5. Inter-annual changes in monsoonal rainfall (JJAS) by direct $\chi(\Delta ET + \Delta IN)$ and indirect effects $\Delta \chi(ET + IN)$ for WET-CTL (upper) and DRY-CTL (lower) in the analysis domains E and W.

Edited by: H. Formayer
Reviewed by: two anonymous referees

References

Collini, E. A., Berbery, E. H., Barros, V. R., and Pyle, M. E.: How does soil moisture influence the early stages of south American Monsoon?, J. Climate, 21, 195–213, 2008.

Cook, B. I., Bonan, G. B., and Levis, S.: Soil moisture feedback to precipitation in southern Africa, J. Climate, 19, 4198–4206, 2006.

Dobler, A. and Ahrens, B.: Precipitation by a regional climate model and bias correction in Europe and South Asia, Meteorol. Z., 17(4), 499–509, 2008.

Kim, Y. J. and Wang, G. L.: Impact of initial soil moisture anomalies on subsequent precipitation over North America in the coupled land-atmosphere model CAM3-CLM3, J. Hydrometeor., 8, 534–550, 2007.

Pielke Sr., R. A., Listan, G. E., Eastman, J. L., and Lu, L.: Seasonal weather prediction as an initial value problem, J. Geophys. Res., 104, 19463–19479, 1999.

Schär, C., Lüthi, D., and Beyerle, U.: The soil-precipitation feedback: A processes study with a regional climate model, J. Climate, 12, 722–741, 1999.

Sellers, P.: Modeling and observing land-surface-atmosphere interaction on large scale, Surv. Geophys., 12, 85–114, 1991.

Zangvil, A., Lamb, P. J., Portis, D. H., Jin, F., and Malka, S.: Comparative study of atmospheric water vapor budget associated with precipitation in Central US and eastern Mediterranean, Adv. Geosci., 23, 3–9, doi:10.5194/adgeo-23-3-2010, 2010.

Zhang, H., McGregor, J. L., Seller, A. H., and Katzfey, J. J.: Impacts of land surface model complexity on a regional simulation of a typical synoptic event, J. Hydrometeorol., 5, 190–198, 2004.

18

The role of nocturnal Low-Level-Jet in nocturnal convection and rainfalls in the west Mediterranean coast: the episode of 14 December 2010 in northeast of Iberian Peninsula

J. Mazón[1] **and D. Pino**[1,2]

[1]Applied Physics Department, BarcelonaTech (UPC), Barcelona, Spain
[2]Institute for Space Studies of Catalonia (IEEC-UPC), Barcelona, Spain

Correspondence to: J. Mazón (jordi.mazon@upc.edu)

Abstract. The night of 14 December 2010 radar images of the Spanish Weather Agency recorded a large rain band that moved offshore at the Northeast coast of the Iberian Peninsula. MM5 mesoscale model is used to study the atmospheric dynamics during that day. A Nocturnal Low Level Jet (NLLJ) generated by an inertial oscillation that brings cold air to the coast from inland has been simulated in the area. This cold air interacts with a warmer air mass some kilometers offshore. According to the MM5 mesoscale model simulation, the cold air enhances upward movements of the warm air producing condensation. Additionally, there is a return flow to the coastline at 600–900 m high. This warm air mass interacts again with the cold air moving downslope, also producing condensation inland. The simulation for the night before this episode shows large drainage winds with a NLLJ profile, but no condensation areas. The night after the 14th the simulation also shows drainage winds but without a NLLJ profile. However, an offshore convergence area was produced with a returned flow, but no condensation inland occurred. This fact is in agreement with radar observations which reported no precipitation for these two days. Consequently, NLLJ in combination with a synoptic wind over the sea could enhance condensation and eventually precipitation rates in the Mediterranean Iberian coast.

1 Motivation

Low-Level-Jet (LLJ) is a quite well known phenomenon that has been detected in many areas of the world (Blackadar, 1957; Banta et al., 2002; Stensrud, 1996; Rife et al., 2010). Nocturnal LLJ (NLLJ) is produced by an inertial oscillation (Van de Wiel et al., 2010) over flat terrain in response to a strong radiative cooling and a rapid stabilization of the boundary layer under relatively dry, cloud-free conditions (Blackadar, 1957). Usually NLLJ reaches a peak intensity in the early morning hours, and then decays shortly after dawn, when the convective mixing begins. The wind maximum of a NLLJ develops typically at levels less than 1 km above ground level, and frequently at levels lower than 500 m. As discussed by Stensrud (1996) and Shapiro and Fedorovich (2010), NLLJ exert significant influence on weather and re-

gional climate. It is important to note that these authors affirm that the NLLJ provide dynamical and thermodynamical support for the development of deep convective storms and heavy rain events, besides advection and thermodynamic processes including water vapor (Mahrt et al., 1998; Acevedo and Fitzjarrald, 2001).

Nocturnal Low-Level-Jet is not a typical phenomenon in the Mediterranean coast (Rife et al., 2010). The aim of this work is, by using MM5 mesoscale simulations, to show and describe the role of this NLLJ caused by inertial oscillation detected in one precipitation event observed close to the coastline during the night of 14 December 2010, characterized by its persistence close to the northeastern coast of the Iberian Peninsula (see Fig. 1).

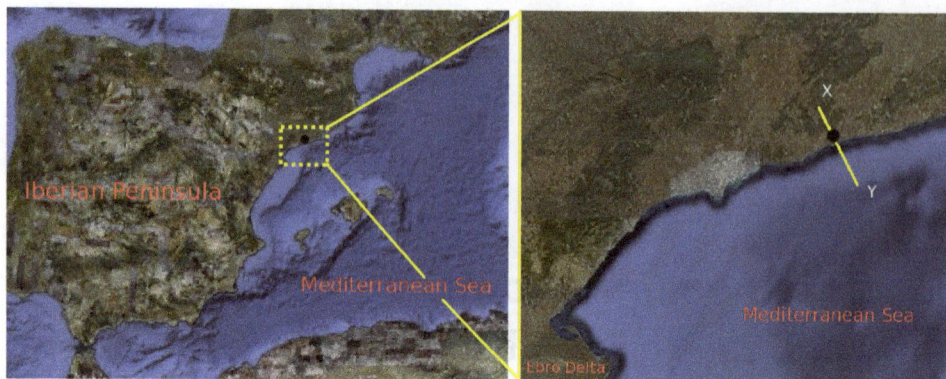

Figure 1. The western Mediterranean area (left) and enlargement of the studied region (right). The yellow line marks the cross section where the wind field is analyzed in Sect. 2.

2　Analyzed episode: 14 December 2010

After some days with temperatures between 5 and 10 °C at 850 hPa, during 13 December a high-pressure system placed at the north of UK, and a low-pressure system extended over the Açores islands (not shown) sent cold and dry air mass from northeast to the Iberian Peninsula, cooling the air temperature to −5 to −8 °C at 850 hPa. The night of 13th and the early morning of 14th rainfall was reported in the northeast of Iberian Peninsula. The radar images from the Spanish Weather Agency for 14 December shows a large precipitation area moving offshore. A small precipitation cell close the coast remained for some hours, from 03:00 to 08:00 UTC. Figure 2 show the precipitation pattern at 04:00 and 06:00 UTC. The black circle indicates the stationary precipitation cell.

The hypothesis proposed here is that such precipitation was caused by the interaction between a cold air mass moving offshore with a LLJ profile and a warmer air mass associated to synoptic winds.

Different wind patterns have been simulated during 13 and 15 December and no precipitation was recorded. Our aim in this work is to analyze the atmospheric dynamics of these three days to study the possible cause of the different precipitation records.

MM5 numerical simulation

The three studied days were simulated by using 5th generation of the PSU/NCAR mesoscale model (MM5, Grell et al., 1995). The simulation setup was as follows. Four nested domains with resolution of 27, 9, 3 and 1 km working in two-way nesting option were defined. The smallest domain is focused on our interested area (see Fig. 1). *Kain-Fritsch* scheme for cumulus (none in the two smallest domains), *Simple Ice* for explicit moisture scheme and *MRF* for the planetary boundary layer parametrization were used. The initial and boundary conditions were updated every six-hours with

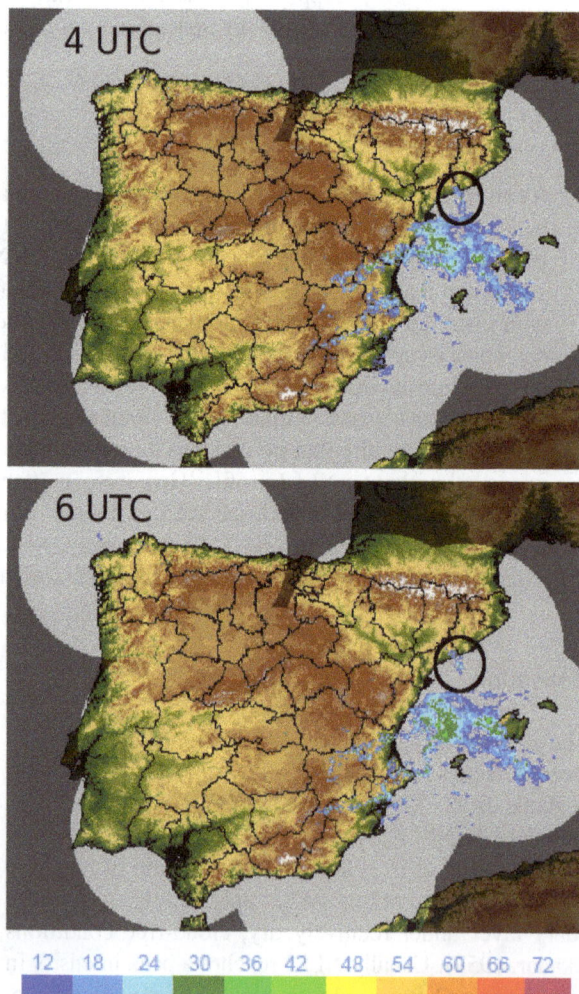

Figure 2. PPI radar reflectivity (in dB) from Spanish Weather Agency at 04:00 and 06:00 UTC on 14 December 2010. The black circle indicates the rainfall area studied.

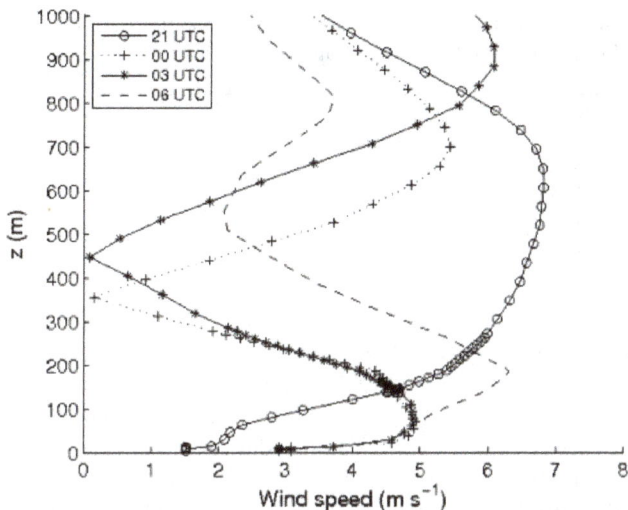

Figure 3. Simulated vertical wind profile at different hours during the night of 13th and early morning of 14th at the point marked by a black dot in Fig. 1.

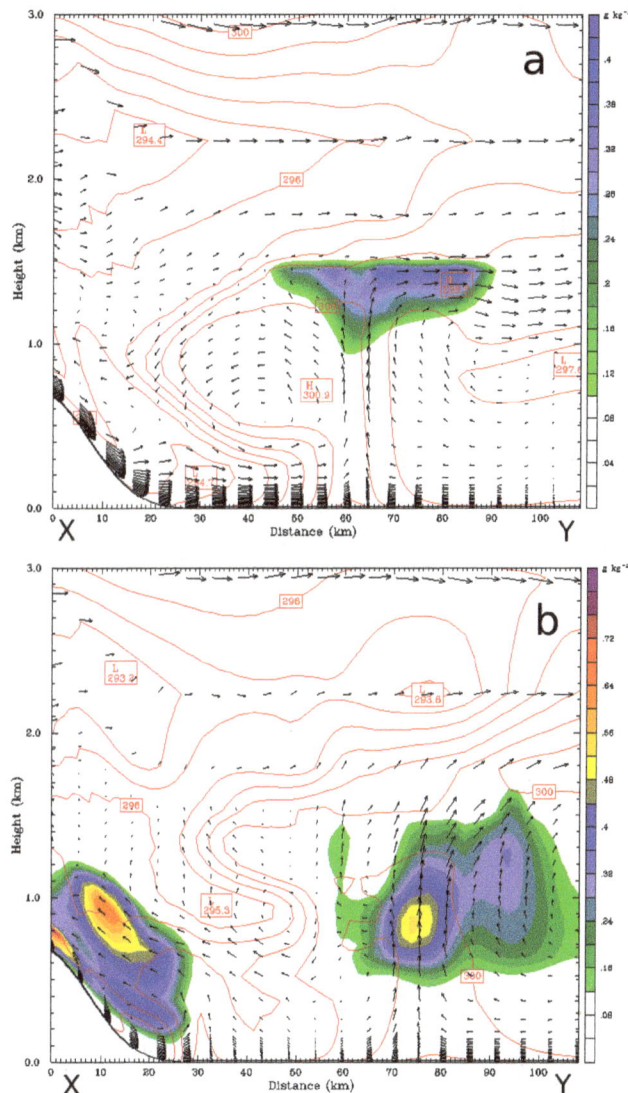

Figure 4. Vertical cross sections along the yellow line of Fig. 1 of the wind (arrows), equivalent potential temperature (red contour lines) and cloud mixing ratio (colored contour) at 00:00 (**a**) and 03:00 UTC (**b**) on 14 December 2010. The maximum wind speed is 9.5 m s^{-1} approximately.

information obtained from the ECMWF analysis model. The model simulation ran during 90 h starting at 00:00 UTC on 12 December 2010. The night of 13th and early morning of 14th a NLLJ appeared as can be observed in Fig. 3. This figure shows the temporal evolution of the simulated wind speed vertical profile at a point marked by a black dot in Fig. 1. At 21:00 UTC of 13th the wind speed increased in height. At 00:00 UTC of 14th the maximum wind speed was located around 100 m high, with values of 5 m s^{-1}. Until 06:00 UTC the maximum velocity of the NLLJ increases, being at this time 6.4 m s^{-1} and located around 200 m high. The hodograph (not shown) confirms that the wind speed profile is a NLLJ, due to an inertial oscillation.

Figure 4 shows the vertical cross section along the yellow line shown in Fig. 1 of the simulated wind field (arrows), equivalent potential temperature (red contour lines) and cloud mixing ratio (colored contour, q_{cl}) at 00:00 (a) and 03:00 UTC (b). It is interesting to note the two condensation areas produced by the simulation in Fig. 4b. One is located offshore, around 65 km far away from the coast and appeared early that night. The second condensation area is located inland, and appeared later. This area of cloudiness is produced at 600–900 m high by the interaction of the return flow that brings warm and moist air inland with the cold air mass moving downslope. The equivalent potential temperature (red lines) is larger in the condensation areas due to a relative warm and moist air mass advection from the sea. Note that in the offshore convection area a larger value of equivalent potential temperature extends vertically, and to the coastline.

These two simulated areas with $q_{cl} > 0$ approximately corresponds to the precipitation areas observed by the radar images indicated by a black circle in Fig. 2. Moreover, the sim-

ulated hourly accumulated precipitation (not shown) has a good correspondence with the precipitation pattern observed in radar images inside the black circle shown in Fig. 2 but the precipitation rates are underestimated. Eventually, the offshore convergence area moved to the coast and the two precipitations areas shown in Fig. 4b joined forming only one area of precipitation from 04:00 to 08:00 UTC (not shown).

The NLLJ drives cold inland air to the coastline. The relative warm and wet air returned to the coast and condensates when interacts with the cold inland air. According to Fig. 3, at early morning the wind speed presented a maximum value, coinciding with an increase of cloud mixing ratio at the mountain slope (not shown). The vertical and temporal

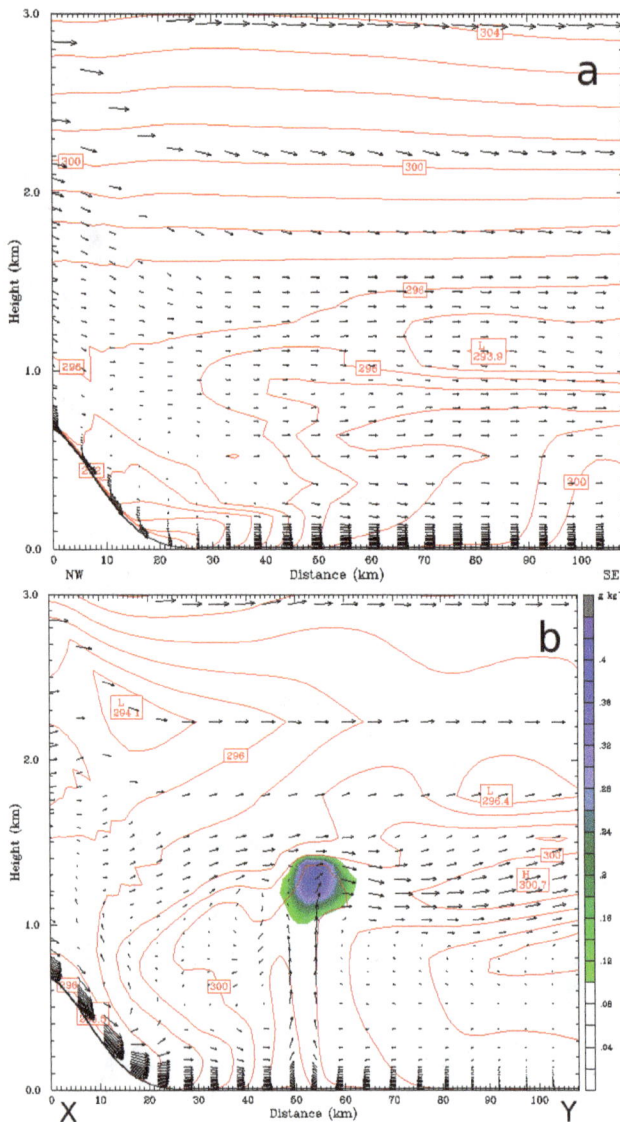

Figure 5. Simulated vertical cross sections along the yellow line of Fig. 1 of the wind field (arrows), equivalent potential temperature (red lines) and cloud mixing ratio (color contour) at 04:00 UTC on (**a**) 13 December and (**b**) 15 December. The maximum wind speed is $8.1\,\mathrm{m\,s^{-1}}$ and $6.5\,\mathrm{m\,s^{-1}}$ in the upper and lower panels, respectively.

evolution of potential temperature has also been analyzed (not shown). The minimum values appear between 00:00 and 01:00 UTC on 14th, with 280.5 K at low levels, below 150 m high, where the wind speed has the maximum values. Above these levels the potential temperature increases 4 K and remains constant at the upper levels. At early morning the potential temperature has a constant value with height of 283 K.

In order to analyze the role of the NLLJ and the return flow on the condensation process inland, the wind field, equivalent potential temperature and cloud mixing ratio obtained with

Table 1. Relation between the atmospheric conditions according MM5 simulation and radar observation.

Night	Sim. div < 0	Sim. Rotor	Sim. NLLJ	Sim. $q_{cl} > 0$	Radar precip.
13th	weak	No	Yes	No	No
14th	Yes	Yes	Yes	Yes	Yes
15th	Yes	Yes	No	No	No

the MM5 simulation for the previous night (13 December) are shown in Fig. 5a. A NLLJ appear at the same place during several hours but not a convergence area offshore because the synoptic wind blows in the same direction. Thus, there is not return flow and condensation inland. The simulated wind field, equivalent potential temperature and cloud mixing ratio for 15 December are shown in Fig. 5b. During this day it is possible to observe downslope winds (in this case without a LLJ profile) and a convergence area about 50 km offshore. A return flow appeared, but there is no condensation inland because the downslope air has an equivalent potential temperature 4 K larger than the air on 14 December.

In order to compare the differences in wind configuration during 13, 14 and 15 December 2010, Table 1 shows the main characteristics obtained from the MM5 simulation and the radar images.

3 Summary and conclusions

Analyzing the radar images during the night of 14 December 2010, we have observed a static precipitation area close to the coastline in the Northeast of Iberian Peninsula. By using MM5 simulation, two different condensation areas have been identified: the first one due to offshore convergence and the second one inland, associated with the interaction between the cold air driven by a NLLJ with the wet air that returns aloft to the coastline. These simulated condensation areas approximately correspond to the precipitation areas observed by the radar images in this area, at the same hours.

By comparing the simulation of the 14th and 15th nights, it can be concluded that when the air advected by the NLLJ is not enough cold, condensation inland is not produced. By comparing the episode on 14 December with the previous night (13 December), a NLLJ appears in both simulations but no offshore convergence during 13 December because the synoptic wind flew in the direction of the drainage winds. Consequently, there is not return flow of wet air to the coast and no condensation inland.

Summarizing, the appearance of a NLLJ that brings cold air to the coast combined with a synoptic offshore wind could enhance not only convergence and precipitation over the sea but also inland precipitation due to the interaction of the NLLJ with the return flow.

Acknowledgements. This project has been carried out by using the resources of the Supercomputing Center of Catalonia (CESCA) and it has been performed under the Spanish MICINN project CGL2009-08609, and the INTERREG EU project FLUXPYR EFA 34/08. The comments of both referees were helpful to improve the manuscript.

Edited by: M. M. Miglietta
Reviewed by: two anonymous referees

References

Acevedo, O. C. and Fitzjarrald, D. R.: The Early Evening Surface-Layer Transition: Temporal and Spatial Variability, J. Atmos. Sci., 58, 2650–2667, 2001.

Banta, R. M., Newsom, R. K., Lundquist, J. K., Pichugina, Y. L., Coulter, R. L., and Mahrt, L.: Nocturnal Low-Level Jet characteristics over Kansas during CASES-99, Bound.-Lay. Meteorol., 105, 221–252, 2002.

Blackadar, A. K.: Boundary layer wind maxima and their significance for the growth of nocturnal inversions, B. Am. Meteorol. Soc., 38, 283–290, 1957.

Grell, G. A., Dudhia, J., and Stauer, D. R.: A description of the fifth-generation Penn State/NCAR mesoscale model (MM5), Tech. Rep. NCAR/TN-398, NCAR Tech. Note, 122 pp., 1995.

Mahrt, L., Sun, J., Blumen, W., Delany, T., and Oncley, S.: Nocturnal boundary-layer regimes, Bound.-Lay. Meteorol., 88, 255–278, 1998.

Rife, D. L., Pinto, J. O., Monaghan, A. S., Davis, C. J., and Hannan, J. R.: Global Distribution and Characteristics of Diurnally Varying Low-Level Jets, J. Climate, 23, 5041–5064, doi:10.1175/2010JCLI3514.1, 2010.

Shapiro, A. and Fedorovich, E.: Analytical description of a nocturnal low-level jet, Q. J. Roy. Meteorol. Soc., 136, 1255–1262, 2010.

Stensrud, D. J.: Importance of low-level jets to climate, J. Climate, 9, 1968–1711, 1996.

Van de Wiel, B. J. H., Moene, A. F., Steeneveld, G. J., Baas, P., Bosveld, F. C., and Holtslag, A. A.: A Conceptual View on Inertial Oscillations and Nocturnal Low-Level Jets, J. Atmos. Sci., 67, 2679–2689, 2010.

An overview of a regional meteorology warning system

S. Gaztelumendi[1,2]**, J. Egaña**[1,2]**, K. Otxoa-de-Alda**[1,2]**, R. Hernandez**[1,2]**, J. Aranda**[3]**, and P. Anitua**[3]

[1]Basque Meteorology Agency (Euskalmet), Vitoria-Gasteiz, Spain
[2]TECNALIA, Meteo Unit, Vitoria-Gasteiz, Spain
[3]Basque Government, Interior Dept., Directorate of Emergencies and Meteorology, Vitoria-Gasteiz, Spain

Correspondence to: S. Gaztelumendi (santiago.gaztelumendi@tecnalia.com)

Abstract. In this work we present a regional meteorology warning system, particularly the operational weather warning system used by the Basque Meteorology Agency (Euskalmet) for Basque Country. System considers different meteorological phenomena capable of generate warnings, and is based on combined thresholds criteria depending on particular weather event and area of territory where is applied. In this work we describe the most significant aspects related with the warning event definition and the warning bulletin. Conclusions from comparison with the former system (prior to 2009) and feedback from different users are presented.

1 Introduction

The Basque Meteorology Agency (Euskalmet) has among its responsibilities severe weather warning issues for the Basque Country area (see Fig. 1). This information is the basis that Basque Government Civil Protection authorities use for action, including recommendations in alert situations to the Basque population. In the beginning of operations (2003), the Euskalmet weather warning system began, as in other meteorological services in the past century (WMO, 1999), like a simple system for a reduced set of meteorological hazards based on a unique threshold criteria (GV, 2004). In the last years, the system has progressively migrated towards a more sophisticated one based on extended meteorological phenomena and a traffic-light colour concept extensively used today by other European meteorological services (WMO, 2010; GV, 2009a).

As is well known from the risk assessment community, developing an effective public warning system is a complex process that requires the integration and management of many different elements. Aspects related with data collection and analysis, decision process, issue format, content, dissemination, public reception, validation and action are crucial. There is a common agreement that a capital factor for success is fluent communication and discussion with civil protection authorities (WMO, 2006; PPW, 2002, 2004).

In order to put into perspective the early severe weather warning system evolution in the case of Basque Country, it is important to consider the Euskalmet history in the Basque Government context. In 2002, the Meteorology and Climatology Directorate (Department of Transportation and Public Work) was founded, assuming regional subjects related to meteorology and climatology. In this context was born Euskalmet. Since its creation in late 2003, it is in charge of different forecast and surveillance operational aspects, including severe weather. In late 2009, the Directorate of Emergencies and Civil Protection merged with the Directorate of Meteorology and Climatology, becoming the Directorate of Emergencies and Meteorology (DEM) in the Interior Department.

In this context, meteorology, climatology, emergencies and civil protection regional and local experts work together with other agents (health, road, infrastructures, municipalities, etc.) in order to design a new severe weather surveillance and prediction protocol (GV, 2010). During past years there were multiple meetings at different levels with different partners to design and implement early warning systems, surveillance and prediction mechanisms, intervention plans and crisis management tools.

In this work we focus only on those aspects related with the meteorological warning event definition (Sect. 2) and warning bulletin aspects (Sect. 3). Finally, some results and

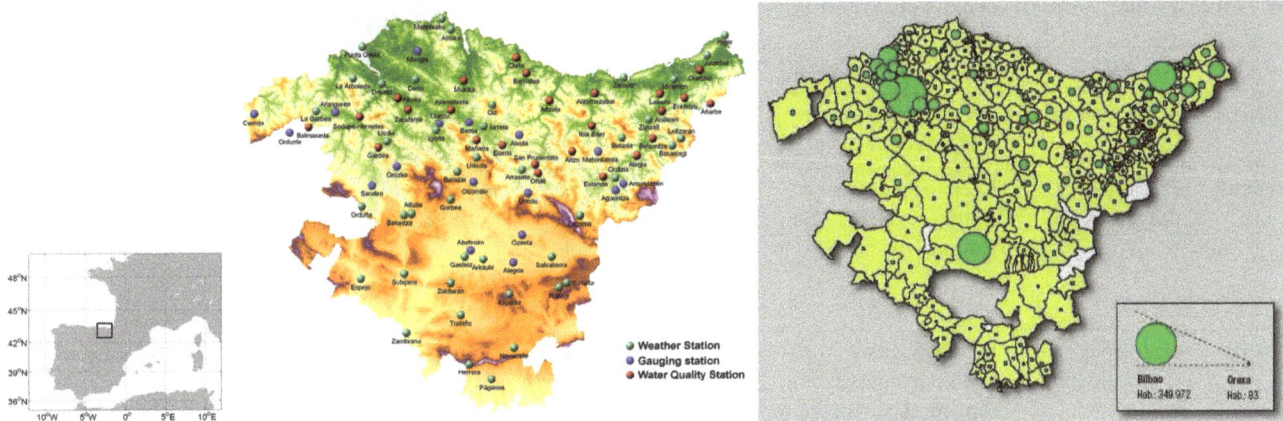

Figure 1. Basque Country Automatic Weather Station Network and population distribution.

conclusions from those years of development and operation are presented.

2 Warning event definition

Criteria and methodologies for defining meteorological warning events and assessing thresholds show a wide variation from one region to another (even in Europe), usually for reasons of climatology, vulnerability and operational aspects. Nevertheless, it is clear that selected thresholds must have an objective and local character, considering statistical studies for representative data in conjunction with risk analysis in a regional/local context. In Europe, the meteoalarm framework uses an extended four level colour code. Awareness colours are assigned according to impact and damage, so certain awareness levels for given phenomena have similar meaning to the public (WMO, 1999, 2002, 2003, 2005, 2010).

In the Euskalmet case, a meteorological warning event is established according to a set of thresholds for different meteorological variables or weather conditions related to potential risk situations caused by rain, snow, wind, temperatures and maritime conditions. Potential risk is evaluated based on previous experience and geographical and climatic characteristics of the area, as well as the distribution of population, properties and infrastructures in the territory (see Fig. 1).

The Basque Country Automatic Station Network (Gaztelumendi et al., 2003) is the main data source used to set up some statistical criteria such as return periods or percentiles. Even so, in the end, operational thresholds are somehow flexible and include practical experiences from historical severe weather events and considerations from the operational prediction capabilities (GV, 2010). During the weather warning process, Euskalmet experts work in cooperation with emergency experts and others agents, depending on the severity of the weather event, similar to regional health authorities during extreme temperature-

related events, responsible harbour and beach authorities in coastal-maritime cases, or roads and municipality authorities responsible in snow cases.

For every warning event case, a colour coded approach scheme philosophy is assumed, considering different aspects like weather severity, weather awareness, risk level, damages and frequency of a given phenomenon in the area of Basque Country (see Table 1). In this context we consider not only all kinds of atmospheric events capable of causing damages to people or properties (red/orange level i.e. alert level), but also those susceptible to affecting in some degree a particular human activity in a given scenario (yellow level, i.e. warning level).

2.1 Rain

In the rain case, two different aspects are considered: intensity and persistence. As in other meteorological services, these two aspects are considered through hourly rain intensities (e.g. Spanish AEMET case) and daily rain accumulation (e.g. Meteofrance case) (see WMO, 1999, 2003; INM, 2007, Stepek et al., 2010a). In the case of local intense rainfall events, problems in Basque Country are mainly related with flash floods of minor rivers, water accumulation on roads and terrain shifts. In persistent rainfall cases, risk comes usually from flood episodes of major rivers. In order to establish the different thresholds, previous studies and expertise from meteorology, emergencies, civil protection and regional river/water authorities are considered (e.g. GV, 1997, 1999a, b, 2004; DFG, 2006; Euskalmet, 2010; Egaña et al., 2005, 2007, 2008a, 2009a; Gaztelumendi et al., 2009a).

At present, for rain warning event definition, we consider hourly precipitation of 15/30/60 mm (limit for strong/very strong/torrential definition) and daily precipitation of 60/80/120 mm for yellow/orange/red level, respectively. Those thresholds are based on a return period estimation of 1/2/5 yr in each case, and are the same for all parts in the territory (see Table 2). Moreover, they are not far from

Table 1. Scenario thresholds and operative procedures for severe weather events.

Level	Weather	Weather Awareness	Risk	Damages	Frequency
Green	Not dangerous	No particular awareness is required.	No risk.	No damages.	Usually
Yellow	Potentially dangerous	Keep informed especially depending on your activity.	No risk for general population but do not take any avoidable risk, depending on your activity.	Some disturbances and very few/occasional damages.	Many times a year/not unusual
Orange	Dangerous	Be very vigilant and keep regularly informed about the detailed expected meteorological conditions.	Moderate/high risk. Follow any advice given by authorities. Be aware of the risks that might be unavoidable.	Moderate and/or localized damages.	Very few times a year
Red	Very dangerous	Keep frequently informed about detailed expected meteorological conditions and risks.	Very high risk. Follow orders and any advice given by authorities under all circumstances. Be prepared for extraordinary measures.	Major and/or generalized damages; casualties are possible.	One time in a few years.

common thresholds used in other countries (e.g. Stepek et al., 2010a; INM, 2007). The former system considered a unique threshold of 30 mm in one hour, and 60 mm in one day (see Table 3).

2.2 Snow

In the snowfall case, we consider an event when meteorological conditions support snow presence during some time (hours) at surface (WMO, 1999). We establish different criteria for different areas in the territory considering population, roads/highways and infrastructures distributions together with some social aspects (see Fig. 1). In our territory, main problems linked with snow episodes are related with transportation, usually from roads/highways users. In order to define procedures and thresholds, previous studies and knowledge in snow episodes from meteorology, civil protection, emergencies, roads maintenance and municipalities experts are considered (Euskalmet, 2009a, 2010; GV, 1997, 2004, 2009a).

In the snow warning case, we divide the territory into four different areas considering altitude. First zone, between 0 and 300 m, covers high-density populated areas in the north part of Basque Country (including Bilbao and Donostia). Second zone (300–700 m) covers the internal basin, Vitoria (Basque capital) area and main communications roads between Alava and the rest of the territory. The third zone (700–1000 m) covers less populated high land areas. No warning events are

considered when snow level is over 1000 m, as no populated areas and no main roads are present (see Fig. 2).

Yellow/orange/red levels are established for different snow accumulation depending on the area considered, as it is shown in Table 2. Note that for 0–300 m area (more than 77 % Basque population, including Bilbao and Donostia), yellow warning level is activated just for snow presence under 1 cm (see Table 2). The former system considered a warning event when snow was present below 1000 m (see Table 3).

2.3 Wind

In the wind case, usually mean wind and/or wind gusts are used in order to characterize a windy meteorological scenario (e.g. WMO, 1999). In our case, we adopt wind gust as the key variable, which proves to be a simple and effective way to take into account problems associated with high wind, as most part of damages are produced by the short duration pushing of wind. Nevertheless, as usual in complex topography, differences are present between mountainous, coastal and inland areas (Stepek et al., 2010b).

In Basque Country cases, main damages in windy scenarios are related with fallen trees, roofs, power cuts, transportation or wildfires. Available studies and expertise from meteorology, emergencies, civil protection and local/regional authorities are considered in order to establish the reference thresholds with a scientific and practical orientation (e.g.

Table 2. Event threshold classification using traffic-lights colour concept for new Euskalmet warning system.

Warning Event	What?	When?	Where?	Green Level	Yellow Level	Orange Level	Red Level	Units
Persistent precipitation risk.	Precipitation	Accumulated rain in 24 h	All the territory	<60	[60–80)	[80–120)	≥120	$l\,m^{-2}$
Intense precipitation risk.		Accumulated rain in 1 h	All the territory	<15	[15–30)	[30–60)	≥60	$l\,m^{-2}$
Snow risk		Snow presence at surface	Altitudes 0–300 m	no snow	(0–1)	[1–5)	≥5	$l\,m^{-2}$ or cm
			Altitudes 300–700 m	<1	[1–5)	[5–20)	≥20	$l\,m^{-2}$ or cm
			Altitudes 700–1000 m	<1	[1–10)	[10–30)	≥30	$l\,m^{-2}$ or cm
Wind risk	Wind	Maximum Wind gust	At exposed zones	<100	[100–120)	[120–140)	≥140	$Km\,h^{-1}$
			At non-exposed zones	<80	[80–100)	[100–120)	≥140	$Km\,h^{-1}$
Extreme low temperatures/ice presence risk	Temperature	Minimum Temperatures	Zone 1	>0	≤0	≤ −2	≤ −4	°C
			Zone 2	>0	≤0	≤ −4	≤ −7	°C
			Zone 3	>0	≤0	≤ −6	≤ −10	°C
			Zone 4	>0	≤0	≤ −5	≤ −8	°C
Extreme high temperatures risk		Maximum Temperatures	Zone 1	>3	≤33	≤35	≤37	°C
			Zone 2	>36	≤36	≤38	≤40	°C
			Zone 3	>35	≤35	≤37	≤39	°C
			Zone 4	>36	≤36	≤38	≤40	°C
Persistent high temperatures risk		Max/Min Temperatures	Zone 1	<30/19	30/9 during 1–2 days	30/19 during 3–4 days	30/19 during ≥5 days	°C
			Zone 2	<35/17	35/17 during 1–2 days	35/17 during 3–4 days	35/17 during ≥5 days	°C
			Zone 3	<35/17	35/17 during 1–2 days	35/17 during 3–4 days	35/17 during ≥5 days	°C
			Zone 4	<36/18	36/18 during 1–2 days	36/18 during 3–4 days	36/18 during ≥5 days	°C
Coastal-maritime risk	Waves/maritime conditions	Significant high wave in summer	0–2 miles	<2	[2–3.5)	[3.5–5.5)	≥5.5	m
		Significant high wave (winter, autumn, spring)	0–2 miles	<3.5	[3.5–5)	[5–7)	≥7	m
	Coastal Trapped Disturbance (CTD) phenomena or similar	Rapid change in Wind direction and module intensification (wind gust)	Coastal area (Zone 1 plus 2 miles offshore)	<60	[60–90)	[90–120)	≥120	$Km\,h^{-1}$

Euskalmet, 2010; GV, 2004; Egaña et al., 2004; Egaña and Gaztelumendi, 2009; Gaztelumendi et al., 2009b).

In the wind case, we consider a territory distribution on exposed and non-exposed zones due to historical wind data series, topography and population/infrastructures allocation. Exposed zones are mountainous, and shoreline zones and non-exposed areas correspond to the rest of the territory (see Fig. 2). We consider wind gust limits of 80/100/120 km h^{-1} and 100/120/140 km h^{-1} for yellow/orange/red level for exposed and non-exposed zones, respectively (see Table 2). Those values are established for return periods of less than one year in the yellow case and one–three years in orange/red case, not far from values used by other services in Europe and particularly for those used by French and Spanish Met services (Stepek, 2010b; INM, 2007). The former system is based on a unique gust threshold of 80/100 km h^{-1} for non-exposed and exposed zones (see Table 3).

2.4 Temperatures

In order to consider human response to heat and cold events, in the context of operational meteorological warning definitions, different approaches are possible based on different event definitions and temperature indexes considerations (see WMO, 1999, 2004; WHO, 2009; InVS, 2005; Robinson, 2001). In our case, we consider extreme high and low temperatures, and persistence of high temperatures (in some sense, heat-waves). This is done in order to consider major temperature-related incidences in our territory, mostly dealing with human health. In extremely high temperature cases, they are mainly due to dehydration or heatstroke usually as results from vigorous physical activity. When excessive heat persists, some chronic affections may worsen, especially in the elderly or very young populations. In the case of very cold temperatures, direct human health impacts come from

Table 3. Event threshold classification for former Euskalmet warning system.

Warning Event		What?	When?	Where?	No Severe Weather	Severe Weather	Units
Persistent precipitation risk		Precipitation	Accumulated rain in 24 h	All the territory	<60	≥60	$1\,m^{-2}$
Intense precipitation risk			Accumulated rain in 1 h	All the territory	<30	≥30	
Snow level risk			Snow at surface	All the territory	no snow below 1000 m	Snow below 1000 m	$1\,m^{-2}$ or cm
Wind risk		Wind	Maximum Wind gust	At exposed zones	<100	≥100	$Km\,h^{-1}$
				At non-exposed zones	<80	≥80	
Extreme temperatures	very low temp	Temperature	Minimum Temperatures	Zone 1	> −3	≤ −3	°C
				Zone 2, 3, 4	> −8	≤ −8	
	cold wave		Maximum Temperatures	Zone 1	>3 or ≤3 during <3 days	≤3 during ≥3 days	
				Zone 2, 3, 4	>3 or ≤3 during <2 days	≤0 during ≥3 days	
	very high temp		Maximum Temperatures	Zone 1	<35	≥35	
				Zone 2	<38	≥38	
				Zone 3	<37	≥37	
				Zone 4	<38	≥38	
	heat wave		Max/Min Temperatures	Zone 1	<31/20 or <3 days	>31/20 during ≥3 days	
				Zone 2	<35/18 or <3 days	>35/18 during ≥3 days	
				Zone 3	<36/17 or <3 days	>35/17 during ≥3 days	
				Zone 4	<36/18 or <3 days	>36/18 during ≥3 days	
Coastal-maritime risk	Waves/maritime conditions	Significant high wave	0–2 miles		<3	≥3	m
		Coastal Trapped Disturbance (CTD) phenomena or similar	Rapid change in Wind direction and module intensification (wind gust)	Coastal-maritime area (Zone 1 plus 2 miles offshore)	No CTD	CTD	

lack of awareness of hypothermia and protection against cold environmental conditions, and indirectly from car accidents due to ice on roads.

In Basque Country cases, we just use maximum and minimum temperatures as meteorological variables for severity definition. We have divided the area into four different zones according to climatology and similar meteorological behavior. These four zones correspond to the coastal area (zone 1) to the Cantabric interior zone (zone 2), the transition zone (zone 3) and the Ebro basin zone (zone 4), as can be seen in Fig. 4.

In order to establish zonification and the different reference thresholds, available studies and knowledge from meteorology, emergencies, civil protection and health authorities are considered (e.g. Euskalmet, 2005, 2010; Egaña et al., 2009b, 2010a, 2005; GV, 2009b, 2004).

In the case of extremely low temperatures, we consider yellow threshold as zero degrees in order to capture any potential problem in roads due to ice formation. For the orange/red alert level, minimum temperature thresholds are established for each zone according to percentile 1/0 for a set of representative temperature data. In the extreme high temperature event case, temperature thresholds are based on

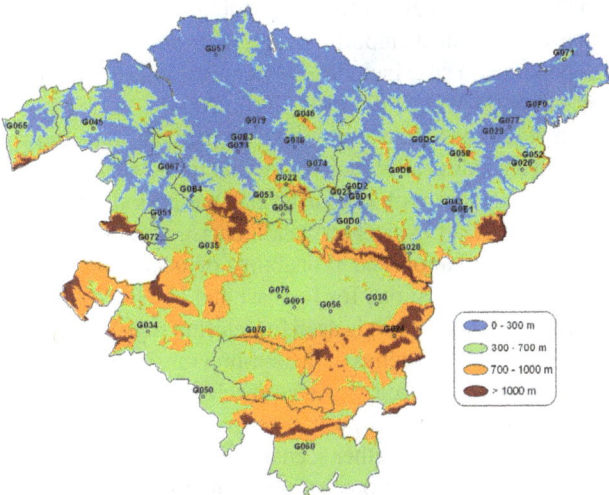

Figure 2. Snow warning areas and heated rain gauge distribution.

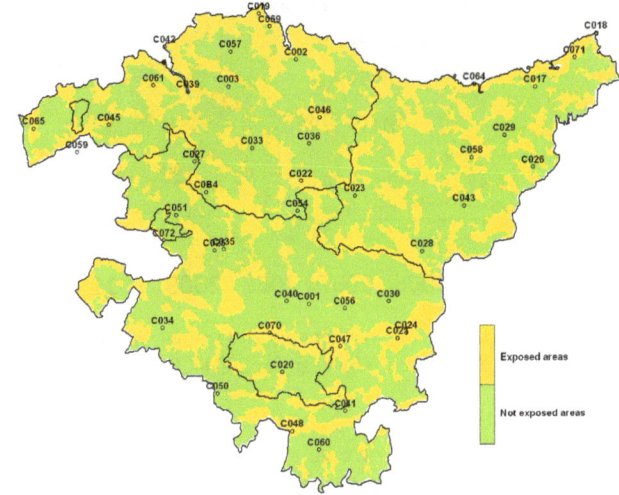

Figure 3. Wind warning areas and wind sensor distribution.

percentile 95/99/100 for yellow/orange/red levels, respectively, using temperature series for the same representative weather stations. Finally, in the high temperature persistence event, a combination of daily max/min temperatures are considered taking into account the 95 percentile; yellow/orange/red level are considered when this situation remains for 1–2 days, 3–4 days or more than 5 days respectively (Table 2).

The former system assumes a heat wave episode when during three or more consecutive days we have daily temperature maximum and minimum down to 95 percentile in each of the four different zones. We define a very high temperature situation when the maximum temperatures will be superior to 99 percentile at any point of each zone. For minimum temperature cases a 1 percentile was used (Table 3).

2.5 Maritime-coastal

For a maritime-coastal warning event, we consider two aspects that usually promote problems for Basque country coastal and maritime areas under Gales, Maritime storm, rapid cyclogenesis or wind reversal ("Galerna") conditions. In order to consider in an easy and effective way these phenomena, we focus on significant wave height in coastal waters and some criteria for sudden wind shift and intensification under *Galerna* conditions (see Table 2).

Usually damages and personal injuries associated with wave events in Basque Country are related with small ships, beaches and coastal-promenade users. In Euskalmet cases, we assume that a good reference variable is the significant wave height considered in the near coastal zone (less than 6 miles). Thresholds for significant wave height are established according to previous experiences in historical maritime-coastal severe events, some statistical studies based on wave data available from the Basque Coast and forecast aspects (e.g. Euskalmet, 2010; GV, 2004; Egaña and Gaztelu-

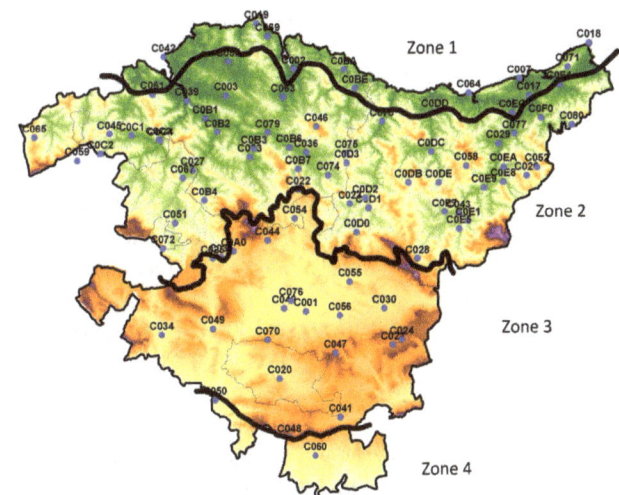

Figure 4. Temperature warnings areas and temperature sensor distribution.

mendi, 2009; Egaña et al., 2010b; Gaztelumendi et al., 2008, 2009, 2010a). Limit values of 3.5/5/7 m and 2/3.5/5.5 m for yellow/orange/red level in winter and summer time, respectively, are used (Table 2). Seasonal criteria are introduced to consider increasing risk in summer time (from beaches and marinas).

On the other hand, problems associated with a *Galerna* event affect beach users, fishermen and small ships users. During a *Galerna* episode, wind turns and intensifies suddenly from S to NW in few minutes, affecting a few kilometers onshore and offshore, and propagates along the coast from west to east (e.g. Euskalmet, 2010; Arasti Barca, 2001; Gaztelumendi et al., 2011). In such cases a combined criteria is used, assigning colours depending on wind gust value (see Table 2) and sea state (GV, 2010). The former system

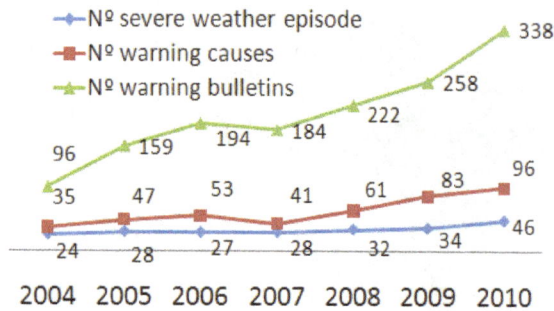

Figure 5. Number of severe weather episodes related with warning causes and number of bulletins by year, for 2004–2010 period.

considered a significant wave height threshold of 3 m, and the *Galerna* yes/no occurrence (see Table 3).

3 Warning bulletins

Thresholds and different criteria we have described are used to distinguish different warning events in a potential severe weather scenario, and to proceed with internal Euskalmet operational routines, including the issuance of warning bulletins. In these cases, actions must be taken according to the Basque Government protocol for prediction and surveillance in severe weather cases (GV, 2010). Implemented procedures, including warnings bulletin elaboration and dissemination, respond to international findings and recommendations for early warning systems development and operations (e.g. WMO, 2002, 2010; PPW, 2002, 2004), and noteworthy are not far from practices in nearby countries such as France or Spain (see WMO, 1999, 2003; INM, 2007).

Under a potential warning event, Euskalmet operational staff (forecast team, senior forecaster and coordinator) analyze the situation (severe weather briefing) and translate recommendations for warning issuance to meteorology responsible within the DEM. When the recommendation from Euskalmet includes orange/red level, briefings and the decision chain extend to emergencies and civil protection parts within the DEM.

It is worth mentioning that a yellow warning is considered for awareness situations so remains at the information level under the meteorological part of the decision chain. Orange and red warnings are considered for alert situations and focus on threats to civil protection, as actions must be taken (see Table 1). Although the meteorological warning bulletin format essentially remains the same format, different dissemination and communication procedures are considered depending on the warning level (GV, 2011).

3.1 Content

Effective warning messages must be short, concise, understandable, and actionable, answering the questions of "what?", "where?", "when?", "why?", and "how to re-

spond?". Usually they include a heading, a headline that summarizes the most important aspects, a descriptive text, and depending on case include non-technical information for public safety (WMO, 2002; PPW, 2002, 2004).

In our study case the warning bulletin includes a title with warning event causes. For each day, we include a sentence for each warning event with colour level, time period, affected area, and finally an explanatory and concise text describing the particular situation and probability of occurrence. The bulletin has two parts: one for forecast and one for observations. The observation part is used for relevant registered data or to incorporate warnings if a non-previously forecasted warning event is observed. We have developed different tools to help in diagnostic and prognostic procedures for warning events and severe weather scenarios identification. In the prognostic part, information coming from different available numerical models and nowcasting systems is the most important information source (e.g. Egaña et al., 2008b; Gaztelumendi et al., 2005, 2007, 2008a; Gelpi et al., 2006a, b). In the observation part, real-time information coming from the AWS mesonet and other data sources available for surveillance purposes is essential (e.g. Aranda and Morais, 2006; Gaztelumendi et al., 2003, 2005, 2006a, b, c, 2009c, 2010b; Hernández et al., 2010; Maruri et al., 2009).

Depending on the warning cause, complementary information is included with the warning bulletin. In snow cases, it consists of a special bulletin containing relevant text, extra graphical and tabular information on the event evolution, including iso0, snow level and precipitation forecast for next four days. In persistent temperature cases, a special bulletin with maximum and minimum temperature tables for each zone and for seven days ahead is incorporated (GV, 2010; Euskalmet, 2009a, b).

Warnings need to be understable, accessible, timely and tied to response actions to be taken by the people (WMO, 2002; Basher, 2006). For this purpose, for red and orange scenarios, information is specially prepared for media (radio, TV, newspapers) focusing on what is happening or going to happen and why/what to do to minimize harm.

3.2 Dissemination

Effective early warnings have to be communicated and disseminated to people to ensure they are warned in advance of impending hazardous events and to facilitate civil protection and others authority activities (WMO, 2002). In Euskalmet cases a supervised system was implemented for easier elaboration of different warning products. Operational personnel must fill out different input forms for the available software. The system formats this information for its dissemination, including e-mail and the web. All the products are elaborated in Spanish and Basque languages.

In Basque Country, case warning information (via mail and web) are routinely updated at 10:30 and 19:30 LT. If new relevant information is available, the update is done at the

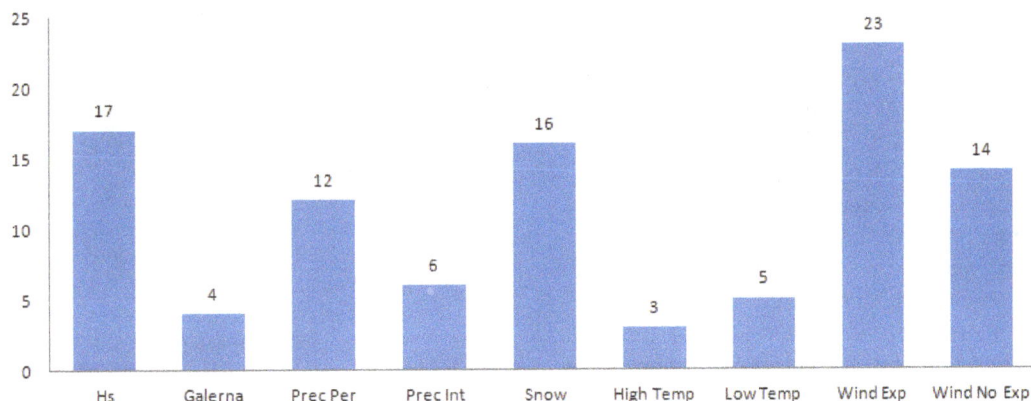

Figure 6. Warning causes frequency (%), for 2004–2010 period.

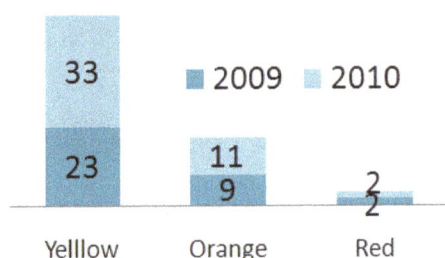

Figure 7. Number of worst warning level colour reached in 2009 and 2010 severe weather episodes.

time it becomes available. Warning information is disseminated to the public 24/48 h previous to the occurrence. The first Euskalmet warning bulletin was issued in May 2004; the first warning bulletin based on traffic light concept was issued in January 2009. In 2011 we started to include Twitter in the warning dissemination procedures.

To be effective, warnings must have not only a scientific/technical basis but also a strong focus on the people exposed to risk (WMO, 2010). Different zonification based on meteorological criteria are translated to political entities (provincial and municipalities) as is needed for civil protection actions and efficient dissemination. Particular communications actions are planned, depending on the warning event, in order to disseminate information effectively to those agents potentially affected or involved in security aspects at the local level (GV, 2010).

4 Results

From 2004 to 2010, 219 weather warning episodes were produced, with a mean ratio of 30 per year. During the same period, 1491 bulletins were issued, for an average rate of more than 200 per year. From the cause perspective, 23 % cases dealt with wind in exposed zones, 17 % with high waves and 16 % with snow events (see Figs. 5 and 6).

The new traffic light-based warning system proves to be an easy way for rapid identification of risk, removing in part difficulties for the general population in understanding meteorological language. Colour levels are an intuitive way to transmit risk and to prepare the population for a proactive response to recommendations, even for extraordinary measures required when a red level is activated, as in the case of the Klaus event (Gaztelumendi et al., 2009b).

In the Basque Country case, warning issues are never activated automatically and never based on a single-man philosophy; the decision chain must be followed and a consensus with emergency part is always needed. In the end, surpassing established meteorological thresholds is important but, in the operational perspective, are just used as guidelines. For instance, yellow level for temperatures below zero degrees is established depending on ice formation probability, or in wave cases some considerations to tides or wave periods are taken into account. Experience during those years of operation showed that a more open and complete vision is needed, especially in red level cases where human losses are feasible.

This new philosophy has had an impact on the increase of warning events during those last years (see Fig. 5), mainly due to new yellow thresholds affecting temperature, precipitation and wave events.

The rise in the number of warning bulletins observed (Fig. 5) is a direct consequence of the increase in warning events (severe weather events), but also reflects a tendency for issuing warning bulletins more than two days in advance. In pre-2009 situations, the number of bulletins issued per warning event was around 6.5, whereas in the new scenario it has increased to 7.5.

During 2009 and 2010, the number of red/orange events stays around 2/10 events per year, whereas in the yellow cases an increase from 23 in 2009 to 33 in 2010 is observed, mainly due to an increase in extreme temperatures and wave related events during 2010 (Fig. 7).

5 Conclusions

In this work we present an overview of a regional meteorology warning system focusing on main aspects related with warning event definition and warning bulletin generation and dissemination, putting this system into perspective with the former system available in Basque Country before 2009, as well as other international practices.

Both meteorology and emergency experts are essential in the formulation of severe weather risk decisions. The first are experts in assigning probability of occurrence of a weather related hazard, and the second are experts in the vulnerability evaluation to a particular hazard. This duality is a key component, not only in the thresholds/criteria for warning event definition, but also at the time of warning issues and dissemination. The new Basque Government context (meteorology and civil protection partners under the same structure) promotes synergies to ensure that users get full benefit of reliable forecasts and warnings.

From an operational meteorological point of view, some problems still remain in the new system. The temperature thresholds used are near expected prognosis error. The new yellow level for minimum temperature is often activated during long winter periods in some areas. Snow events according to established criteria are very difficult to predict and to validate.

In the future, new concepts will be introduced in the system, like taking into account storms in a clearer way, improving dissemination; or considering radar capabilities for rain episodes under 1 h duration, among others.

Acknowledgements. The authors would like to thank the Emergencies and Meteorology Directorate – Interior Department – Basque Government for public provision of data and operational service financial support. We also would like to thank all our colleagues from Euskalmet for their daily effort in promoting valuable services for the Basque community. Finally we would like to thank to all partners that participate in the Severe Weather Surveillance and Prediction Plan elaboration for the Basque Country.

Edited by: B. Reichert
Reviewed by: two anonymous referees

References

Aranda, J. A. and Morais, A.: The new weather radar of Euskalmet. Site selection, construction and installation, Proceedings 4th ERAD Conference, 356–359, 2006.

Arasti Barca, E.: Estudio de la galerna típica del Cantábrico, INM, ISBN: 84-8320-175-5, 2001.

Basher, R.: Global early warning systems for natural hazards: systematic and people-centred, Phil. Trans. R. Soc. A, 364, 2167–2182, 2006.

DFG (Diputacion Foral de Gipuzkoa): Caracterización de las situaciones hidrológicas extremas en Gipuzkoa y situación frente a inundaciones en Gipuzkoa, 2006.

Egaña, J. and Gaztelumendi, S.: Klaus overview and comparison with other cases affecting Basque Country area, Proceedings of the 5thECSS Conference, 6 pp., 2009.

Egaña, J., Gaztelumendi, S., Gelpi, I. R., and Otxoa de Alda, K.: Synoptic characteristics of extreme wind events in the Basque Country, 6th EMS Conference, 2004.

Egaña, J., Gaztelumendi, S., Mugerza, I., and Gelpi, I. R.: Synoptic patterns associated to very heavy precipitation events in the Basque Country, 5th EMS Conference, 2005.

Egaña, J., Gaztelumendi, S., Gelpi, I. R., and Otxoa de Alda, K.: A preliminary analysis of summer severe storms in the Basque Country area: synoptic characteristics, 4th ECSS Conference, 2007.

Egaña, J., Gaztelumendi, S., Gelpi, I. R., Otxoa de Alda, K., Maruri, M., and Hernández, R.: Radar Analysis of Different Meteorological Situations in the Basque Country Area, Proceedings 5th ERAD Conference, Helsinki, 4 pp., 2008a.

Egaña, J., Gaztelumendi, S., Otxoa de Alda, K., Gelpi, I. R., and Hernandez, R.: Synoptical and mesoscale information for forecast purposes, 8thEMS/7thECAC Conference, 2008b.

Egaña, J., Gaztelumendi, S., Pierna, D., Gelpi, I. R., and Otxoa de Alda, K.: "Convective storms over Basque Country: June 2008 cases study", Proceeding of the 5th ECSS Conference, 291–292, 2009a.

Egaña, J., Gaztelumendi, S., Otxoa de Alda, K., and Gelpi, I. R.: Synoptic characteristics of extreme heat episodes in the Basque Country, 9th EMS/9th ECAM Conference 2009b.

Egaña, J., Gaztelumendi, S., Otxoa de Alda, K., and Gelpi, I. R.: Synoptic characteristics of extreme low temperatures episodes in the Basque Country, 10thEMS/8thECAC Conference, 2010a.

Egaña, J., Gaztelumendi, S., Gelpi, I. R., and Otxoa de Alda, K.: Analysis of oceano-meteorological conditions during Klaus episode on Basque Country area, 10thEMS/8th ECAC Conference, 2010b.

Euskalmet: Estudio de umbrales de temperatura para la CAPV, 2005.

Euskalmet: Procedimiento operativo ante situaciones de nieve, 2009a.

Euskalmet: Procedimiento operativo ante situaciones de temperaturas altas extremas, 2009b.

Euskalmet: Compendio de informes meteorológicos anuales, 2010.

Gaztelumendi, S., Otxoa de Alda, K., and Hernandez, R.: Some aspects on the operative use of the automatic stations network of the basque country, 3th ICEAWS Conference, 2003.

Gaztelumendi, S., Egaña, J., Gelpi, I. R., and Mugerza, I.: An Automatic Warning and Alert system: Design and validation, 5th EMS Conference, 2005.

Gaztelumendi, S., Otxoa de Alda, K., Gelpi, I. R., and Egaña, J.: An Automatic Surveillance System for Severe Weather Real Time Control in Basque Country Area, 4th ICEAWS Conference, 2006a.

Gaztelumendi, S., Egaña, J., Gelpi, I. R., Otxoa de Alda, K., Maruri, M., and Hernández, R.: The new radar of Basque Meteorology Agency: Configuration and some considerations for its operative use, Proceedings 4th ERAD Conference, 363–366, 2006b.

Gaztelumendi, S., Otxoa de Alda, K., Egaña, J., and Gelpi, I. R.: Inclusion of Radar information in the surveillance panel of the Basque Meteorology Agency, Proceedings 4th ERAD Conference, 352–355, 2006c.

Gaztelumendi, S., Gelpi, I. R., Egaña, J., and Otxoa de Alda, K.: Mesoscale numerical weather prediction in Basque Country Area: present and future, EMS7/ECAM8 Conference, 2007.

Gaztelumendi, S., Egaña, J, Gelpi, I. R., and Otxoa de Alda, K.: The Euskalmet wave forecast system – preliminary results and validation, Proceedings 5th EuroGOOS Conference, 168–176, 2008a.

Gaztelumendi, S., Hernandez, R., Otxoa de Alda, K., Egaña, J., and Gelpi, I. R.: Basque Meteorology Agency annual meteorological bulletins, 8thEMS/7thECAC Conference, 2008b.

Gaztelumendi, S., Otxoa de Alda, K., Egaña, J., Gelpi, I. R., Pierna, D., and Carreño, S: Summer showers characterization in the Basque Country, Proceedings of the 5th ECSS, 81–82, 2009a.

Gaztelumendi, S. and Egaña, J.: Klaus over Basque Country: Local characteristics and Euskalmet operational aspects, Proceedings of 5th ECSS, 255–256, 2009b.

Gaztelumendi, S., Lopez, J., Egaña, J., and Aranda, J. A.: Preliminary results from lightning detection in Basque Country, 5th ECSS Conference, 2009c.

Gaztelumendi, S., González, M., Egaña, J., Rubio, A., Gelpi, I. R., Fontán, A., Otxoa de Alda, K., Ferrer, L., Alchaarani, N., Mader, J., and Uriarte, Ad.: Implementation of an operational oceanometeorological system for the Basque Country, Thalassas, 26, 151–167, 2010a.

Gaztelumendi, S., Otxoa de Alda, K., Hernández, R., Egaña, J., and Gelpi, I. R.: Meteosat products for surveillance: the Euskalmet case, Proceedings 2010 EUMETSAT Conference, p. 57, 6 pp., 2010b.

Gaztelumendi, S., Egaña, J., Ruiz, M., Pierna, D., Otxoa de Alda, K., and Gelpi, I. R.: An analysis of Cantabric coastal trapped disturbances, 6th EuroGOOS Conference, 2011.

Gelpi, I. R., Gaztelumendi, S., Egaña, J., and Otxoa de Alda, K.: Implementing a Data Assimilation System: Preliminary Results, 6th EMS Conference, 2006a.

Gelpi, I. R., Gaztelumendi, S., Otxoa de Alda, K., and Egaña, J.: Some results from assimilation on Kapildui radar wind information in the Basque Country, Proceedings of the 4th ERAD Conference, 550–552, 2006b.

GV (Gobierno Vasco): Plan de Protección Civil de Euskadi, 1997.

GV: Plan Integral de Prevención de Inundaciones (PIPI), 1999a.

GV: Plan Especial de Emergencias ante el Riesgo de Inundaciones de la CAPV, 1999b.

GV: Protocolo para la predicción y vigilancia de fenómenos meteorológicos adversos, 2004.

GV: Plan de predicción y vigilancia de fenómenos meteorológicos adversos, Dirección de Meteorología y Climatología, 2009a.

GV: Procedimiento operativo en situaciones de temperaturas altas persistentes y temperaturas altas extremas, Dirección de Meteorología y Climatología 2009b.

GV: Protocolo para la predicción, vigilancia y actuación ante fenómenos meteorológicos adversos, Departamento de interior, 2010.

Hernández, R., Gaztelumendi, S., Otxoa de Alda, K., Gelpi, I. R., and Egaña, J.: Combining Meteosat data and weather radar products to improve the meteorological surveillance and nowcasting, Proceedings 2010 EUMETSAT Conference, p. 57, 6 pp., 2010.

INM: Plan nacional de predicción y vigilancia de meteorología adversa – meteoalerta, Ministerio de Medio Ambiente, Gobierno de España, 2007.

InVS (Institut de Vielle Sanitaire): Système d'alerte canicule et santé (SACS) Raport, June 2005.

Maruri, M., Gaztelumendi, S., Gelpi, I. R., and Egaña, J.: Some quality aspects related with Punta Galea wind profiler and Kapildui weather radar, 8th International Symposium on Tropospheric Profiling, Delft, 2009.

PPW (Partnership for Public Warning): Developing a unified all-hazard public warning system, PPW-Report, 2002.

PPW: An introduction to public alert and warning, PPW-Report, 2004.

Robinson, P. J.: On the definition of heat waves, J. Appl. Meteorol., 40, 762–765, 2001.

Stepek, A., Wijnant, I. L., and van der Schrier, G.: Analysis of daily precipitation thresholds in Meteoalarm, ECA&D Report, 2010a.

Stepek, A., Wijnant, I. L., and van der Schrier, G.: Analysis of wind gust thresholds in Meteoalarm, ECA&D Report, 2010b.

WHO: Improving public health responses to extreme weather/heatwaves, EuroHEAT: Technical Summary, World Health Organization Europe, 2009.

WMO: Guide to public weather services practices, TD No. 834, 1999.

WMO: Guide on improving public understanding of and response to warnings, TD No. 1139, 2002.

WMO: Guidelines on cross-border exchange of warnings, TD No. 1179, 2003.

WMO: Proceedings of the Meeting of Experts to Develop Guidelines on Heat/Health Warning Systems, WCASP No. 63, WMOTD No. 1212, 2004.

WMO: Guidelines on integrating severe weather warnings into disaster risk management, TD No. 1292, 2005.

WMO: Comprehensive risk assessment for natural hazards, TD No. 966. Rep, 2006.

WMO: Guidelines on early warning systems and application of nowcasting and warning operations, TD No. 1559, 2010.

20

An operational forecasting system for the meteorological and marine conditions in Mediterranean regional and coastal areas

M. Casaioli[1], F. Catini[2], R. Inghilesi[1], P. Lanucara[2], P. Malguzzi[3], S. Mariani[1], and A. Orasi[1]

[1]Italian National Institute for Environmental Protection and Research (ISPRA), Rome, Italy
[2]Italian Interuniversity Consortium High Performance Systems (CINECA), Rome, Italy
[3]Institute of Atmospheric Sciences and Climate-Italian National Research Council (ISAC-CNR), Bologna, Italy

Correspondence to: R. Inghilesi (roberto.inghilesi@isprambiente.it)

Abstract. The coupling of a suite of meteorological limited area models with a wave prediction system based on the nesting of different wave models provides for medium-range sea state forecasts at the Mediterranean, regional and coastal scale. The new system has been operational at ISPRA since September 2012, after the upgrade of both the meteorological BOLAM model and large-scale marine components of the original SIMM forecasting system and the implementation of the new regional and coastal (WAM-SWAN coupling) chain of models. The coastal system is composed of nine regional-scale high-resolution grids, covering all Italian seas and six coastal grids at very high resolution, capable of accounting for the effects of the interaction between the incoming waves and the bathymetry. A preliminary analysis of the performance of the system is discussed here focusing on the ability of the system to simulate the mean features of the wave climate at the regional and sub-regional scale. The results refer to two different verification studies. The first is the comparison of the directional distribution of almost one year of wave forecasts against the known wave climate in northwestern Sardinia and central Adriatic Sea. The second is a sensitivity test on the effect on wave forecasts of the spatial resolution of the wind forcing, being the comparison between wave forecast and buoy data at two locations in the northern Adriatic and Ligurian Sea during several storm episodes in the period autumn 2012–winter 2013.

1 Introduction

In the last two decades there has been an increasing interest in applied meteorology to provide marine forecasts as by-product of the traditional meteorological forecasts. Events like Katrina have compelled the improvement of integrated systems capable of correctly simulate and forecast extreme storm-surge events (see, e.g., Wang and Oey, 2008; Sampson et al., 2010), while economic factors associated with both relatively new meteorological applications like weather-routing (Saetra, 2004) and more traditional coastal dynamics applications, have encouraged the meteorological centers to expand in a field which was previously considered as pertaining mostly to oceanography and to coastal engineering. The process has been clearly favored by the rapid development of reliable numerical models (e.g., The WISE Group, 2007).

At the oceanic scale, the scientific and economic interest in improving forecasting systems is to extend the range of a useful forecast in time, that is, implementing Ensemble Prediction Systems (EPS) to get probabilistic forecasts (Saetra and Bidlot, 2004) and couplings between atmospheric, wave and ocean models (see, e.g., Janssen et al., 2013). At the scale of relatively small, almost enclosed seas like the Mediterranean, it becomes likewise critical to improve the quality of the short-term forecasts in the coastal zones. In fact, since a freighter usually traverses the Mediterranean Sea from east to west in only a few days, it follows that the internal distances put a limit to the practical value of extended medium-range wave forecasts. Moving towards the shore, the interactions with the sea-bottom and with morphological features as islands, gulfs or artificial structures become critical factors in determining the time evolution of the wave fields along their path. The wind-wave dynamics assumes an increasingly "local" nature, until waves reach a point where the solution of the stochastic wave-propagation problem can not satisfy the conditions typical of the surf-zone, and the waves break.

Given the complexity of the processes involved, there is considerable scientific interest in the problem of wave propagation on many different spatial-temporal scales, from the global scale to the regional, local and small scale. Numerical models have been developed specialising in some particular range, that is, deep water – global scale models, near-shore models, surf-zone, small-scale models. While the range of scales on which the models are applied are widening in time, the most common approach to deal with the problem is still to nest different models from the large to the smaller scale. If the models are similar in the formulation, as it happens from large scale models to regional and perhaps local scale, a chain of numerical models can be implemented operationally to produce forecasts. The principal advantage of this approach, when applied to closed or almost-closed basins like the Mediterranean Sea, is that there are no lateral boundary conditions to fix because all boundaries are closed. Then, every nested model inherits boundary conditions from the parent model and the important aspect to be dealt with care is the blending of the different grid geometries, the parametrizations of the models and the external forcings.

The meteorological forcing, operating itself at the synoptic, regional and at local scales, is the cause of the swell and the local wave generation, this making wave forecast to depend almost completely on the accuracy of surface wind field. Unfortunately, the wind is hardly the easiest quantity to forecast, especially at the surface. The two main problems with surface wind forecast are the parametrisation of the boundary layer processes and the spatial resolution of the meteorological model. There are many studies in literature (for Mediterranean Sea see, e.g., Signell et al., 2005; Bertotti et al., 2006, 2009; Cavaleri and Bertotti, 2003; Cavaleri and Sclavo, 2006; Bellafiore et al., 2012) indicating that the use of meteorological hydrostatic limited area models (LAM) can effectively improve wave forecast at regional scale, in particular where the the topography affects significantly the mesoscale flow.

Due to its peculiar weather, some of the mentioned studies were focused on the Adriatic Sea, which is a small, almost-enclosed sea bounded by several mountain ranges along most of its borders. The recurrent flooding of Venice, located inside the Venice Lagoon on the north-western border of the sea, is an example of how the meteorological forcing can affect both the regional and local scales. In fact, the floodings (named aqua alta) are related to south-easterly winds (Sirocco) on the entire Adriatic Sea in connection with high tides generating storm surges and high sea waves directed towards the Lagoon. The bora, which is an almost-katabatic wind forced through the Dinaric Alps at the north-eastern border of the Adriatic Sea, is instead an example of strong wind dominated by highly local structures. The two wind regimes, bora and Sirocco, almost completely define the wave climate in the northern Adriatic Sea.

Clearly, bora is a kind of meteorological feature which can hardly be reproduced by low-resolution general circulation models, and even by LAMs. Non-hydrostatic, very high resolution LAMs (VHRLAM) are traditionally aimed to deal with this kind of local scale (loosely defined as on the lower end of the meteorological meso-β scale) phenomena. With the availability of VHRLAMs the expectation was that they would lead to a definite improvement of local wave forecast, like in the case of bora episodes. On the contrary, even though wind fields produced by VHRLAM are more realistic in reproducing complex-terrain situations, some studies (see, in particular, Signell et al., 2005; Bertotti et al., 2006, 2009) indicated that their use as input of wave models have seldom performed better than the simple LAMs. To be fair, it must also be mentioned that forecast verification at very small spatial and temporal scales presents a real challenge to the traditional methods of analysis (the double penalty issue, see, e.g., Mass et al., 2002). A very well resolved field (in terms of spatial distribution, gradients, etc.) with an active dynamics on the short time scale could be outclassed by a field having a reduced spatial and temporal variability when compared with a small number (or even a single) time series. It has been suggested, at least, that in the long term the use of VHRLAMs would provide a better wave climate statistics due to the enhanced variability of the local fields (Signell et al., 2005).

In pondering the advantages of the downscaling against the complexity of the procedures involved, it is important to consider that the main processes related to wave propagation occur in the coastal zone. For example, the formation of wave-induced currents in the region of wave bottom-induced breaking (the surf zone), and the interaction between breaking waves and currents are crucial for the coastal dynamics and the evolution of the shoreline in the long term. The dispersion of pollutants in coastal areas can also be significantly affected by the presence of waves interacting with currents and riverine plumes or jets. Unfortunately, not all the dynamical processes, like bottom-induced breaking or diffraction induced by obstacles, can be easily taken into account by numerical models.

As far as the surf zone is concerned, the traditional wave propagation models, that is, models based on the the solution of the action balance equation (see Whitham, 1999; Lavrenov, 2003), are not adequate to simulate the coastal dynamics properly, and a completely different approach must be considered, like 3-D Navier-Stokes (NS) equations or possibly Boussinesq or non-hydrostatic shallow water equations (Zijlema and Stelling, 2005). Due to the severe time and computational requirements of the 3-D NS models, the idea of considering a complete operational coverage of the coastal processes up to the shore seems definitely premature at the moment. On the contrary, it is completely possible to determine the sea conditions operationally up to the offshore boundary of the surf zone in a large part of the coastal areas. Along these boundaries, wave spectra generated by the operational models can be stored in order to provide boundary conditions for the application of NS surf-zone models in case studies or in climate analyses. An operational chain of wave models from the Mediterranean to the coastal scale is then aimed at not only providing the numerical forecast at the regional and sub-regional scale but also defining the surf zones and determining the boundary conditions to apply for NS models to resolve the individual waves, the alongshore and the rip currents (see, e.g., Lavrenov, 2003).

These considerations were the rationale for the set-up in the late nineties of the ISPRA Sistema Idro-Meteo-Mare (SIMM) and for its further, recent improvement, including the use of higher-resolution models and the development of the Mediterranean-embedded Costal WAve Forecasting system (Mc-WAF). After one year from the Mc-WAF start-up (September 2012), a first preliminary evaluation of the system performance is presented here.

The paper is organized as follows. The SIMM chain and the advances in the meteorological and marine segments are outlined in Sect. 2. Section 3 illustrates the results from the comparison between Mc-WAF forecast and the wave climate at two different locations in the Tyrrhenian and Adriatic Seas. The use of LAM's and VHRLAM's wind during three case studies at La Spezia, Ancona and Venice is also presented in the section. Finally, conclusions and final remarks are presented in Sect. 4.

2 The SIMM forecasting system

2.1 The original SIMM

SIMM is an integrated forecasting system, based on a cascade of several one-way nested numerical models, which has been providing since 2000 hydro-meteorological and marine forecasts over the Mediterranean basin and storm surge (acqua alta) forecasts in the Northern Adriatic Sea (Speranza et al., 2004, 2007).

The SIMM forecasting system is continually updated to incorporate the latest results of research. In particular, attention has been paid to the improvement of the hydrostatic BOlogna Limited Area Model (BOLAM, developed by ISAC-CNR: Buzzi et al., 1994; Malguzzi and Tartaglione, 1999), since it not only provides weather forecasts over the Mediterranean basin, but also gives the input to the marine models of the SIMM chain. A fully updated, parallelized BOLAM version was implemented in 2009 (Mariani et al., 2014b) even keeping the original 0.1° horizontal grid step and domain extension (Fig. 1).

2.2 The new BOLAM-MOLOCH suite

A new higher-resolution BOLAM configuration was implemented in late 2012 after a sensitivity study based on a

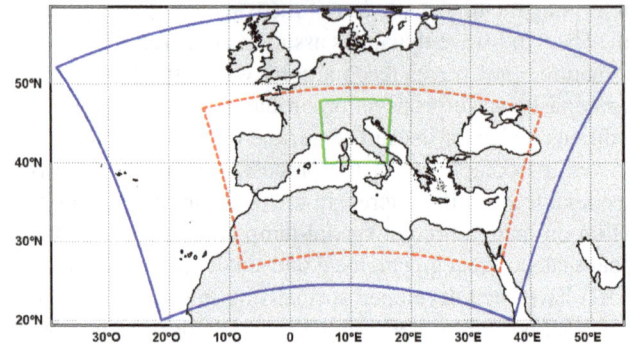

Figure 1. SIMM: model domain for the 0.1° BOLAM (red dashed line), the 0.07° BOLAM (blue solid line) and the 0.0225° MOLOCH (green solid line).

massive "reforecast" campaign (Casaioli et al., 2013). In that study, five different BOLAM configurations were intercompared on a densely-instrumented verification area. The comparison was made up by combining different settings, namely horizontal grid spacing, domain extension, initial and boundary condition, nesting design, and code version. The optimal configuration found had 0.25° ECMWF initial and boundary conditions and a 0.07° resolution grid (7.8 km) over an extended domain (Fig. 1).

This configuration was hence tested operationally (Ferretti et al., 2013) in the framework of the Special Observation Periods (SOPs) of the international initiative HYdrological cycle in Mediterranean EXperiment (HyMeX, http://www.hymex.org/). In addition, a very-high, 0.0225° version of the convection-permitting ISAC-CNR MOLOCH model (Malguzzi et al., 2006) was implemented in cascade to the hydrostatic BOLAM for SOPs. As shown in Fig. 1, the MOLOCH domain covered Central and Northern Italy. This new configuration will be completely operational within SIMM in 2014.

2.3 Mc-WAF

Mc-WAF is a complex operational tool designed to merge different scales for the generation and propagation of the wave energy in the Mediterranean Sea. The system is effective in connecting the Mediterranean scale with the coastal scale using an intermediate nesting at regional scale. In principle, it allows the use of altimeter data assimilation, regional and coastal surface currents, and also different surface wind fields, that is, LAM's winds at regional scale and VHRLAMS's winds at coastal scale. A high level of modularity in the chain implementation makes also possible to improve the operational system step by step, adding and testing new areas, independently from what was implemented before.

The marine forecast system is the operational implementation of a wind wave hindcasting system which was extensively verified on a series of test-cases in several Italian locations (Inghilesi et al., 2012). The system works on three

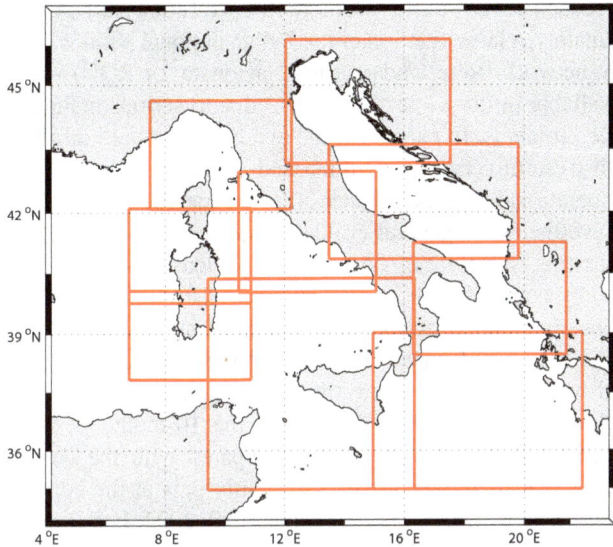

Figure 2. Mc-WAF: implemented regional areas.

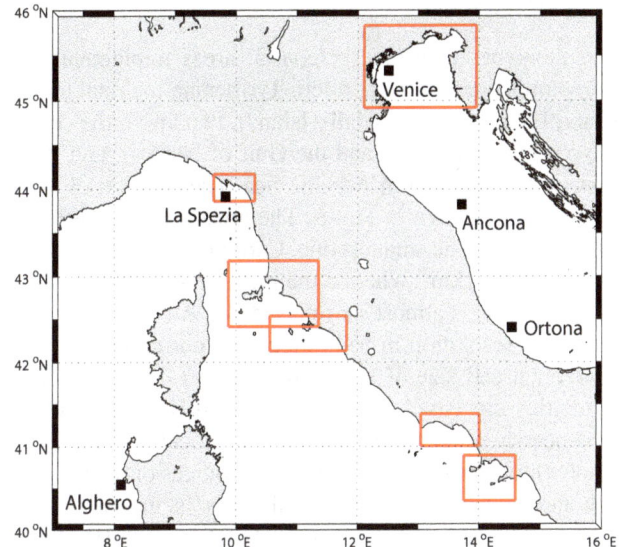

Figure 3. Mc-WAF: implemented coastal areas.

levels of nesting: the Mediterranean model passes the boundary conditions to the regional runs (Fig. 2), each of which, in turn, creates the boundary conditions for all the coastal runs present in the particular regional grid (Fig. 3).

Currently, the wind input is provided by the 0.1° BOLAM model: the same hourly wind fields are used by all wave models in the operational version of the system. The forecast range is 84 h, with 1 h frequency output. The system will be upgraded to use the higher resolution meteorological input tested in the present work in the late 2014, after the complete implementation of the refined meteorological segment of the system.

2.3.1 Mediterranean scale

The marine forecast system at the Mediterranean scale is the first step of the wave operational forecast. The model used is the WAve Model (WAM) cycle 4.5.3 implemented on a grid extending from 5.5° W to 35.73° W in longitude and from 30.0° N to 46.0° N in latitude at 1/30 degree resolution. WAM is a third-generation wave model, that is, it integrates the wave energy-balance equation:

$$\frac{\partial E}{\partial t} + \vec{c}_\mathrm{g} \cdot \nabla_{\mathrm{x}E} = S_\mathrm{wind} + S_\mathrm{nl} + S_\mathrm{ds}, \tag{1}$$

where $E(f,\theta)$ is the variance density spectrum of surface elevation, $f = \omega/2\pi$, is the spectral frequency, θ is the direction, $c_\mathrm{g} = \partial\omega/\partial\kappa$ is the group velocity, and ω and κ are related by means of the dispersion relationship. The S functions on the right hand side of the equation represent the sources for wind-wave interaction, resonant nonlinear wave-wave interaction and dissipation, respectively. In the WAM cycle 4.5.3 the wind-generation function and dissipation terms implement the Janssen's formulation, and the nonlinear interaction

source function is evaluated using the discrete interaction approximation (see, e.g., Janssen, 2008).

It is perhaps worth mentioning that the use of the Hamiltonian formulation for the time evolution of wave packets implied in the equation relies on the hypothesis formulated by Whitham about the averaged Lagrangian (described in Whitham, 1999). The consequence is that any information about the phase of the waves which constitute the actual sea state is lost. The class of wave models based on the approach described are sometimes named stochastic wave models as opposed to deterministic wave models, which can reproduce the evolution of individual waves (used mostly in the surf zone).

In WAM, the number of directions considered is 24, whereas the number of frequencies is 25 ranging from 0.04177 Hz to 0.4114 Hz. The bathymetry used is the general Bathymetric Chart of the Ocean (GEBCO) at 30 arc second grid resolution.

2.3.2 Regional scale

In all regional areas, WAM is implemented as in the Mediterranean grid, except for the position and dimension of the grids and for the resolution, which is 1/60° in latitude and longitude. As shown in Fig. 2, the nine regional areas covering all the Italian Seas are: Ligurian-North Tyrrhenian Sea, North Sardinia, South Sardinia, Central Tyrrhenian Sea-Sicily Channel, Ionian Sea, Gulf of Taranto-Otranto Channel, Central Adriatic Sea, Northern Adriatic Sea. The GEBCO bathymetry was locally corrected in each area using the Istituto Idrografico della Marina (IIM) digital maps.

2.3.3 Coastal scale

At the moment, there are six coastal areas implemented: three are imbedded in the northern Tyrrhenian regional area (Carrara, Elba Island and Giglio Island), two are in the central Tyrrhenian (Terracina and the Gulf of Naples), and the last one is in the northern Adriatic Sea. The position of the coastal areas is shown in Fig. 3. The areas are very different in extension. The smallest one, Carrara, covers a surface of about $120 \times 60\,km^2$, whereas the bigger one, in the northern Adriatic Sea, is almost six times larger. All coastal grids have $1/240°$ resolution in both directions, corresponding to an horizontal cell size of approximately $400 \times 400\,m^2$. The bathymetries are based on the regional GEBCO-IIM sets, refined nearshore by the inclusion of all local information available (multi-beam cruises and other sources of data).

The model in use in all coastal areas is the Simulating WAves Nearshore (SWAN) model 40.91, cycle III (The SWAN Team, 1993). SWAN is also a third-generation wave model, differing from WAM mostly in the numerical methods used and in the presence of additional source functions in Eq. (1) for shallow water applications.

The forecasts in coastal areas are very sensitive to the correct parametrization of the physical processes, that is, wind generation-whitecapping, shoaling and bottom refraction. Consequently, it is very important to determine in each case which combination of numerical schemes gives the better results and how smooth is the transition from the regional to the coastal scale. Tests in all areas indicated that, for wind generation and white-capping dissipation, the nonlinear saturation-based scheme described in Van der Westhuysen et al. (2007) gives the best results and a good coupling between WAM and SWAN at different scales (see Inghilesi et al., 2012).

3 Test and upgrade of the system

Forecast verification is an important component of the system. SIMM meteorological forecasts have been monitored and verified for over a decade and there is a well-established methodology (see, e.g., Casaioli et al., 2013; Mariani et al., 2014b), which drives the evolution of the meteorological chain of models. Marine verification, implemented originally for deep water forecasts (Speranza et al., 2007), is now being extended to check also the new coastal system. Due to the nesting procedure, accurate forecasts in deep sea are a necessary condition for good forecasts in coastal areas, thus marine forecast verification is still a key factor in the assessment of the reliability of the whole Mc-WAF system. Several aspects have to be considered: the climatological, long-term reliability of the forecasts, the accuracy in the forecast of single storms as compared to buoy or satellite data, and the reliability of the forecast for a given event in terms of probability of occurrence. In this paper, only the first issue is considered, being a more thorough analysis of the problem at the regional

and coastal scale the aim of a more focused future study. Unfortunately, a large fraction of the Italian national wind wave buoy network (Rete Ondametrica Nazionale, or RON) was unavailable in the winter 2012–2013 due to serious maintenance problems. In particular, the two buoys moored off the north-western coasts of Sardinia and central Adriatic were un-operational from September 2012 to June 2013.

Nevertheless, a general evaluation of the behavior of the forecasting system can be assessed, at least on the long term, by comparing the joint frequency distributions (JFD) of observed and simulated time series of significant wave height (H_{m0}) and wave direction (or observed and simulated wave climate). Hence, the distribution obtained from the hourly series of H_{m0} forecast in the 10-month period September 2012–June 2013 was compared with the wave climate evaluated over 12 years of hourly data at the corresponding buoy location in the period 1989–2002. For all the analyses presented in this study, forecast time series were assembled using the first 24 h of every daily forecast, starting at 00:00 UTC. A preliminary analysis was aimed at extracting a set of independent values of H_{m0} from both the time series. As a first step, individual storms (continuous periods for which H_{m0} is over a given threshold) were identified on the assumption that different storms are separated by at least 48 h of "calm" sea ($H_{m0} <$ a given threshold). Then, for each storm the maximum value was taken as representative of the event. The procedure was exactly the same used to statistically analyse the extreme wave events in the Mediterranean Sea; a full description of the methodology can be found in Inghilesi et al. (2001).

The hypothesis that the distribution of independent events extracted from the forecast series was identical to the distribution of the independent events extracted from the buoy time series was consequently tested. In order to test the hypothesis, the U Mann-Whitney-Wilcoxon test (hereafter U test) was applied on both the distributions of independent events measured and forecast at Alghero (NW Sardinia) and Ortona (central Adriatic). This statistical non-parametric test is commonly used to decide whether two population distributions are identical without assuming them to follow any known distribution (for more information and applications see, e.g., Mood et al., 1974; Wilks, 2011). In particular, given two samples of independent data and without assuming the data to have normal distribution, it allows to decide at the a-priori significance level $p = 0.05$ whether the data have identical distribution or not. If the result of the test, p value, does not exceed 0.05 the null hypothesis – that the difference is due to random sampling – is rejected, and it is concluded that the two populations are distinct. If the p value > 0.05, the test do not support the rejection of the null hypothesis.

3.1 Alghero

Time series of data and forecasts refer to the position 40.545° N, 8.011° E, corresponding to a water depth of 90 m.

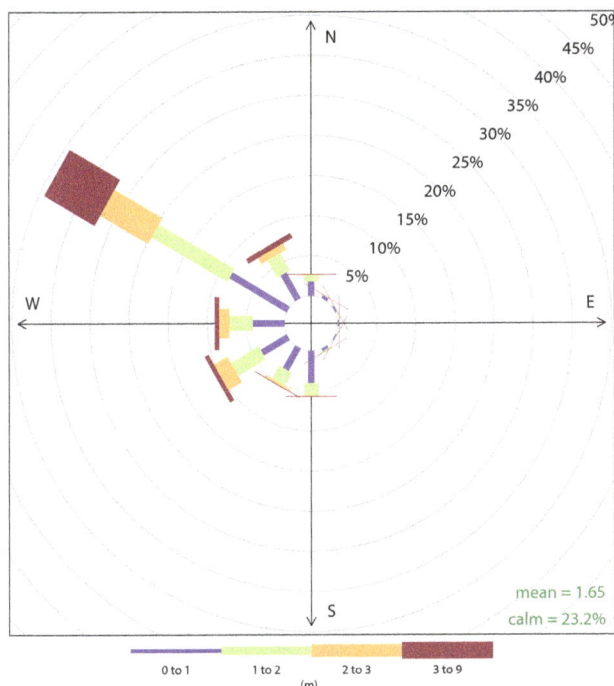

Figure 4. Observed wave climate over 12 years of hourly time series at Alghero (1989–2002).

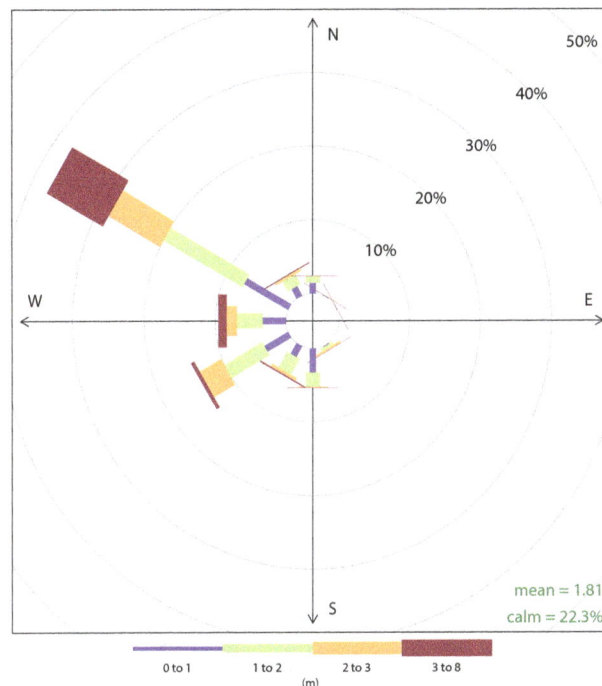

Figure 5. Simulated wave climate over 10 months of hourly time series at Alghero.

The buoy is off the north-western coast of Sardinia (see Fig. 2). Being directly exposed to mistral wind generated in the Gulf of Lyon, the area has experienced the highest storms recorded in the Italian seas since 1989, with maximum observed H_{m0} close to 10 m. The buoy wave climate (Fig. 4), is mostly unimodal, with the higher waves and the higher frequency occurrence of waves over the $H_{m0} = 2$ m threshold directed toward south-east (see Inghilesi et al., 2001; Franco et al., 2004). The JFD relative to 10 months of forecasts are illustrated in Fig. 5. The two distributions look very similar, both indicating that the higher waves come mainly from north-west, with smaller contributions from west and south-west. The set of independent values extracted from 12 years of buoy data have 403 records, the dimension of the set of independent events extracted from the 10 months of forecasts is 28.

The U test was applied to test the null hypothesis that the measured and forecast independent H_{m0} are identical populations. The p value turns out to be 0.8439, exceeding the 0.05 significance level. As a consequence the null hypothesis cannot be rejected and the distribution of the forecast wave climate is identical to the buoy wave climate.

3.2 Ortona

Time series of data and forecasts refer to the position 42.406° N, 14.537° E, corresponding to a water depth of 72 m. The location is representative of the climate of the central and southern Adriatic Sea. The buoy wave climate (see Fig. 6) is mostly bimodal, corresponding to the two main regimes of northerly and easterly winds. The higher waves have been recorded in the northern sector, with maximum waves around $H_{m0} = 7$ m. This is also the sector with the highest frequency of occurrence of episodes with $H_{m0} > 1.5$ m. Maximum waves exceeding $H_{m0} = 4$ m in the eastern sector have been occasionally recorded since 1989 (see Franco et al., 2004). The polar diagrams of the JFD of H_{m0} and mean wave direction relative to 10 months of forecasts are shown in Fig.7.

In the forecast wave climate, it is clear the presence of the same directional sectors found in the observed one, one from north and the other from east. However, the angular distribution of the northern sector is much more narrow and the directional separation between the sectors is much more definite. The set of independent values extracted from 12 years of buoy data has 336 records, whereas the dimension of the set extracted from 10 months of forecast is 22 records. The U test applied to Ortona gives a p value < 0.03, which is less than the 0.05 significance level. Consequently, the null hypothesis of identical distributions is then rejected. The result of the test is that the distribution of the forecast is not identical to the buoy distribution, the two populations are distinct and the differences cannot be explained purely in terms of random sampling.

The consequence of the application of the U test is that, while it is expected that the forecast system will reproduce correctly the wave climate of the western Mediterranean Sea

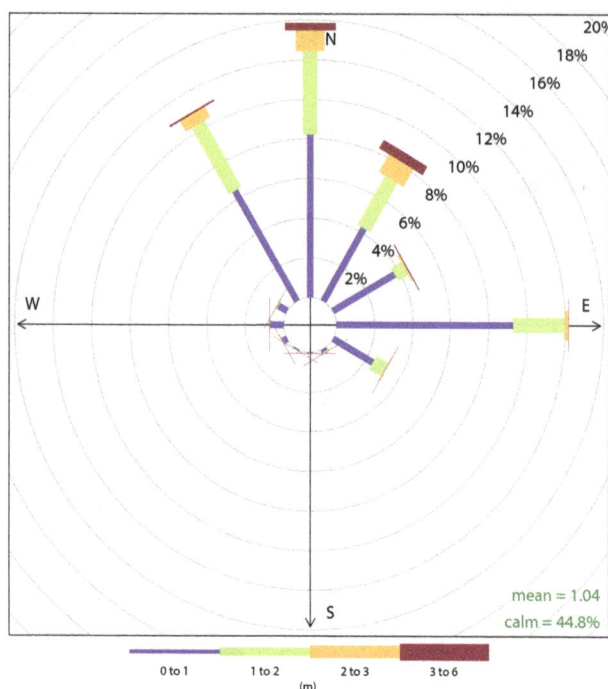

Figure 6. Observed wave climate over 12 years of hourly time series at Ortona (1989–2002).

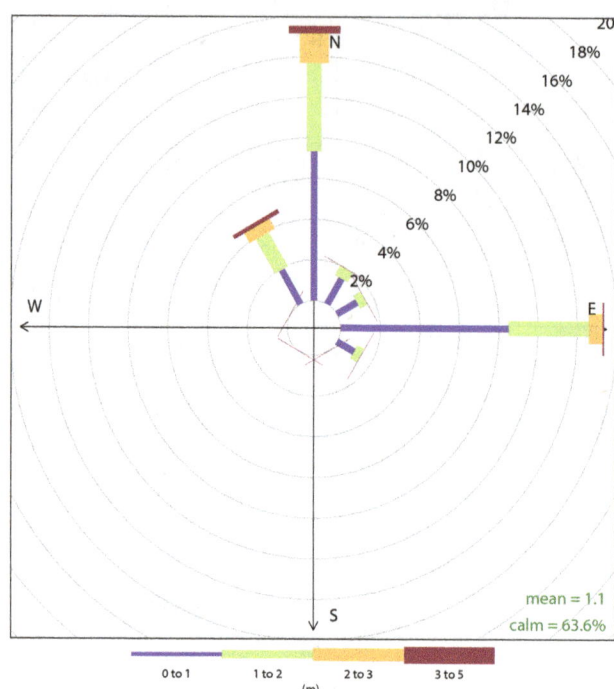

Figure 7. Simulated wave climate over 10 months of hourly time series at Ortona.

related to the mistral wind, there is the possibility that the wave climate in the Adriatic Sea will not properly reproduced by the system. The differences between the directional distributions are mostly in the east sector, like if some of the northeastern episodes that should have been forecasted were erroneously placed here. This is an indication that the problem is in the correct prediction of the bora-easterly winds. This problem has been recently discussed from the meteorological point of view in Bellafiore et al. (2012) for the northern Adriatic Sea. In view of the complex orography of the area, given that the operational BOLAM runs at 0.1° resolution, an improvement in the wind accuracy can be expected by simply increasing the BOLAM resolution or, alternatively, using a more complex model as MOLOCH. The effects of the two methods of downscaling are discussed in the next section.

3.3   Use of high-resolution wind

During the HyMeX SOPs, a common platform was implemented to compare products from different numerical weather prediction models and efficiently support the planning of the observing strategy of high-impact hydrometeorological and marine events. Given the positive performance of the new BOLAM-MOLOCH suite during this forecasting activity (Ferretti et al., 2013; Mariani et al., 2014a), it was decided to test and quantify the forecast skill of the Mc-WAF marine component when fed by the high-resolution wind fields generated with both the 0.07° BOLAM and the 0.0225° MOLOCH.

The test was made in two regional areas, that is, the Ligurian Gulf and the northern Adriatic Sea, and in one coastal area, in the upper Adriatic.

Given the availability of forecast and observations during the period September 2012–March 2013 (which includes the two HyMeX SOP campaigns), three different test-cases were selected: the first was in early Autumn, from 25 October to 10 November 2012; the second was in late autumn, from 4 to 18 December 2012, and the last was in late winter, from 12 to 26 March 2013. The test was organized as a comparison between forecast and observed data at the buoy locations of La Spezia, Ancona and Venezia.

For Ancona and La Spezia, which are in reasonably deep waters, the forecast time series considered were those obtained from the outputs of WAM at regional scale. The model was driven by three different wind fields generated with: (i) the operational 0.1° BOLAM (wave forecasts indicated as WAM_{OP}); (ii) the new 0.07° BOLAM configuration (wave forecasts indicated as WAM_{B78}); and, (iii) the 0.0225° MOLOCH (wave forecasts indicated as WAM_{MO}). Since the Venice buoy is moored at only 20 m depth, the four H_{m0} forecast series considered in the comparison were obtained using (i) the operational WAM_{OP} as a reference, (ii) the WAM model fed by the new 0.07° BOLAM (i.e., WAM_{B78}), (iii) the WAM model fed by the 0.0225° MOLOCH (i.e., WAM_{MO}), and, finally, iv) the SWAN model fed by the new 0.07° BOLAM (i.e., $SWAN_{B78}$).

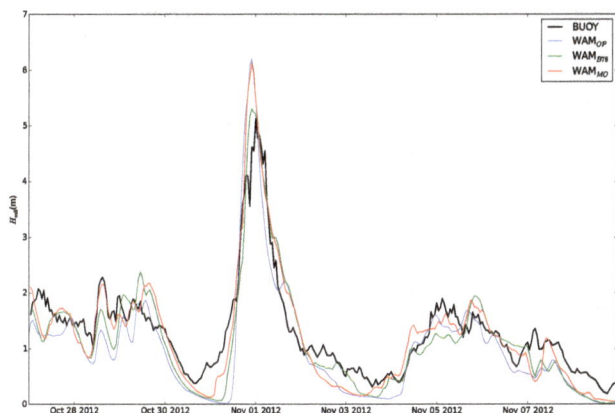

Figure 8. Test-case 1: comparison between forecast and buoy data at Ancona location.

Table 1. Statistics for the Ancona forecast verification.

model	ρ	bias	MSE	MAE	mean	case
Buoy	–	–	–	–	1.19	1
WAM$_{OP}$	0.92	0.79	0.25	0.40	0.92	1
WAM$_{B78}$	0.93	0.89	0.16	0.32	1.12	1
WAM$_{MO}$	0.94	0.95	0.15	0.29	1.08	1
Buoy	–	–	–	–	1.21	2
WAM$_{OP}$	0.82	0.73	0.37	0.47	0.87	2
WAM$_{B78}$	0.83	0.75	0.37	0.45	0.97	2
WAM$_{MO}$	0.82	0.97	0.43	0.46	1.08	2
Buoy	–	–	–	–	1.29	3
WAM$_{OP}$	0.95	1.01	0.13	0.28	1.31	3
WAM$_{B78}$	0.95	0.95	0.11	0.26	1.19	3
WAM$_{MO}$	0.95	1.14	0.18	0.32	1.27	3

Several statistics were taken into account for the test: the correlation coefficient (ρ), the mean square error (MSE), the mean absolute error (MAE), the fractional bias expressed as the ratio between the the mean of the forecast H_{m0} and the mean of the observed H_{m0}, and finally the mean value of H_{m0} in the considered time period.

3.3.1   Deep water – Ancona

The Ancona buoy is located in the north-central Adriatic Sea at the position 43.821° N, 13.717° E (see Fig. 2), and its depth is 70 m.

In the first test-case (see Fig. 8), a single 5 m storm was recorded between 31 October and 2 November 2012. The episode is related to the passage of a small but intense depression over the northern part of Italy. The pressure low was blocked for some days resulting in southerly winds carrying moist, warm water interacting with the dry and cold air to the north of the Adriatic Sea (see Fig. 9). There were intense rainfall episodes, aqua alta at Venice and wind-waves exceeding 4 m in both the Tyrrhenian and the Adriatic Seas.

As shown in Fig. 8, all forecast time series closely follow the development of the storm, the main difference being in the magnitude of the storm peak. Both WAM$_{OP}$ and WAM$_{MO}$ forecast a maximum H_{m0} exceeding 6 m, while WAM$_{B78}$ forecast a value around 5 m that is close to the observed maximum. The lag between the forecast and observed peaks is only one hour. The statistics, shown on top of Table 1, indicate indeed a high correlation for all the series. WAM coupled with the new meteorological suite has a better performance in terms of correlation and bias than those operational. In addition, WAM$_{MO}$ gives a slightly better results in terms of MSE and MAE.

The second test-case occurred with several small events. It was observed a maximum of H_{m0} around 2 m in the period 2–19 December, and a peak reaching 4 m of H_{m0} between 8 and 10 December 2012 (not shown). The statistics, reported in the middle part of Table 1, indicate results similar to the

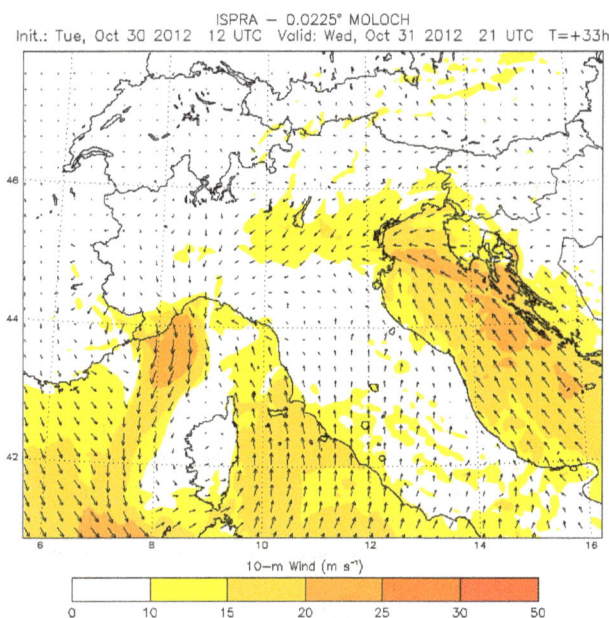

Figure 9. Test-case 1: MOLOCH wind forecast at 21:00 UTC 31 October 2012, initialized at 12:00 UTC 30 October 2012 using 0.07° BOLAM forecasts

first for all the series, but with lower correlations. The best correlation is given by WAM$_{B78}$ with $\rho = 0.83$. The peak of the bigger storm was slightly overpredicted by WAM$_{OP}$ and largely overpredicted (more than 2 m) by WAM$_{MO}$. The peak given by WAM$_{B78}$ was close to the observed one. The high value of the fractional bias obtained for WAM$_{MO}$ indicates that the large overprediction of the main event acts to compensate the underprediction in the remaining part of the time period considered.

In the third test-case (not shown), there were three small episodes associated with south-easterly winds around 2.5–3.0 m. The correlation, shown at the bottom of Table 1, is high for all series ($\rho = 95\,\%$). In all episodes, WAM$_{MO}$

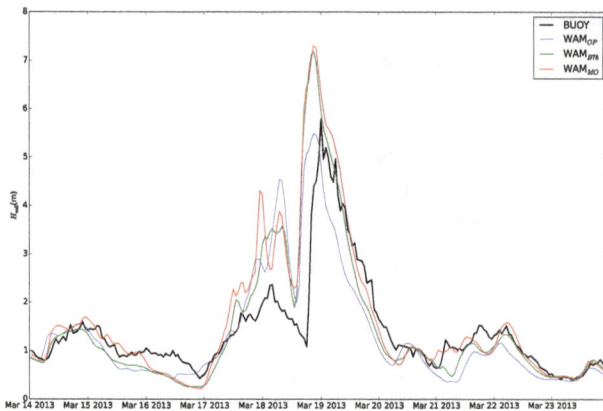

Figure 10. Test-case 3: comparison between forecast and buoy data at La Spezia buoy location.

Table 2. Statistics for the La Spezia forecast verification.

model	ρ	bias	MSE	MAE	mean	case
Buoy	–	–	–	–	1.64	1
WAM$_{OP}$	0.86	0.91	0.35	0.46	1.46	1
WAM$_{B78}$	0.92	1.23	0.64	0.6	1.91	1
WAM$_{MO}$	0.92	1.32	0.95	0.75	2.03	1
Buoy	–	–	–	–	1.49	2
WAM$_{OP}$	0.85	0.91	0.21	0.33	1.35	2
WAM$_{B78}$	0.89	1.1	0.37	0.42	1.30	2
WAM$_{MO}$	0.88	1.2	0.41	0.45	1.33	2
Buoy	–	–	–	–	1.31	3
WAM$_{OP}$	0.78	0.93	0.45	0.42	1.24	3
WAM$_{B78}$	0.84	1.01	0.53	0.38	1.29	3
WAM$_{MO}$	0.86	1.12	0.56	0.37	1.30	3

overpredicted the peak of the storm by more than 0.5 m, and the WAM$_{OP}$ overpredicted two out of three peaks. WAM$_{B78}$ was reasonably close to the observed series during all the period. For this event, the statistics clearly indicate that WAM$_{B78}$ performs better than its competitors.

3.3.2 Deep water – La Spezia

The location of La Spezia buoy is in the Ligurian Gulf, position 43.914° N, 9.827° E (see Fig. 2); it is moored at 90 m depth. The wave climate of the area is mostly unimodal, with waves coming from south-west.

In the first test-case considered (25 October–8 November 2012; Fig. 9), a 5 m H_{m0} event was observed between 27 and 29 October 2012, followed by two smaller events between 3 and 4 m H_{m0} (not shown). The peak of the first event was underpredicted by WAM$_{OP}$ by more than 1 m, while both WAM$_{B78}$ and WAM$_{MO}$ overpredicted the peak by respectively 1 m and more than 1.5 m. The two smaller events were both also largely overpredicted by the new suite of meteorological models. The statistics at the top of Table 2 indicate that the new chain (0.07° BOLAM and 0.0225° MOLOCH + Mc-WAF) performed better in terms of correlation than the operational run (0.1° BOLAM + Mc-WAF). On the contrary, they showed a larger fractional bias and MSE.

In the second test-case, an event exceeding 4.5 m was recorded between 4 and 5 December, followed by two smaller events reaching 2–3 m at the peak of the storm. All models overestimated the most important storm, the operational run by more than 1 m, the others by more than 2 m. The smaller events were more or less closely reproduced by all the models. The statistics in the middle of Table 2 indicate a relatively low correlation for all models, similar bias (WAM$_{OP}$ underestimating, the others overestimating), and shown that the errors are significantly larger for WAM$_{B78}$ and WAM$_{MO}$ than for the operational model.

In the third test-case, a single event exceeding 5 m was observed between 19 and 20 March 2013 (see Fig. 10). The event had an abrupt growth after a small peak occurred the day before, and a short duration. The peak of the storm was anticipated by a few hours by the forecasts, WAM$_{OP}$ being very close to the recorded value. WAM$_{B87}$ and WAM$_{MO}$ both largely overpredicted the peak by approximately 2 m, and all models overpredicted the small storm precursor by more than 2 m. As a consequence, the statistics at the bottom of Table 2 indicate that the correlation is relatively low, with WAM$_{MO}$ and WAM$_{B78}$ being better than the operational run. The fractional bias indicates the overprediction tendency of the new model chain (0.07° BOLAM and 0.0225° MOLOCH + Mc-WAF) opposite to the underprediction tendency of the operational model chain (0.1° BOLAM + Mc-WAF). The MSE of WAM$_{MO}$ and WAM$_{B78}$ are significantly larger than the operational run.

3.3.3 Intermediate-shallow water – Venice

The location of the Venice buoy is in the north-eastern part of the Adriatic Sea (marked in Figs. 3 and 11), at latitude 45.330° N, longitude 12.516° E, and it is moored at 20 m depth.

Two events with H_{m0} around 3–4 m were observed during the first episode, in the period 28 October–8 November 2012 (Fig. 12). In the second test-case, only one small event reaching 2.5 m at the top of the storm was recorded between 7 and 9 December (not shown). In the third test-case, two similar events with H_{m0} not exceeding 2.5 m were observed between 17 and 27 March 2013 (not shown). Since the focus of the study is mainly on wave peaks exceeding 2.5 m, only the results of the first Venice test-case are illustrated in the present study.

The case considered is the evolution at regional and local scale of a complex meteorological condition of blocking. It began with a bora wind, with strong north-easterly winds

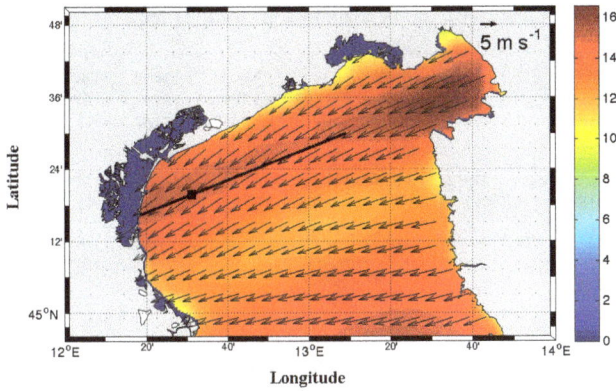

Figure 11. Test-case 1: the 0.07° BOLAM wind field (black arrow), the WAM_B78 wave height field (color filled contour plot) and a transect at the peak of the storm on 29 October. The Venice buoy is also indicated as a black square.

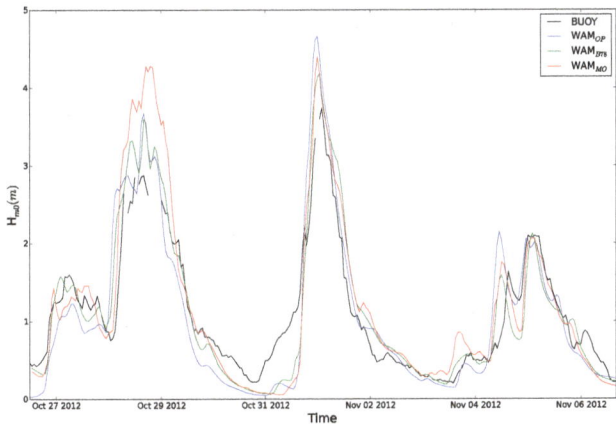

Figure 13. Test-case 1: Comparison between wind forecast and anemometer data at Venice during the first test case.

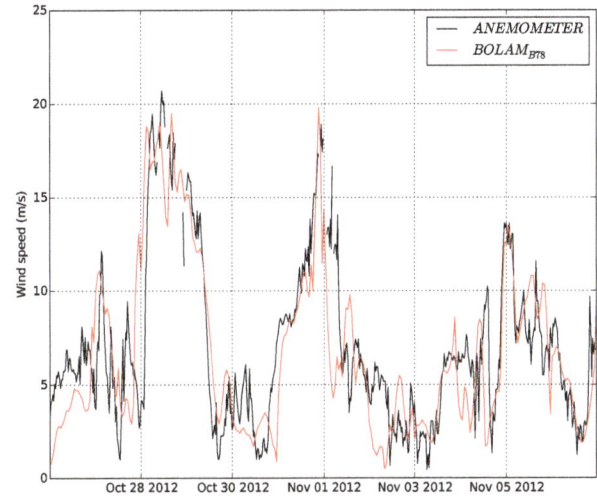

Figure 12. Test-case 1: comparison between forecast and buoy data at Venice buoy location.

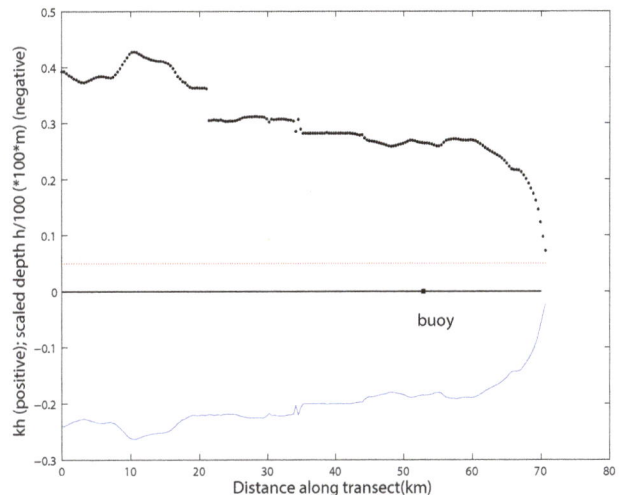

Figure 14. Test-case 1: kh and depth h along the transect in Fig. 11.

in the north Adriatic Sea and wind speed up to $20\,\mathrm{m\,s^{-1}}$ occurring on 29 October. The wind field forecast by the 0.07° BOLAM at 00:00 GMT, 29 October is illustrated in Fig. 11. In the same figure is marked in black the position of a 70 km long transect traced along the direction of the waves crossing the buoy position. Few days after the first case, the wind was strong and south-easterly in all the Adriatic Sea except in the very upper part of the basin, where it rotated locally counter-clockwise in the north-easterly direction (see Fig. 9). It was the result of a typical blocking situation. The wind in this circumstances is referred to as "dark bora", because the southerly flux of warm and moist air which impinges against the blocking cold air to the north of the Adriatic Sea carries dark clouds and rain. This complex situation produced severe weather conditions on 1 November.

The comparison between buoy data and the regional scale models, shown in Fig. 12, indicates that all the models reproduced the evolution of the two storms both in the growth and in the decay of the events. In the first bora episode, the

regional WAM_B78 and the WAM_OP overestimated the peak of the storm by approximately 1 m, while the WAM_MO overestimated the peak by more than 2 m. In the second "dark bora" episode, all models overestimated the peak by approximately 1 m. The regional WAM_B78 implementation is better correlated, but WAM_OP has smaller bias and MSE. The WAM_MO statistics reflect the fact that the 29 October peak was largely overestimated. Wind data collected on the buoy is available for the time period considered, even though the data quality is not ideal. The comparison between 0.07° BOLAM and the anemometer data corrected for the reference 10 m height indicates that the agreement between the forecast and the anemometer data is acceptable (see Fig. 13). The comparison with MOLOCH and the operational BOLAM produces very similar results, with MOLOCH wind speed just $1\,\mathrm{m\,s^{-1}}$ stronger than the other forecasts during the first episode.

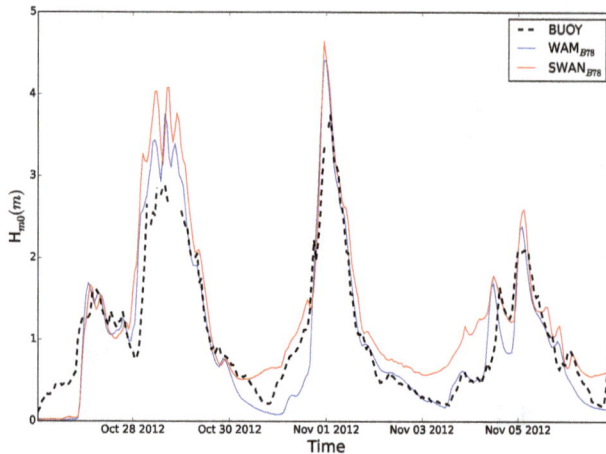

Figure 15. Test-case 1: comparison between $SWAN_{B78}$ and WAM_{B78} at Venice buoy location.

Table 3. Statistics for the Venezia forecast verification.

model	ρ	bias	MSE	MAE	mean	case
Buoy	–	–	–	–	0.89	1
WAM_{OP}	0.91	0.95	0.16	0.28	0.82	1
WAM_{B78}	0.92	1.03	0.12	0.23	0.96	1
WAM_{MO}	0.91	1.09	0.15	0.26	1.01	1
$SWAN_{B78}$	0.91	1.31	0.20	0.30	0.95	1

sion of the regions where the friction and breaking induced by the bathymetry are clearly visible in the figure, in particular to the north of the Venice Lagoon and to the south, in the region of the Po river delta.

4 Conclusions

An operational forecasting system for the Mediterranean Sea working at regional and coastal scale has been implemented coupling several meteorological and wave models. The system is aimed to forecast meteorological events at regional scale in complex orography, and is able to predict the generation and propagation of wind waves in marine and coastal areas. The coastal-scale simulations provide not only operational forecast but also the mapping of the shallow water areas and the extension of the surf zones in relation to the local wave climate. The information gathered at coastal scale is the basis for the investigation of coastal scale processes like wave breaking induced currents and interactions with jets and riverine plumes.

In order to verify whether the waves produced during a major bora storm propagate in deep, intermediate or shallow waters up to the buoy location, the wavelength k along the 70 km long transect x indicated in Fig. 11 has been evaluated using $SWAN_{B78}$ data for the predicted 4 m H_{m0} peak related to the 29 October event. Figure 14 shows, on the negative y axis, the bathymetry $h(x)$ along the transect x scaled as $1/100$ m. The product $k(x)h(x)$ is shown on the positive y axis. The bathymetry along the transect up to the position of the buoy rises very gently up to $h = 20$ m. From this point up to the shoreline the depth climbs much more rapidly. The value of $kh < 0.5$ along all the transect indicates that, up to the buoy, the propagation of the wave is not in "deep water", meaning that the phase velocity of the waves is not independent of h, but neither is equal to the group velocity, as it would be if it was $kh < 0.05$ somewhere. It means that, even when the waves are relatively high, the propagation is affected mainly by the effect of the shoaling. The dissipative effects directly related to the bottom (friction, breaking, triads nonlinear interaction), which are typical of shallow waters, are not very significant in intermediate water. In this situation both WAM and SWAN are applicable, WAM becoming more and more overpredicting as the waves move to the region $kh < 0.1$ close to the shoreline.

The comparison between H_{m0} buoy data, $SWAN_{B78}$ and WAM_{B78} is shown in Fig. 15. For $H_{m0} > 1$ m there are small differences between the two models, while for small values of H_{m0} it seems that $SWAN_{B78}$ overestimates systematically. It is probably a symptom that the Mc-WAF implementation of the nesting WAM/SWAN in the northern Adriatic area is not adequate in very low wind conditions. The statistics in Table 3 indicate that all series are well-correlated with the buoy data.

The SWAN results in the coastal areas directly affected by the north-easterly winds, corresponding to the bora peak on 29 October, are shown in Fig. 16. The position and the exten-

A preliminary test on the capability of the system to reproduce the wave climate features in the west Mediterranean and in the Adriatic Sea has been performed. The U test was aimed to assess if the JFD of the events predicted in the first year of Mc-WAF operativity has identical distribution as the distribution extracted from 12 years time series of buoy data. The results indicate substantial differences in the reliability of the forecasts in the two areas. In the north-west Mediterranean Sea the simulated wave distribution is in close agreement with the observed wave climate, indicating that the mistral regime is well-reproduced by the system. In the central Adriatic Sea the test concluded that the distribution of the forecast is not identical to the known wave climate, indicating a possible systematic error in the reproduction of the weather at regional scale. In particular, the analysis of the differences of the JFD of H_{m0} an direction suggests that a significant fraction of the bora events were not correctly reproduced. In order to investigate the methods to improve the forecasts at sub-regional scale, several tests were made to assess the benefit of using a downscaling in areas bordered by complex topography. In this work two different methods for meteorological downscaling were applied: the refinement of the BOLAM grid in order to better resolve the orography and the use of a more complex, non-hydrostatic model, MOLOCH.

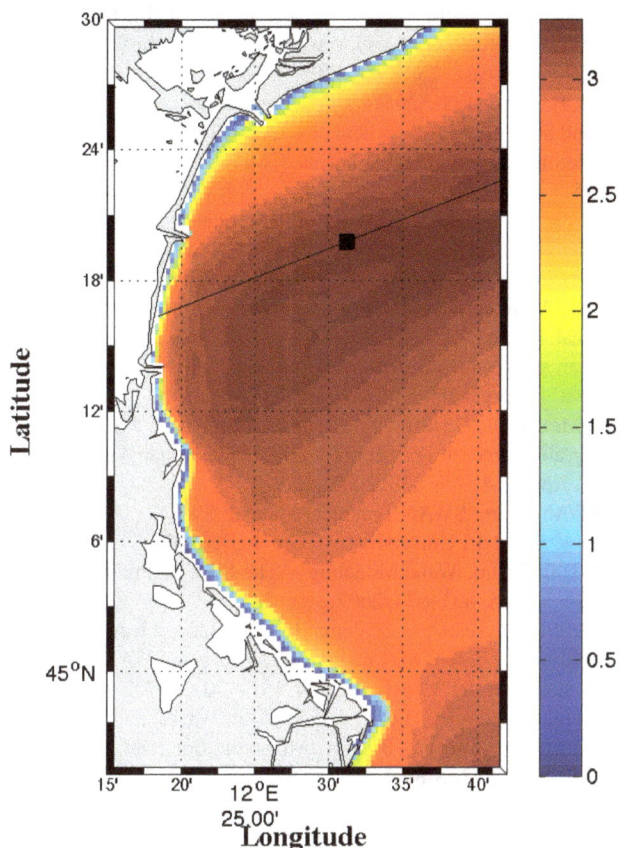

Figure 16. Test-case 1: field of H_{m0} produced by SWAN$_{B78}$ at the peak of the storm on 29 October.

The downscaling has been applied to three test-cases in order to test the limitations and the benefits of the use of more intensive and time-consuming operational procedures. The first test-case was particularly severe for the system, in that the particular meteorological blocking considered proved to be dynamically complex with strong spatial gradients and rapidly varying local conditions. Nevertheless, the marine conditions were reasonably well forecasted both in the Tyrrhenian and the Adriatic Sea.

The results of the comparisons between forecast and buoy data during several high-impact meteo-marine events in the Ligurian and Adriatic Seas are encouraging and support the operational implementation of the new high-resolution BO-LAM in all the regional areas. The improvement is due to the increase in spatial resolution of the model but also to an optimization in the nesting with the ECMWF IFS model. The use of VHRLAM MOLOCH did not provide better performances than the BOLAM in the test-cases considered. It was seen that often the wave forecasts overpredicted the buoy data more than the other implementations, and the MSE and MAE were generally higher. On the contrary, the test-cases considered were relatively few, especially in the northern Adriatic Sea where, due to the complex orography, the potential benefits of the use of VHRLAM are larger.

Acknowledgements. The present study was carried out within the scope of the Directive 2008/56/EC of the European Parliament and of the Council of 17 June 2008 establishing a framework for community action in the field of marine environmental policy (Marine Strategy Framework Directive), the authors are grateful to the Italian Ministry for Environment, Territory and Sea for supporting the research on the evaluation of wave exposure in coastal areas. The authors are also indebted to Heinz Guenter and Arno Behrens (GKSS) for making the version cycle 4.5 of WAM available and to Stefano Tagliaventi (CINECA) for strenuously collaborating on the update of the SIMM meteorological component. We are in debt with Luigi Cavaleri and Luciana Bertotti (CNR-ISMAR) for their patient explaining the complexity of the Northern Adriatic weather and for making the forecasts of the "Nettuno" project available.

Edited by: M. M. Miglietta
Reviewed by: two anonymous referees

References

Bellafiore, D., Bucchignani, E., Gualdi, S., Carniel, S., Djurdjevi'c, V., and Umgiesser, G.: Assessment of meteorological climate models as inputs for coastal studies. Ocean Dynam., 62, 555–568, 2012.

Bertotti, L. and Cavaleri, L.: On the influence of resolution on wave modeled results in the Mediterranean Sea, Nuovo Cimento, 29, 411–419, doi:10.1393/ncc/i2005-10210-6, 2006.

Bertotti, L. and Cavaleri, L.: Large and small scale wave forecast in the Mediterranean Sea, Nat. Hazards Earth Syst. Sci., 9, 779–788, doi:10.5194/nhess-9-779-2009, 2009.

Buzzi, A., Fantini, M., Malguzzi, P., and Nerozzi, F.: Validation of a limited area model in cases of Mediterranean cyclogenesis: Surface fields and precipitation scores, Meteorol. Atmos. Phys., 53, 53–67, 1994.

Casaioli, M., Mariani, S., Malguzzi, P., and Speranza, A.: Factors affecting the quality of QPF: a multi-method verification of multi-configuration BOLAM reforecasts against MAP D-PHASE observations, Meteorol. Appl., 20, 150–163, 2013.

Cavaleri, L. and Bertotti, L.: The accuracy of modelled wind and waves fields in enclosed seas, Technical Memorandum n. 409, ECMWF Technical Memoranda, 15 pp., 2003.

Cavaleri, L. and Sclavo, M.: The calibration of wind and wave model data in the Mediterranean Sea, Coast. Eng., 53, 613–627, 2006.

Ferretti, R., Pichelli, E., Gentile, S., Maiello, I., Cimini, D., Davolio, S., Miglietta, M. M., Panegrossi, G., Baldini, L., Pasi, F., Marzano, F. S., Zinzi, A., Mariani, S., Casaioli, M., Bartolini, G., Loglisci, N., Montani, A., Marsigli, C., Manzato, A., Pucillo, A., Ferrario, M. E., Colaiuda, V., and Rotunno, R.: Overview of the first HyMeX Special Observation Period over Italy: observations and model results, Hydrol. Earth Syst. Sci. Discuss., 10, 11643–11710, doi:10.5194/hessd-10-11643-2013, 2013.

Franco, L., Piscopia, R., Corsini, S., and Inghilesi, R.: L'Atlante delle onde nei mari italiani – Italian Wave Atlas, Full Final Report, University of Roma TRE – APAT, 2004.

Inghilesi, R., Corsini, S., Guiducci, F., and Arseni, A.: Statistical analysis of extreme waves on the Italian coasts from 1989 to 1999, Bollettino di Geofisica Teorica ed Applicata, 41, 3–4, September–December 2000, 315–337, 2000.

Inghilesi, R., Catini, F., Bellotti, G., Franco, L., Orasi, A., and Corsini, S.: Implementation and validation of a coastal forecasting system for wind waves in the Mediterranean Sea, Nat. Hazards Earth Syst. Sci., 12, 485–494, doi:10.5194/nhess-12-485-2012, 2012.

Lavrenov, I. V.: Wind Waves in Oceans Dynamics and Numerical Simulations, 12–34, Springer, 2003.

Janssen, P. A. E. M.: Progress in ocean wave forecasting, J. Comput. Phys., 227, 3572–3594, 2008.

Janssen, P. A. E. M., Breivik, Ø., Mogensen, K., Vitart, F., Balmaseda, M., Bidlot, J., Keeley, S., Leutbecher, M., Magnusson, L., and Molteni, F.: Air–Sea Interaction and Surface Waves, Technical Memorandum n. 712, ECMWF Technical Memoranda, 34 pp., 2013.

Malguzzi, P. and Tartaglione, N.: An economical second-order advection scheme for numerical weather prediction, Q. J. Roy. Meteor. Soc., 125, 2291–2303, 1999.

Malguzzi, P., Grossi, G., Buzzi, A., Ranzi, R. and Buizza, R.: The 1996 "century" flood in Italy. A meteorological and hydrological revisitation, J. Geophys. Res., 111, D24106, doi:10.1029/2006JD007111, 2006.

Mariani, S., Casaioli, M., and Malguzzi, P.: Towards a new BOLAM-MOLOCH suite for the SIMM forecasting system: implementation of an optimised configuration for the HyMeX Special Observation Periods, Nat. Hazards Earth Syst. Sci. Discuss., 2, 649–680, doi:10.5194/nhessd-2-649-2014, 2014.

Mariani, S., Casaioli, M., Lanciani, A., Flavoni, S., and Accadia, C.: QPF performance of the updated SIMM forecasting system using reforecasts, Meteor. Appl., in print, doi:10.1002/met.1453, 2014.

Mass, C. F., Ovens, D., Westrick, K., and Colle, B. A.: Does Increasing Horizontal Resolution Produce More Skillful Forecasts?, B. Am. Meteorol. Soc., 83, 407–430, 2002.

Mood, A. M., Graybill, F. A., and Boes D. C.: Introduction to the theory of statistics, Mcgraw-Hill Ed., 1974.

Saetra, Ø.: Ensamble Shiprouting, ECMWF Technical Memorandum n. 435, ECMWF Technical Memoranda, 10 pp., 2004.

Saetra, Ø. and Bidlot, J. R.: Potential benefits of using probabilistic forecasts for waves and marine winds based on the ECMWF Ensemble Prediction System, Weather Forecast., 19, 673–689, 2004.

Sampson, C. R., Wittmann, P. A., and Tolman, H. L.: Consistent Tropical Cyclone Wind and Wave Forecasts for the U.S. Navy, Weather Forecast., 25, 1293–1306, 2010.

Signell, R., Carniel, S., Cavaleri, L., Chiggiato, J., Doyle, J., Pullen, J., and Sclavo, M.: Assessment of wind quality for oceanographic modelling in semi-enclosed basins, J. Marine Syst., 53, 217–233, 2005.

Speranza, A., Accadia, C., Casaioli, M., Mariani, S., Monacelli, G., Inghilesi, R., Tartaglione, N., Ruti, P. M., Carillo, A., Bargagli, A., Pisacane, G., Valentinotti, F., and Lavagnini, A.: POSEIDON: An integrated system for analysis and forecast of hydrological, meteorological and surface marine fields in the Mediterranean area, Nuovo Cimento, 27, 329–345, 2004.

Speranza, A., Accadia, C., Mariani, S., Casaioli, M., Tartaglione, N., Monacelli, G., Ruti, P. M., and Lavagnini, A.: SIMM: An integrated forecasting system for the Mediterranean Area, Meteorol. Appl., 14, 337–350, 2007.

The SWAN Team: SWAN Technical Manual, SWAN Cycle III version 40.81, Delft University of Technology, 1993.

The WISE Group: Wave Modeling – The State of the Art, Prog. Oceanogr., 75, 603–674, 2007.

Van der Westhuysen, A. J., Zijlema, M., and Battjes, J. A.: Nonlinear saturation based white capping dissipation in SWAN for deep and shallow water, Coast. Eng., 54, 151–170, 2007.

Wang, D.-P. and Oey, L.-Y.: Hindcast of Waves and Currents in Hurricane Katrina, B. Am. Meteorol. Soc., 89, 487–495, doi:10.1175/BAMS-89-4-487, 2008.

Whitham G. B.: Linear and Nonlinear Waves, Wiley-Interscience, 390–402, 1999.

Wilks, D. S.: Statistical Methods in the Atmospheric Sciences, Academic Press, 467 pp., 1995.

Zijlema, M. and Stelling, G. S.: Further experiences with computing non-hydrostatic free-surface flows involving water waves, Int. J. Numer. Meth. Fl., 48, 169–197, doi:10.1002/fld.821, 2005.

21

Evening transitions of the atmospheric boundary layer: characterization, case studies and WRF simulations

M. Sastre[1], **C. Yagüe**[1], **C. Román-Cascón**[1], **G. Maqueda**[2], **F. Salamanca**[3], and **S. Viana**[4]

[1]Dept. de Geofísica y Meteorología, Universidad Complutense de Madrid, Spain
[2]Dept. de Astrofísica y Ciencias de la Atmósfera, Universidad Complutense de Madrid, Spain
[3]Lawrence Berkeley National Laboratory (LBNL), Berkeley (CA), USA
[4]Agencia Estatal de Meteorología (AEMET), Delegación Territorial de Cataluña, Barcelona, Spain

Correspondence to: M. Sastre (msastrem@fis.ucm.es)

Abstract. Micrometeorological observations from two months (July–August 2009) at the CIBA site (Northern Spanish plateau) have been used to evaluate the evolution of atmospheric stability and turbulence parameters along the evening transition to a Nocturnal Boundary Layer. Turbulent Kinetic Energy thresholds have been established to distinguish between diverse case studies. Three different types of transitions are found, whose distinctive characteristics are shown. Simulations with the Weather Research and Forecasting-Advanced Research WRF (WRF-ARW) mesoscale model of selected transitions, using three different PBL parameterizations, have been carried out for comparison with observed data. Depending on the atmospheric conditions, different PBL schemes appear to be advantageous over others in forecasting the transitions.

1 Introduction

The Planetary Boundary Layer (PBL) goes through different dynamical and thermal situations throughout a single day. Knowing how these changes are reached will be helpful to improve our understanding on various topics of the PBL, especially the transport of scalars – pollutants, water vapor, heat, etc. – in the lower troposphere and Earth-atmosphere exchanges and interactions (Baklanov et al., 2010). Having an enhanced comprehension of this subject will be very helpful for improvements on practical applications, such as atmospherically induced health alerts or agricultural topics.

This atmospheric layer is strongly influenced by the diurnal solar cycle, having a direct impact on surface heating and cooling and usually driven by turbulent processes. This fact leads us to look for the mechanisms that trigger the evolution from a convective PBL to a stable one at times around sunset. Changes that occur near the Earth's surface, namely the decay of the turbulence or the crossover of the sensible heat flux, mark the beginning of the evening transition from a certain time before sunset – which can vary between a few minutes and around one hour – Fernando et al. (2004). Here we study the following temporal interval: from two hours before sunset until four hours after sunset. In this way, conditions preceding the transition can be explored and the turbulence decay may also be studied from the time it starts. Addition-

ally, the first hours of the night are investigated in order to explore how the different transitions can affect the development of the subsequent Nocturnal Boundary Layer (NBL).

The main aim of this work is to offer a framework for classifying turbulence decay in terms of TKE during the evening transition and to connect each class to some other phenomena in the NBL, like possible gravity waves or katabatic winds. To achieve this aim we studied some thermal and dynamical issues of the PBL. The study has two parts: the analysis of the experimental data and their comparison to WRF model simulations. Firstly, by using observations, it was investigated how rapidly the turbulence decays when the input solar radiation is reduced. In the second part, transitions corresponding to different situations were simulated with the WRF mesoscale model. The main goal of these simulations is to learn if a mesoscale model can adequately reproduce this turbulence decay. Furthermore, it can be interesting for future improvements in PBL transition modelling by Numerical Weather Prediction (NWP) models. It is also an objective to find relationships between characteristics of the transitions and different model settings, to find out which elements favour that the atmospheric behaviour is properly reflected by the model. Moreover, the comparison simulations-observations provides to us a wider point of view of the evening transition questions.

2   Experimental data

Data employed for this study were obtained during July and August 2009 with some of the permanent instrumentation placed at a 10 m height mast in the Research Centre for the Lower Atmosphere – CIBA, for the Spanish acronym. These devices are sonic anemometers at 10 m height (working at a frequency of 20 Hz), temperature sensors at 1.5 (Z_1 level) and 10 m (Z_2 level) (1 Hz) and cup anemometers and vanes at 1.5 and 10 m (1 Hz). A picture of the mast can be found in Supplement 1. Additionally used, was a GRIMM 365 Monitor for measurements of particulate matter smaller than 1, 2.5 or 10 μm (PM_1, $PM_{2.5}$, PM_{10}, respectively) at the surface (1/6 Hz).

The CIBA site is located on the Northern Spanish plateau (41°49′ N, 4°56′ W), at 840 m above sea level and over a quite flat terrain. Nevertheless, two small slopes can be found: one in the NW-SE direction (1:6000) and the other one in the NE-SW direction (1:1660). These slopes should be taken into account for drainage flows. Some topographic maps of the location and the slopes scheme are provided in Supplement 1 (Viana, 2011). More details on the experimental site can be found by looking up Cuxart et al. (2000), and Yagüe et al. (2009) for the last instrumentation setup.

Five minute means were used for calculations of Turbulent Kinetic Energy (TKE=$\frac{1}{2}(\overline{u'^2} + \overline{v'^2} + \overline{w'^2})$), friction velocity ($U_* = \left[(\overline{u'w'})^2 + (\overline{v'w'})^2\right]^{1/4}$) and vertical heat flux ($H = \rho c_p \overline{\theta'w'}$) from sonic anemometer records, considering the 3-D-wind components variances and covariances. Wind speed (U) and potential temperature (θ) were also evaluated in five minute means. The Bulk Richardson number (from Z_2 and Z_1 measurements) were used to look up stability and it is calculated as (Arya, 2001):

$$Ri_B = \frac{\frac{g}{T_0} \sqrt{Z_1 Z_2} \ln\left(\frac{Z_2}{Z_1}\right) \Delta\theta}{(\Delta U)^2}. \tag{1}$$

3   Model configuration

WRF-ARW numerical model version 3.3 was adopted for this study. This model is, at present, widely used for different kinds of simulations (Shin and Hong, 2011; García-Díez et al., 2012), both for operational and research goals, given that it can provide an important range of possibilities (i.e., different PBL or surface layer schemes and physical options). Here we briefly explain some aspects of the model configuration chosen to carry out our simulations. Four nested domains were configured, whose grids are, respectively, 27 km, 9 km, 3 km and 1 km, keeping its central point just on the CIBA coordinates. According to the number of grid points used for each domain, the smallest one is a 120 km-side square and the largest one has a side of 2700 km. For the vertical resolution, the model considers 50 eta levels, from which twenty-eight are located under the first kilometre, and

also eight of them are under the first 100 m. The spin up used was 12 h and the time step was configured to be 90 s. For all the simulations the Noah Land Surface Model (LSM) was chosen, which is the unified NCEP/NCAR/AFWA scheme with soil temperature and moisture in four layers. Among the different options of PBL parameterizations, we used three: Mellor-Yamada-Janjic (MYJ) (Janjic, 1990), Mellor-Yamada-Nakanishi-Niino (MYNN) (Nakanishi et al., 2004) and Quasi-Normal-Scale-Elimination (QNSE) (Sukorianski et al., 2005). MYJ is basically the Eta operational scheme, which uses a one-dimensional prognostic TKE scheme allowing local vertical mixing. On the other hand, QNSE assumes a prognostic TKE equation, which is obtained from a theory for stably stratified regions, and its diffusivity allows for anisotropy. MYNN goes a bit further and can predict second order moments besides TKE. Every PBL parameterization uses a specific surface layer scheme: MYJ is run with the Monin-Obukhov (Janjic Eta) scheme while MYNN and QNSE use their own schemes (named also MYNN and QNSE). The long wave radiation (RRTM), short wave radiation (Dudhia) and microphysics package (WSM3) have been the same for all the simulations. Skamarock et al. (2008) can be checked for further details on the parameterizations and the model characteristics. Finally, the initial and boundary conditions were taken from NCEP-NCAR, whose horizontal resolution is 1° and the boundary conditions are forced every 6 h.

4   Results

4.1   Observational data

Thermodynamic and dynamic variables were studied for a six-hour interval, which takes sunset as the reference time ($t_{sunset} = 0$ h). With this normalization, the time interval studied went from −2 to +4 h. First of all, in order to get an overview of the transition, a brief analysis of mean values was done. You can find wind speed (at 1.5 and 10 m), temperature difference between 10 m and 1.5 m, particulate matter concentration, the Bulk Richardson number, Turbulent Kinetic Energy, friction velocity and vertical heat flux mean values in Table 1. They are shown separately in three time sub-intervals lasting two hours each, as far as the latter can be considered as different sub-periods dynamically and thermodynamically. We can generally see that stability increases as time goes on within the transition, with smaller values of turbulence parameters and a significant increase in particulate matter concentration, which does not diffuse to higher levels.

Turbulent Kinetic Energy, 1.5 m temperature, 10 m wind speed and temperature difference between 10 m and 1.5 m were calculated for the temporal interval previously mentioned. Figure 1 shows the 10 m Turbulent Kinetic Energy evolution of four evening transitions corresponding to the same week of August 2009. The same days are presented

Table 1. Mean values of wind speed, temperature difference between 10 and 1.5 m, the Bulk Richardson number, particle concentrations and turbulent parameters for the time ranges used at the CIBA site (July–August 2009). The sunset time is the reference: $t_{\text{sunset}} = 0\,\text{h}$.

	$t=[-2,0]\,\text{h}$	$t=[0,2]\,\text{h}$	$t=[2,4]\,\text{h}$
$U_{1.5}$ (m s^{-1})	2.74	2.10	2.08
U_{10} (m s^{-1})	4.35	3.74	3.72
$\Delta T_{10-1.5}$ (°C)	−0.27	1.17	1.01
Ri_B	−0.09	0.24	0.18
PM$_1$ (μg m^{-3})	3.02	4.76	6.90
PM$_{2.5}$ (μg m^{-3})	5.05	7.22	9.04
PM$_{10}$ (μg m^{-3})	14.36	20.19	17.66
TKE (m^2 s^{-2})	0.94	0.55	0.42
U_* (m s^{-1})	0.36	0.26	0.24
H (W m^{-2})	19.02	−16.20	−17.56

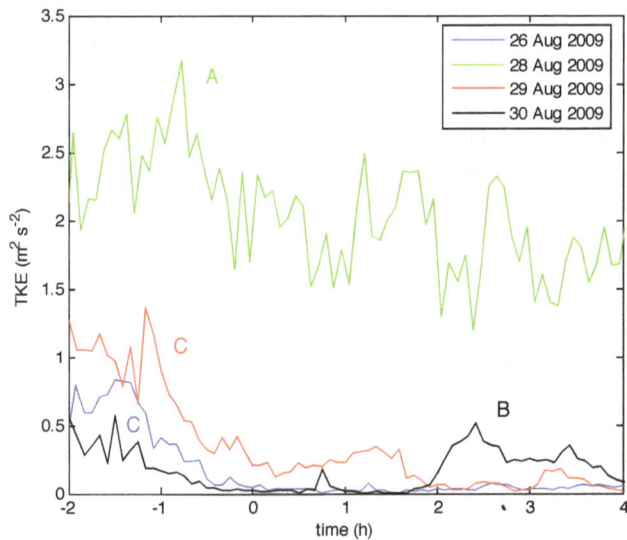

Figure 2. Observed 10–1.5 m temperature difference evolution for different types of evening transitions (A, B, C). Times are normalized around sunset for each day ($t_{\text{sunset}} = 0\,\text{h}$).

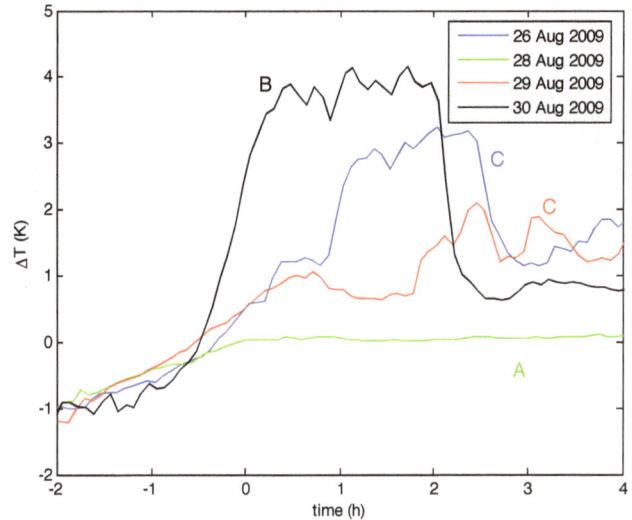

Figure 1. Observed TKE evolution at 10 m for different types of evening transitions (A, B, C). Times are normalized around sunset for each day ($t_{\text{sunset}} = 0\,\text{h}$).

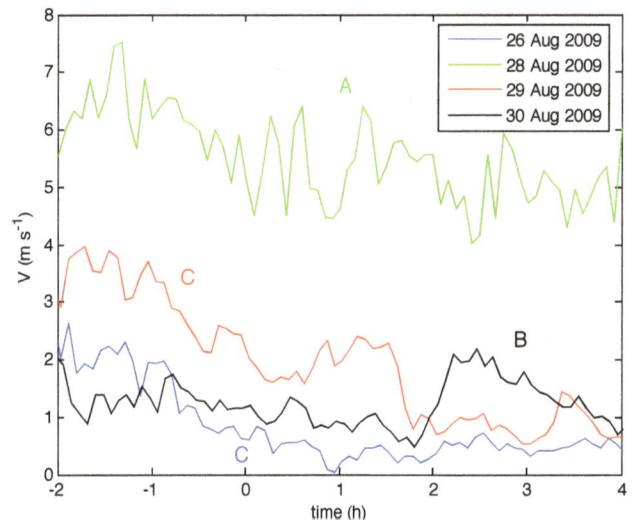

Figure 3. Observed 1.5 m wind evolution for different types of evening transitions (A, B, C). Times are normalized around sunset for each day ($t_{\text{sunset}} = 0\,\text{h}$).

for the temperature difference (Fig. 2) and 1.5 m wind speed (Fig. 3). Actually, days plotted on Figs. 1–3 are examples of the three types of transitions we found during the two months of data analysed. First of all, we have the ones that are controlled by moderate to high synoptic winds (labelled in Figs. 1–3 with A). These were quite turbulent evenings, with no surface-based inversion temperature or a very weak one, and where TKE kept reaching values higher than 1.5 m^2 s^{-2}, sometimes not very different from diurnal ones. On the other hand, there were some transitions with very small values of TKE (<0.5 m^2 s^{-2}) and wind speed before sunset, so that an early and strong surface-based inversion developed (B). This strong stability is very likely to the occurrence of katabatic winds, which can erode the stability (see increase of TKE

two hours after sunset) and are sometimes related to the generation of gravity waves (Viana et al., 2010). Finally, a third group of transitions would consist of those ones with light to moderate winds before sunset, developing a soft and continuous inversion during the night without important katabatic events (C). TKE values between 0.5 and 1.5 m^2 s^{-2} were characteristic of the latter group. The two months of data collected for this work show that in this period the most common transitions were type C (39 %), followed by type B (32 %), while type A (18 %) was the least frequent to occur. There are still some cases (11 %) that cannot be easily classified as any of these three types.

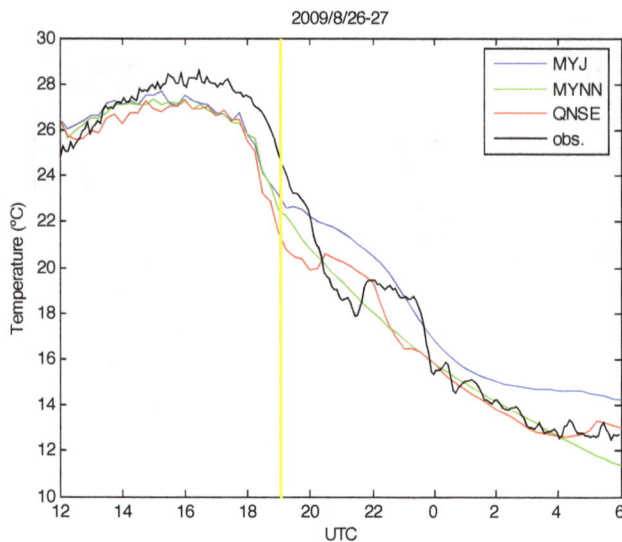

Figure 4. Simulations of 2 m-air temperature and comparison with 5-min averaged observations for the 26–27 August 2009 transition (type C).

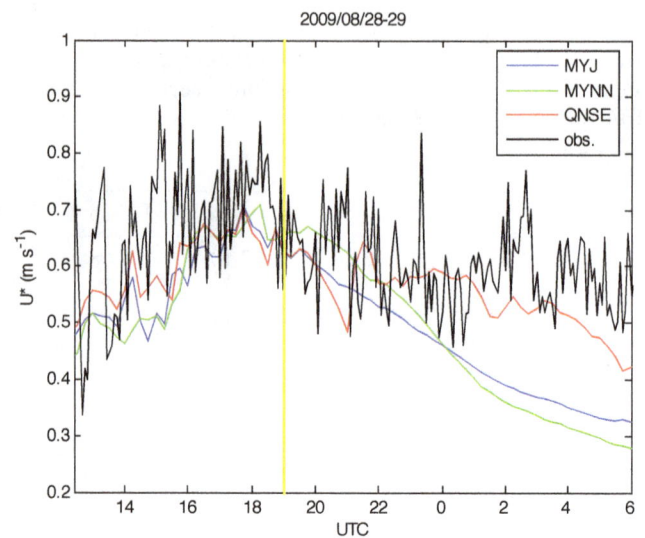

Figure 6. Simulated and observed (5-min averaged) friction velocity for 28–29 August transition (type A).

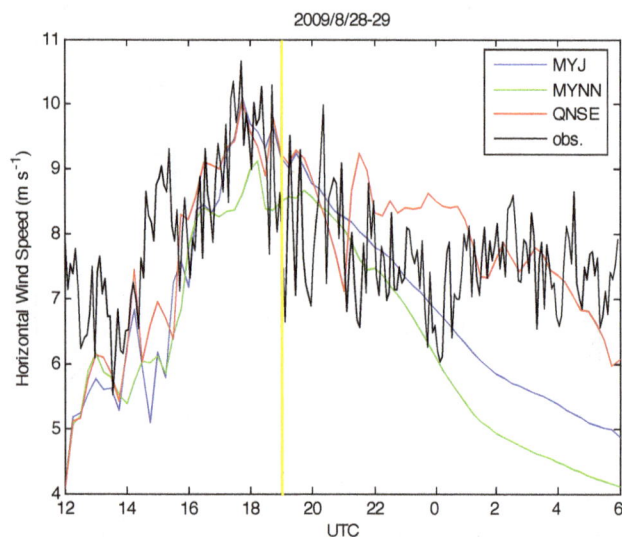

Figure 5. Simulated and observed (5-min averaged) wind speed at 10 m for 28–29 August transition (type A).

4.2   WRF simulations

Wind speed, temperature and friction velocity were simulated for the evening transitions of certain days which are representative of different situations, for comparison with observed data.

The selected PBL parameterizations tend to smooth the observed behaviour of the magnitudes represented. Temperature was, as a whole, well-simulated both qualitatively and quantitatively and with a correct timing by the three parameterizations, although high frequency peaks were not captured

(see day 26 August, Fig. 4). In Fig. 4 it is remarkable that during the temperature decay there was an observed upturn, which seems to be captured only by QNSE parameterization, although one or two hours in advance. Windy transitions (type A) are usually better simulated by QNSE, while MYJ and MYNN fail to reproduce the beginning of the night, providing a continuous decay of the wind and friction velocity when it does not really happen (see wind speed and friction velocity for day 28 August, Figs. 5 and 6). Regarding transitions with early-developed inversions and katabatic winds, (type B) not very good agreement between experimental data and simulations has been had at times for big changes in the PBL structure (see friction velocity for day 30 August, Fig. 7). Nevertheless, the observed decay and later fast rise might have been simulated by QNSE and MYJ, but a couple of hours before it happened, and reaching significantly smaller maximum and minimum values than the observed ones. For type C, QNSE gave better results when the inversion was already developed, probably because QNSE is especially designed for stable situations.

To evaluate simulations a bit more deeply than from a visual inspection, two parameters were calculated to compare the model's outputs with observations: bias and root-mean-square error (RMSE), which are respectively defined as:

$$\text{BIAS} = \frac{1}{N} \sum_{i=1}^{N} (\phi_{\text{m}_i} - \phi_{\text{o}_i}) \tag{2}$$

$$\text{RMSE} = \left[\frac{1}{N} \sum_{i=1}^{N} (\phi_{\text{m}_i} - \phi_{\text{o}_i})^2 \right]^{1/2} \tag{3}$$

where "N" is the number of data considered to calculations, "ϕ" indicates the variable being evaluated and de sub-indexes

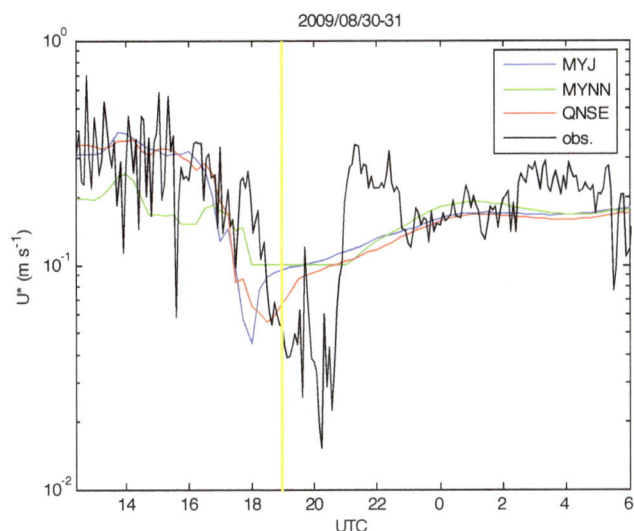

Figure 7. Simulated and observed (5-min averaged) friction velocity for the 30–31 August 2009 transition (type B).

"m" and "o" refer to modelled or observed data, respectively. A table of bias and RMSE values associated to the data for Figs. 4 to 7 can be found in Supplement 2.

According to the bias, the friction velocity is usually underestimated by the parameterizations – opposite to the overestimation that is frequently seen –, except for one particular situation: in the interval of two hours after sunset during day 30 August, which is type B.

MYNN parameterization is most of the time the one that obtains lower values of RMSE. However, although QNSE is rarely the best one at RMSE, sometimes it is the only one that captures some particular events, such as a short climb in temperature during a descending trend.

5   Conclusions

Three different types of observed PBL evening transitions were found for the summer 2009 at CIBA and some TKE thresholds may be used to classify them: the windy and with nearly no temperature inversion ones have TKE > 1.5 m^2 s^{-2}; the ones with early strong inversions (B) correspond to TKE < 0.5 m^2 s^{-2} and intermediate cases (C) take place when 0.5 < TKE < 1.5 m^2 s^{-2}.

Considering WRF model simulations, we found no clear evidence to conclude which one of the three PBL parameterizations tested is the best at simulating evening transitions, as far as all the three are able to reproduce the observed behaviour in certain circumstances. QNSE seems to simulate some events while the other ones do not, although not with the correct timing or intensity. Further research is required in order to improve the simulation results, especially for difficult events such as katabatic winds. Moreover, a new theoretical framework might be necessary, as suggested by Nadeau

et al. (2011), to describe the TKE during the evening transition when winds are very light and the mechanical turbulence production decreases.

Acknowledgements. The authors wish to thank Javier Peláez (CIBA) for his technical support and help as well as J. L. Casanova, Director of the CIBA. We are also very grateful to the editor and two anonymous referees for their constructive suggestions, which helped to improve this paper. This research has been funded by the Spanish Ministry of Science and Innovation (projects CGL 2006-12474-C03-03 and CGL2009-12797-C03-03). The GR58/08 program (supported by BSCH and UCM) has also partially financed this work through the Research Group "Micrometeorology and Climate Variability" (No 910437). M. Sastre is supported by a FPI-UCM fellowship (reference BE45/10).

Edited by: G.-J. Steeneveld
Reviewed by: two anonymous referees

References

Arya, S. P. S.: Introduction to Micrometeorology, 2nd Edn., International Geophysics Series, Academic Press, London, 307 pp., 2001.

Baklanov, A., Grisogono, B., Bornstein, R., Mahrt, L., Zilitinkevich, S., Taylor, P., Larsen, S., Rotach, M., and Fernando, H. J. S.: On the nature, theory, and modelling of atmospheric planetary boundary layers, B. Am. Meteorol. Soc., 92, 123–128, 2010.

Cuxart, J., Yagüe, C., Morales, G., Terradellas, E., Orbe, J., Calvo, J., Fernández, A., Soler, M. R., Infante, C., Buenestado, P., Espinalt, A., Joergensen, H. E., Rees, J. M., Vilà, J., Redondo, J. M., Cantalapiedra, I. R., and Conangla, L.: Stable Atmospheric Boundary-Layer Experiment in Spain (SABLES 98): a report, Bound.-Lay. Meteorol., 96, 337–370, 2000.

Fernando, H. J. S., Princevac, M., Pardyjak, E. R., and Dato, A.: The decay of convective turbulence during evening transition period. Paper 10.3, 11th Conf. on Mountain Meteorology and MAP Meeting, Bartlett, NH, Amer. Meteor. Soc., 4 pp., 2004.

García-Díez, M., Fernández, J., Fita, L., and Yagüe, C.: WRF sensitivity to PBL parametrizations in Europe over an annual cycle, Q. J. Roy. Meteor. Soc., under review, 2012.

Janjic, Z. A.: The step-mountain coordinate: physics package, Mon. Weather Rev., 118, 1429–1443, 1990.

Nadeau, D. F., Pardyjak, E. R., Higgins, C. W., Fernando, H. J. S., and Parlange, M. B.: A simple model for the afternoon and early evening decay of convective turbulence over different land surfaces, Bound.-Lay. Meteorol., 141, 301–324, 2011.

Nakanishi, M. and Niino, H.: An improved Mellor-Yamada level-3 model with condensation physics: its design and verification, Bound.-Lay. Meteorol., 112, 1–31, 2004.

Shin, H. H. and Hong, S.-Y.: Intercomparison of Planetary Boundary-Layer parametrizations in the WRF model for a single day from CASES-99, Bound.-Lay. Meteorol., 139, 261–281, 2011.

Skamarock, W. C., Klemp, J. B., Dudhia, J., Gill, D. O., Barker, D. M., Duda, M. G., Huang, X.-Y., Wang, W., and Powers, J. G.: A description of the advanced research WRF version 3. NCAR Technical note, NCAR/TN-475+STR, 113 pp., 2008.

Sukorianski, S., Galperin, B., and Perov, V.: Application of a new spectral theory of stable stratified turbulence to the atmospheric boundary layer over sea ice, Bound.-Lay. Meteorol., 117, 231–257, 2005.

Viana, S.: Estudio de los procesos físicos que tienen lugar en la capa límite atmosférica nocturna a partir de campañas experimentales de campo, Ph.D. thesis, Faculty of Physical Sciences, University Complutense of Madrid, Spain, 238 pp., 2011.

Viana, S., Terradellas, E., and Yagüe, C.: Analysis of gravity waves generated at the top of a drainage flow, J. Atmos. Sci., 67, 3949–3966, 2010.

Yagüe, C., Sastre, M., Maqueda, G., Viana, S., Ramos, D., Vindel, J. M., and Morales, G.: CIBA2008, an experimental campaign on the atmospheric boundary layer: preliminary nocturnal results, Física de la Tierra, 21, 13–26, 2009.

22

A modified drought index for WMO RA VI

S. Pietzsch and P. Bissolli

Deutscher Wetterdienst, Offenbach, Germany

Abstract. Drought is a phenomenon which can cause large economical impact even in Europe. To assess the magnitude and the spatial extension of drought events, it is important to have a standardized drought index which is applicable for a large climatically heterogeneous region like Europe or the WMO RA VI Region (Europe and the Middle East). Such an index should describe the drought phenomenon adequately, but it should also be derivable from meteorological quantities which are easily and timely available in whole Europe.

In a first investigation, some candidates for drought indices were chosen, compared and assessed for applicability in whole Europe. The most appropriate one seems to be the widely known Standardized Precipitation Index (SPI) which is a standardized and handy measurement of drought for any location and requires nothing but precipitation data. However, it has turned out that for some places in the RA VI Region, notably in arid regions in summer, the SPI does not always provide reasonable or easily interpretable results.

For that reason, some modifications of the SPI have been tried out and tested statistically. It seems that the gamma distribution of precipitation which is used for computation of the SPI is in fact the most appropriate one and other distributions have not improved the results substantially. On the other hand a so called zero correction, which sets very small precipitation totals to dry values, only dependent on the precipitation distribution, but independent on the individual location delivers more reasonable results.

Maps of the new modified drought index and its anomalies from the climate normal are produced quasi-operationally and distributed via the Internet each month. The drought monitoring is part of the monitoring programme of the WMO RA VI Pilot Regional Climate Centre on Climate Monitoring (RCC-CM) hosted by the German Meteorological Service (Deutscher Wetterdienst, DWD), and the maps can be found on its present RCC-CM platform (http://www.dwd.de/rcc-cm).

1 Introduction

Drought is a phenomenon which can appear in many parts of the globe (Wilhite, 2000) and can cause large economic impacts even in Europe. The spatio-temporal evolution and characteristics of large-scale European droughts have been investigated recently, e.g. by Parry et al. (2010) and Hannaford et al. (2011), including linking droughts to weather types (Fleig et al., 2011) and analysing drought trends for the Mediterranean e.g. by Sousa et al. (2011); further numerous studies have been carried out on a national or subregional scale.

To assess the magnitude and the spatial extent of drought events, it is important to have a standardized drought index which is applicable for a large climatically heterogeneous region like Europe or the WMO RA VI Region (Europe and the Middle East). Such an index should describe the drought phenomenon adequately, but it should also be derivable from meteorological quantities which are easily and timely available for the whole of Europe. It is desirable to have one simple index to show the drought characteristics of the wide and disparate region. The aim of the investigation is to produce an operational system to monitor drought conditions, showing their variability in time and space. It should be an index which is easy to calculate and apply to the whole region so that individual sub-regions can be easily compared to each other. This is possible for standardized indices which are generally applicable for each location within the Region and do not contain any local or sub-regional specifics or, in case there are any, they need to be adapted or modified. There were many previous approaches for deriving and using drought indices, e.g. by Parry et al. (2010). Here, the approach is to create a drought index by choosing an already existing and widely used index which is applicable for most of the WMO RA VI Region, and by introducing modifications for those parts of the Region where improvement of the index is needed due to their special climatic characteristics. Such an index can provide an overview of drought conditions and its variability in space and time operationally.

As agricultural and hydrological droughts need relatively long timescales to be indexed, McKee et al. (1993) originally calculated a drought index, the SPI (see Sect. 2), for 3-, 6-, 12-, 24- and 48-month timescales. This paper concentrates on a monthly timescale, which can be used to characterise meteorological drought. Another reason for having chosen that time scale is that the aim of this paper is the construction of a drought index for monthly climate monitoring, to enable a comparison with other monthly climate monitoring products such as temperature and precipitation anomalies.

2 Comparison of drought indices

In a first investigation (Pietzsch, 2009), three drought indices were chosen, compared and assessed for applicability to the whole of Europe. These were the Aridity Index according to de Martonne (1950) (Precipitation/[Temperature + 10]), the Climatic Water Balance (Müller-Westermeier, 2005) (Precipitation Evaporation according to Wendling, 1995) and the Standardized Precipitation Index (McKee et al., 1993) (precipitation anomalies shown as multiples of the standard deviation of an adapted theoretical distribution; here the gamma distribution and the reference period 1971–2007). The indices were calculated for every month of the years 2002 to 2006 based on various data sources (see Pietzsch, 2009 for more details) and mapped using a GIS (Geographical Information System) on a regular grid with a resolution of 0.5×0.5 degrees latitude/longitude. The results were that the Aridity Index according to de Martonne does not give any useful results for temperatures near or below −10 degrees Celsius and so cannot be applied to the whole of WMO RA VI. The Climatic Water Balance does not show extremes in an adequate way, because it expresses drought events too weakly. The most appropriate drought index seems to be the widely known Standardized Precipitation Index (SPI) which is a standardized and handy measurement of drought for any location and only requires precipitation data. However, for some places in the RA VI Region, notably in arid regions in summer such as southern Spain and the Middle East, the SPI does not always provide reasonable or easily interpretable results. It is problematic to calculate the SPI for very arid regions because precipitation near zero can cause misleading high or low SPI values (Lloyd-Hughes and Saunders,

2002). These problems appear especially for the time scale of 1 month when trying to describe regions in which aridity is normal during certain months (NDMC, 2011). In the case of zero precipitation during the whole period considered it is not possible to calculate any SPI value. And in regions with very little precipitation, the SPI misleadingly shows very high (humid) values.

3 SPI modifications

For that reason, some modifications of the SPI have been tried out and tested statistically using the gridded precipitation data of the Global Precipitation Climatology Centre (GPCC, full data reanalysis product version 4, spatial resolution 0.5×0.5°; Schneider et al., 2008). First a choice of theoretical distributions (the binomial distribution, the exponential distribution, the gamma distribution, the Poisson distribution and the Weibull distribution) was applied to precipitation data from the problematic arid regions. Therefore 51 grid points in Portugal, southern Italy, Greece, Cyprus, Turkey, Lebanon, Syria, Israel and Jordan were chosen and the corresponding empirical precipitation distributions of August in years 1971 to 2007 were adapted to each of the theoretical distributions. Then the theoretical distributions were compared to the real precipitation distributions using the Chi-square test. It appeared reasonable to test a variety of distributions to find out whether there is a distribution function beside the widely accepted gamma distribution (McKee et al., 1993) that better fits the precipitation distribution in the arid regions. In that case the distribution function of the SPI could have been modified. In regions with values of only 0 or 1 mm precipitation the binomial distribution fitted very well. In the wetter regions the gamma distribution provided the best results but also the exponential distribution showed quite good results. All in all, however, the gamma distribution was the most suitable one with the lowest Chi-square values, which means the highest similarity to the real precipitation distributions. In a second step the distributions of the calculated SPI-values for the grid points mentioned above were compared with the normal distribution via the Chi-square test. This examination showed that none of these SPI distributions was normally distributed. Based on these findings, various correction approaches for the SPI were created and tested. One version was changing the distribution function and recalculating the SPI. Here a combination of the exponential distribution and the gamma distribution was used. Another model was a weighted addition of another distribution function to the existing SPI. Furthermore a so-called zero correction, in terms of setting very small precipitation totals to dry SPI values, was used to solve the problem that the present SPI does not adopt dry values in cases of no precipitation and an inappropriate distribution as in arid regions. Every variation of modification, with and without zero correction, was validated via comparison of Chi-square

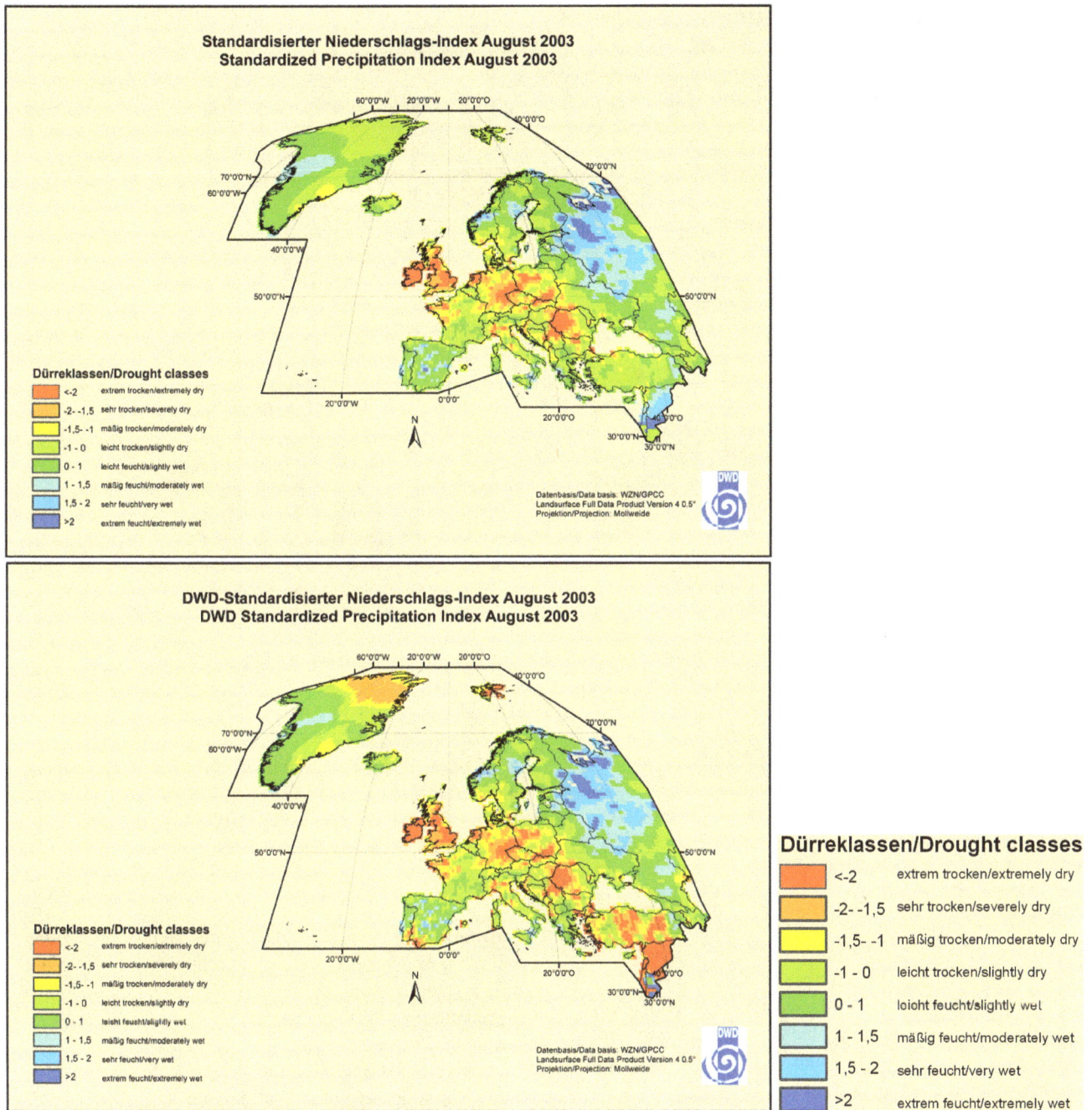

Figure 1. Comparison of the SPI (upper map) and the SPI DWD (lower map).

values. Here again, the Chi-square test was applied to find out whether the different correction approaches were normally distributed (by comparing the drought index distributions with a theoretical normal distribution); or rather which of them was the most normally distributed one. It seems that the gamma distribution of precipitation which is used for computation of the SPI is in fact the most appropriate one and other distributions have not improved the results substantially. On the other hand, the zero correction, only dependent

on the precipitation distribution, but not explicitly dependent on the individual location delivers more reasonable results by making sure that zero precipitation yields low SPI-values for each location. To have a comparison and to get the long-term mean monthly parameters of the gamma distribution (α, β, and $p0$ = the relative frequency for zero precipitation) for further calculations, first the original SPI was computed for the years 1901–2009 for every grid point of WMO RA VI using the function:

Table 1. Comparison of the SPI and the SPI DWD for several grid points.

Grid Point	Precipitation August (mm)		Drought Index August 2003	
	1971–2007	2003	SPI	SPI DWD
02.5° W 39.5° N (Spain)	14.7	31.0	1.2	1.2
07.5° W 40.5° N (Spain)	12.9	30.0	1.2	1.0
21.0° E 38.0° N (Greece)	10.4	19.0	0.9	0.7
35.5° E 33.5° N (Middle East)	0.3	0.0	1.2	−1.9

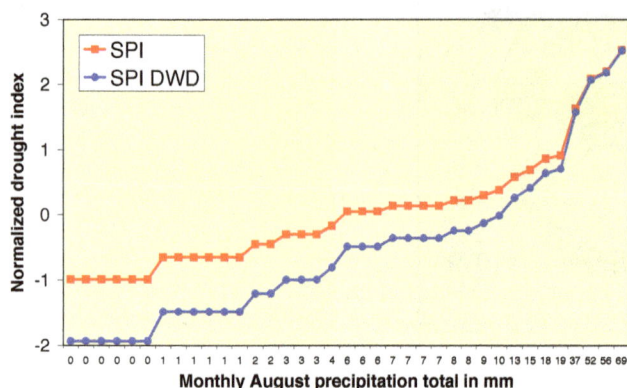

Figure 2. SPI and SPI DWD as function of monthly precipitation in August 1971–2007 at a grid point over Greece (21.0° E 38.0° N). Precipitation totals for each year of that period are displayed in ascending order. Parameters in this example: number of years $n = 37$, relative frequency of zero precipitation $p0 = 6/n = 0.162$.

$$F(x) = (1 - p0) \cdot \Gamma(x, \alpha, \beta) \text{ for } x > 0$$
$$F(x) = p0 \text{ for } x = 0 \tag{1}$$

x = precipitation total over the chosen time interval (e.g. 1 month); $p0$ = relative frequency for zero precipitation total $(x = 0)$ = number of years with zero precipitation in the given time span divided by the total number of evaluated years.

The cumulative probability, $F(x)$, is then transformed to the standard normal random variable Z with mean zero and variance of one, which is the value of the SPI (http://ccc.atmos.colostate.edu/pub/spi.pdf).

Afterwards the most reasonable correction approaches for the SPI were calculated. By successive trying, correcting and modifying the various solutions, the final solution, we call it SPI DWD, was developed:

$$\text{SPI DWD} = \text{SPI} + (\text{NORMINV}(1/n)\,\text{NORMINV}(p0)) \cdot$$
$$(1 - p(\text{SPI}))/(1 - p0) \tag{2}$$

n = Number of evaluated years; NORMINV( ) = Inverse function of the standardized normal distribution; $p0$ = Relative frequency for zero precipitation as in the original SPI; $p(\text{SPI})$ = Value of the distribution function of the standardized normal distribution for z = SPI; for $x = 0$ mm: NORMINV $(1/n)$ (e.g. for $n = 107 = -2.3$ = extremely dry);

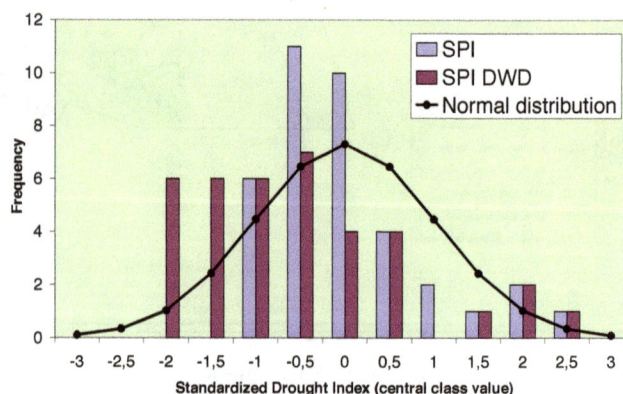

Figure 3. Frequency distribution of the SPI and the SPI DWD for the same data as in Fig. 2, compared with the corresponding normal distribution.

for $x > 0$ mm: the higher the rainfall, the greater $p(\text{SPI})$, the smaller the difference SPI DWD – SPI, and the smaller the correction; for $p0 = 0$: SPI DWD = SPI.

4   Results

The advantages of the new SPI DWD are that the arid regions in summer now can be correctly represented (see Fig. 1), that there is a smooth value transition from zero precipitation to little precipitation and that there are hardly any differences to the original SPI in case of higher precipitation.

The correction is only high for the extremely dry regions. In areas where no zero precipitation values appear, the SPI DWD is equal to the SPI and no correction is made. The smaller the precipitation is in a region, the higher is the difference between the SPI DWD and the SPI, so that now the regions with very little precipitation correctly show low (dry) SPI-values. In case of zero precipitation NORMINV $(1/n)$ is set (e.g. for 107 yr $(n = 107)$ a SPI value of -2.3 is the outcome). The complexity of the correction results from the aim of achieving the smooth value transition from zero precipitation to little precipitation.

The table above (Table 1) shows a comparison of the SPI and the SPI DWD for several grid points. It becomes evident that the difference between the two indices is quite small for the grid points in Spain and Greece, which do not have an extremely small long-term average (here August 1971–2007) precipitation total. The correction via the SPI DWD becomes clear, looking at the very dry grid point in the Middle East. Here the long-term precipitation average is 0.3 mm and the August 2003 value, with 0 mm precipitation, is even smaller. Therefore, a dry SPI value should be expected, but the original SPI index shows the "moderately wet" value of 1.2. The SPI DWD calculation, on the contrary, yields a more reasonable value of −1.9, which means "severely dry".

The graph in Fig. 2 shows how the SPI and the SPI DWD behave over the years 1971 to 2007 at one of the 51 tested grid points mentioned in Sect. 3 with 6 of the 37 yr having zero precipitation in August. In contrast to SPI, SPI DWD shows larger negative values close to −2 (severely/extremely dry) for zero precipitation and small totals, whereas for high precipitation the difference is close to zero since the original SPI already yields high values >2 which means extremely wet.

The disadvantage of the SPI DWD is that the values of its distribution are still not normally distributed in arid regions (see Fig. 3). However, in this case even the original SPI is not normally distributed, because the distribution of the precipitation values is inappropriate, since there are numerous small precipitation values. To achieve the aim of correctly representing very small precipitation values via very small SPI values we had to drop the normal distribution.

5 Production of time series and operational maps

The SPI DWD was computed for every month of the years 1901–2010 and the whole WMO RA VI. Since the GPCC full data reanalysis product presently was only available until 2007, the GPCC monitoring product had been used instead for the years since 2008. The monitoring product has a lower data density and a coarser resolution ($1 \times 1°$) compared to the full data product, but the gridded data are available two months after the actual month. For near real time monitoring, another GPCC product (first guess product) is available around 5 days after completion of the month, but with even lower data density and lower quality control level compared to the GPCC monitoring product (Schneider et al., 2008).

The index was mapped using a GIS from 2005 onwards. Maps of the new modified drought index and its anomalies from the climate normal are produced quasi-operationally and distributed via the Internet each month. The drought monitoring is part of the monitoring programme of the WMO RA VI Pilot Regional Climate Centre on Climate Monitoring (RCC-CM) hosted by the German Meteorological Service (Deutscher Wetterdienst, DWD), and the maps can be found on its present platform http://www.dwd.de/rcc-cm.

Acknowledgements. Sincere thanks are given to the GPCC for providing the gridded precipitation data. The guest editor and three anonymous reviewers contributed some very constructive comments on the paper.

Edited by: J. Prior
Reviewed by: three anonymous referees

References

de Martonne, E.: Traite de géographie physique. Notions générales, climat, hydrogeographie, 8th edn., Colin, Paris, 1950 (in French).

Fleig, A. K., Tallaksen, L. M., Hisdal, H., and Hannah, D. M.: Regional hydrological drought in north-western Europe: linking a new Regional Drought Area Index with weather types, Hydrol. Process., 25, 1163–1179, doi:10.1002/hyp.7644, 2011.

Hannaford, J., Lloyd-Hughes, B., Keef, C., Parry, S., and Prudhomme, C.: Examining the large scale spatial coherence of European drought using regional indicators of precipitation and streamflow deficit, Hydrol. Process., 25, 1146–1162, doi:10.1002/hyp.7725, 2011.

Lloyd-Hughes, B. and Saunders, M. A.: A drought climatology for Europe, Int. J. Climatol., 22, 1571–1592, 2002.

McKee, T. B., Doesken, N. J., and Kleist, J.: The relationship of drought frequency and duration to time scales, Proceedings of the Eighth Conference on Applied Climatology, Anaheim, California, 1993.

Müller-Westermeier, G.: Die Klimatische Wasserbilanz, in: Deutscher Wetterdienst (2005): Klimaatlas Bundesrepublik Deutschland 4, Offenbach am Main, 2005 (in German).

National Drought Mitigation Center (NDMC): Interpretation of 1-Month Standardized Precipitation Index Map, http://drought.unl.edu/MonitoringTools/ClimateDivisionSPI/Interpretation/1month.aspx, 2011.

Parry, S., Prudhomme, C., Hannaford, J., and Lloyd-Hughes, B.: Examining the spatio-temporal evolution and characteristics of large-scale European droughts, in: Role of Hydrology in Managing Consequences of a Changing Global Environment, Proceedings of the BHS Third International Symposium, edited by: Kirby, C., Newcastle, British Hydrological Society, 135–142, 2010.

Pietzsch, S.: Raum-zeitliche Analyse von Dürreperioden in der WMO Region VI (Europa und Naher Osten), Diploma thesis, Johannes Gutenberg University, Mainz, 2009 (in German).

Schneider, U., Fuchs, T., Meyer-Christoffer, A., and Rudolf, B.: Global Precipitation Climatology Centre (GPCC): Global Precipitation Analysis Products of the GPCC, Deutscher Wetterdienst Offenbach am Main, internet publication: http://gpcc.dwd.de, 2008.

Sousa, P. M., Trigo, R. M., Aizpurua, P., Nieto, R., Gimeno, L., and Garcia-Herrera, R.: Trends and extremes of drought indices throughout the 20th century in the Mediterranean, Nat. Hazards Earth Syst. Sci., 11, 33–51, doi:10.5194/nhess-11-33-2011, 2011.

Wendling, U.: Berechnung der Gras-Referenz-Verdunstung mit der FAO Penman-Monteith-Beziehung, Wasserwirtschaft 85E, H.12, 602–660, 1995 (in German).

Wilhite, D. A.: Drought a global assessment, Vol. I, London, 2000.

23

Comparing different meteorological ensemble approaches: hydrological predictions for a flood episode in Northern Italy

S. Davolio[1], T. Diomede[2], C. Marsigli[2], M. M. Miglietta[3,4], A. Montani[2], and A. Morgillo[2]

[1]Institute of Atmospheric Sciences and Climate, National Research Council, Bologna, Italy
[2]HydroMeteoClimate Regional Service of ARPA Emilia Romagna, Bologna, Italy
[3]Institute of Atmospheric Sciences and Climate, National Research Council, Lecce, Italy
[4]Institute of Ecosystem Study, National Research Council, Verbania Pallanza, Italy

Correspondence to: S. Davolio (s.davolio@isac.cnr.it)

Abstract. Within the framework of coupled meteorological-hydrological predictions, this study aims at comparing two high-resolution meteorological ensembles, covering short and medium range. The two modelling systems have similar characteristics, as almost the same number of members, the model resolution (about 7 km), the driving ECMWF global ensemble prediction system, but are obtained through different methodologies: the former is a multi-model ensemble, based on three mesoscale models (BOLAM, COSMO, and WRF), while the latter follows a single-model approach, based on COSMO-LEPS (Limited-area Ensemble Prediction System), the operational ensemble forecasting system developed within the COSMO consortium.

Precipitation forecasts are evaluated in terms of hydrological response, after coupling the meteorological models with a distributed rainfall-runoff model (TOPKAPI) to simulate the discharge of the Reno river (Northern Italy), for a severe weather episode.

Although a single case study does not allow for robust and definite conclusions, the comparison among different predictions points out a remarkably better performance of mesoscale model ensemble forecasts compared to global ones. Moreover, the multi-model ensemble outperforms the single model approach.

1 Introduction

Prediction of the hydrological response of a watershed to rainfall can be handled by coupling meteorological and hydrological numerical weather prediction (NWP) models. This is especially true for small and medium-sized catchments (smaller than $10\,000\,\mathrm{km}^2$), characterized by complex orography and short response times, where the sole observed precipitation is not suitable to drive hydrological models for timely forecasts and adequate emergency planning. The reliability and the practical use of a coupled discharge forecasting system are tightly dependent on the accuracy of the forecast precipitation data. However, increased NWP model resolution and improved rainfall forecast skill do not ensure a positive impact on hydrological predictions, since quantitative precipitation forecasts (QPFs) issued by meteorological models are still affected by errors at the small scales that are particularly relevant for hydrological applications.

In the forecasting process, it is therefore necessary to acknowledge the different sources of errors affecting QPF (initial condition, model structure), which represent the largest source of uncertainty in discharge prediction. An ensemble prediction system (EPS) can quantify at least part of these uncertainties by producing probabilistic forecasts, thus representing an attractive product to be used for flood predictions (for a review, see Cloke and Pappenberger, 2009). Probabilistic forecasts are considered much more valuable than a single deterministic forecast for weather prediction (Buizza, 2008); recently there is a general agreement on the usefulness of ensemble forecasting for early flood warning application

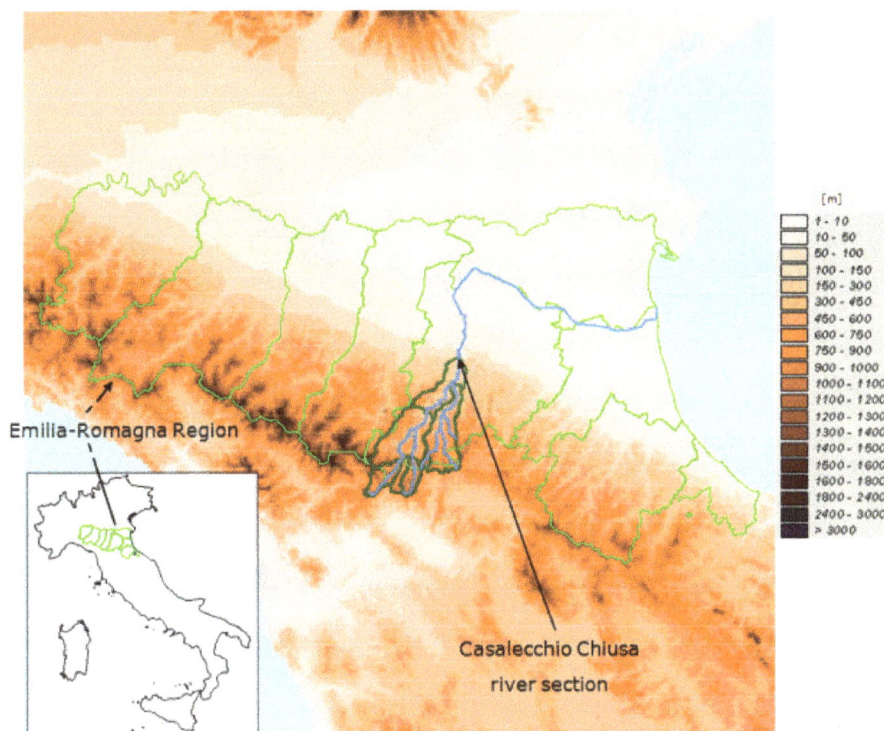

Figure 1. Localisation of the Reno river basin. The main river is showed in cyan. The upper part of the basin, closed at Casalecchio Chiusa, is evidenced with dark green lines.

too. EPS forecasts can be used as an input for hydrological models, thus propagating the uncertainty along the flood forecasting system, in order to provide a probabilistic and hopefully more informative hydrological prediction.

Although providing positive results, global model ensemble predictions suffer from their coarse resolution and often proved not to be accurate enough for application at basin scale, especially in region with complex orography. Thus, during the last decade, different ensemble approaches based on limited area models (LAMs) have been developed (e.g. Montani et al., 2011). These local ensemble prediction systems (LEPSs) basically perform a dynamical downscaling of global EPS and represent the state-of-the-art in meteo-hydrological forecasting (Cuo et al., 2011).

In this study, two different ensemble approaches, both focused on the short/medium range, are compared. In order to allow a fair comparison, the two ensembles have been implemented using almost the same set up in terms of integration domain, horizontal resolution, number of members. The difference resides in the relative importance which has been attributed to the representation of the boundary condition error with respect to that of the LAM error. For the single-model ensemble, the same LAM has been run 16 times receiving initial and boundary conditions from 16 selected members of the ECMWF EPS, while for the multi-model ensemble, only 5 EPS members have been selected out of the EPS but 3 different LAMs have been run on each EPS member, thus

yielding a 15 member ensemble. Both the ensembles have been used to generate probabilistic precipitation maps and to provide the input fields to the same hydrological model. The results, in terms of discharge prediction, allow to evaluate the ensembles performance in a recent severe weather episode affecting the Reno river basin, located in Northern Italy (Fig. 1).

2 Case study and ensemble generation

The analysed severe weather period, between 29 November and 2 December 2008, was characterized by the presence of a deep trough over the Mediterranean Sea, driving several frontal systems towards the Italian peninsula. Persistent and intense moist south-westerly flow impinging on the northern Apennines (where the Reno river basin is located) was responsible for severe weather. Two period of intense precipitation, during the nights of 29 November (Fig. 2) and between 30 November and 1 December, respectively, produced two relevant discharge peaks, both exceeding the warning threshold at the closure section of the mountain portion of the Reno catchment.

Discharge forecasts for this event were produced by the distributed rainfall-runoff model TOPKAPI (Todini and Ciarapica, 2002), whose input rainfall fields have been provided by the following EPSs:

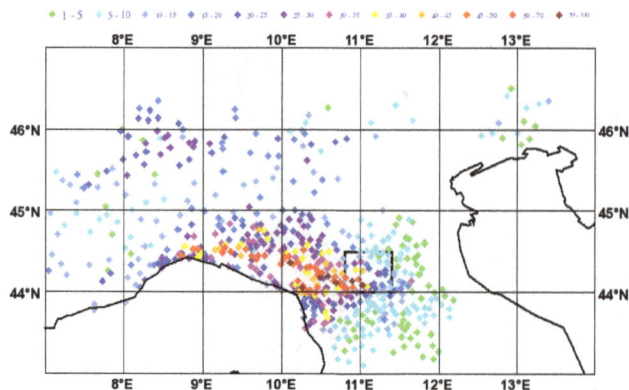

Figure 2. Observed 6-h accumulated precipitation at 00:00 UTC, 30 November 2008. Dark blue diamonds correspond to 20–25 mm/6 h. Black rectangle indicates approximately the Reno river basin.

1. A 15-member multi-model ensemble, based on three mesoscale models, each of them initialized by 5 representative members of ECMWF EPS: BOLAM (Malguzzi et al., 2006) has been developed and implemented by the Institute of Atmospheric Sciences and Climate (ISAC); COSMO (Steppeler et al., 2003) has been developed within the COSMO consortium and implemented by ARPA-SIMC; WRF (Skamarock et al., 2005) is the USA community model and it is implemented at ISAC.

2. A single-model approach, based on COSMO-LEPS (COSMO Limited-area Ensemble Prediction System), the operational forecasting system of the COSMO Consortium, driven by 16 representative members of the ECMWF EPS.

The same cluster analysis (Montani et al., 2011) has been used in both EPSs for selecting the representative members within the 102-members generated by the ECMWF EPS initialized both at 00:00 and 12:00 UTC. The two EPSs have been initialized at three different instants 24-h apart, starting from 26 November at 12:00 UTC, in order to evaluate the forecasting system at different lead times.

In order to have a reference prediction, forecasts provided by ECMWF EPS (51 members), initialized at 12:00 UTC on the same days, were also used as input for the TOPKAPI model.

3   Results: meteorological perspective

The evaluation of the ensemble systems is first performed from a meteorological perspective over a larger area than the single catchment (e.g. entire Northern Italy). For sake of brevity, only the first period of intense precipitation, shown in Fig. 2, will be presented. Intense precipitation affected the

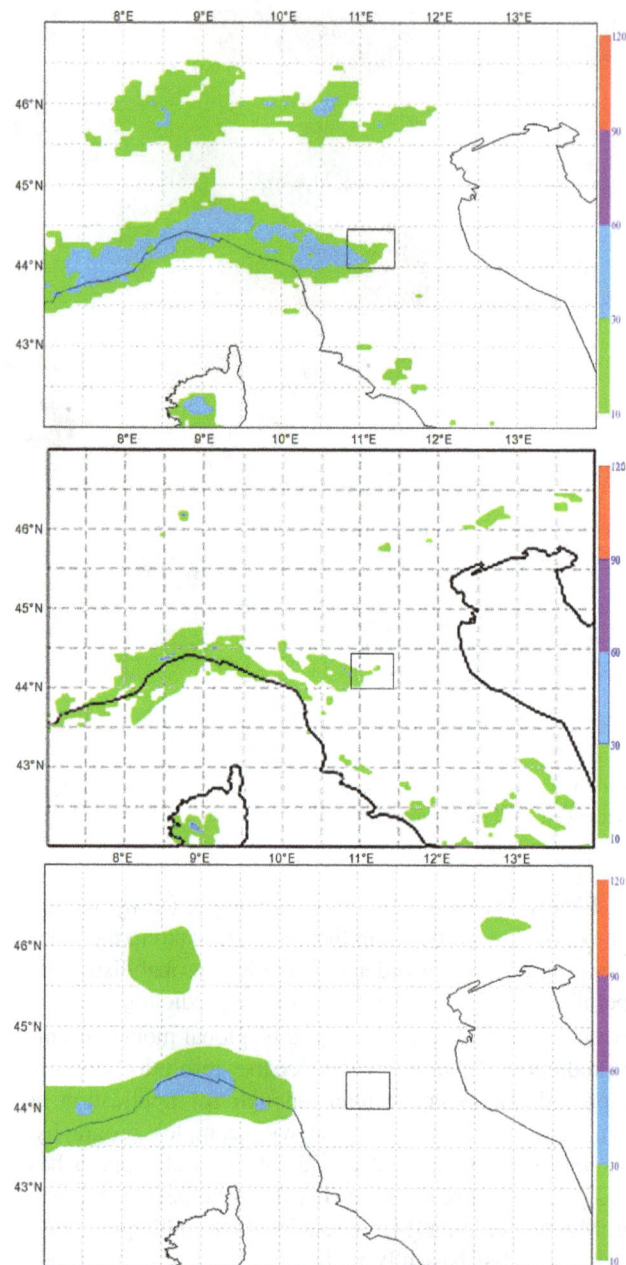

Figure 3. Probability of precipitation exceeding 20 mm/6 h forecast by (from top to bottom) multi-model, COSMO-LEPS and ECMWF EPS, at 00:00 UTC, 30 November 2008. Simulations are initialized at 12:00 UTC, 26 November 2008.

whole northern Apennines and also some Alpine areas. Results of the two LEPSs and of the global EPS are compared for two different forecast lead times, in terms of probability maps of occurrence of precipitation exceeding 20 mm/6 h. At 78–84 h range (Fig. 3, initialization time 12:00 UTC, 26 November), the global EPS does not provide any indication of intense precipitation over the Reno basin, but only over western Apennines (probability up to 60 %). On the other hand, both LEPSs forecast some probability of rainfall (up to 60 % for the multi-model) over the Reno river basin. Moreover, the multi-model provides a signal also over the central Alps, where precipitation did occur.

Similarly, for a shorter forecast range (not shown, initialization time 12:00 UTC, 28 November), only the two LEPSs are able to forecast the possible occurrence of precipitation over the target basin. Very high probability is assigned to rainfall over western Apennines and the Alpine chain by all the systems, with a progressively increasing probability for shorter lead times, thus improving the confidence in the prediction as the event approaches.

Similar results have been obtained for the second period of intense precipitation (not shown).

4　Results: hydrological perspective

The two intense precipitation events generated two relevant discharge peaks in the Reno basin (Fig. 4). The ensemble discharge forecasts are the expected consequence of the result shown by the maps of probability of precipitation. Indeed, while the discharge prediction driven by the global EPS fails to generate any relevant peak, the discharge predictions driven by both LEPSs are remarkably better. Although underestimated in intensity, the possible occurrence of high discharge peaks is forecast almost 4 day ahead by both LEPSs, thus providing a useful indication of the event for the civil protection authorities. In particular, at this long forecast range, some members of the multi-model exceed the warning threshold. At a less extent, also the COSMO-LEPS displays some relevant peaks, although the timing is affected by large uncertainty.

Even at shorter forecast ranges, up to 2 days in advance, LEPSs outperform the global EPS (not shown). The multi-model ensemble displays a large spread among the members and a more accurate discharge prediction, especially concerning the second peak.

Looking into detail at the discharge forecasts generated by every single model of the multi-model ensemble, it is possible to recognize that for longer lead times (more than 3 days) the behaviour of the different members is dominated by the boundary conditions effect, since the higher discharge peaks are associated with mesoscale forecasts driven by the same global ensemble members. This is not true for short forecast ranges, where the impact of boundary conditions is weaker and the spread is reasonably ascribable to the characteristics of the models.

5　Conclusions

Although limited to a single event, the comparison among EPSs provided some interesting results, in particular highlighting the added value of mesoscale models for ensemble forecasting with respect to a global ensemble. At vari-

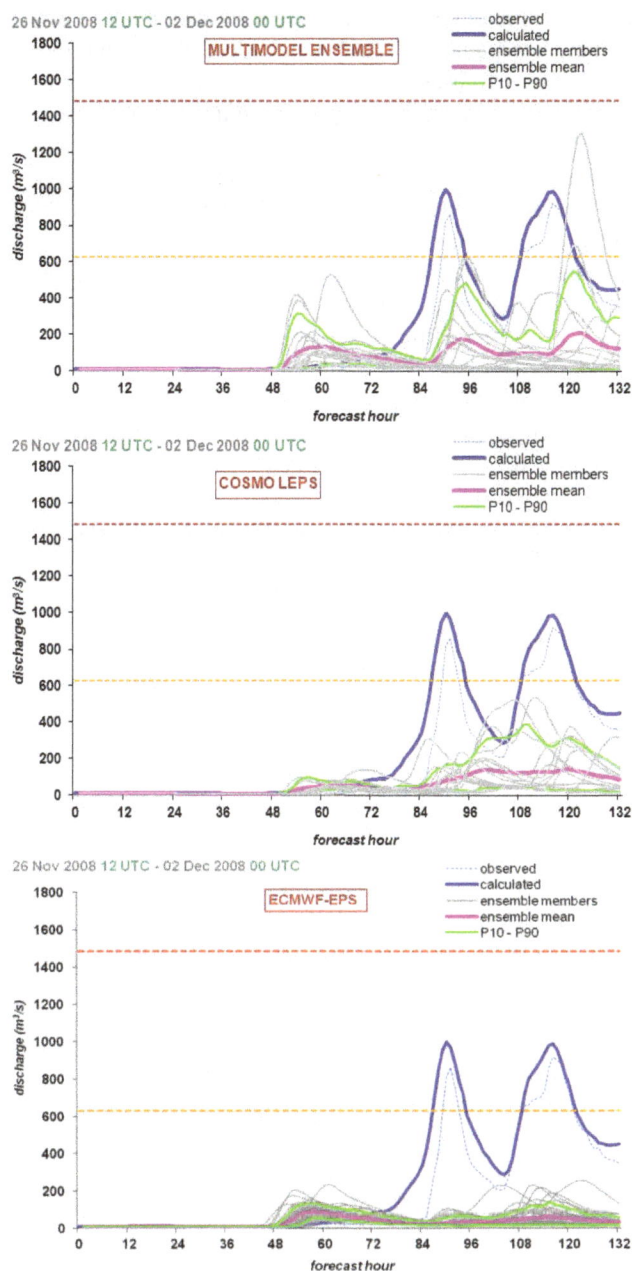

Figure 4. Discharge forecast $(m^3 \, s^{-1})$ vs. forecast range (hours) for multi-model, COSMO-LEPS and ECMWF EPS (51 members), respectively, initialized at 12:00 UTC, 26 November 2008. Each grey line corresponds to an ensemble member. For reference, the observed discharge (dotted blue) and the discharge computed using rainfall observations are provided. Ensemble mean (pink) and the P10-P90 curves (green) are also plotted. Horizontal dashed orange and red lines indicate warning and alarm thresholds, respectively.

ance with LEPS, the global EPS forecasts do not provide evidence of any relevant probability of intense precipitation over the Reno river basin, even at short forecast ranges. This points out that structural model deficiencies (i.e. low resolution, coarse orography representation) cannot be accounted for by this kind of ensemble approach. Instead, higher resolution models are needed. LAMs are indeed able to improve remarkably the forecast quality, also in terms of hydrological response of the basin. Looking in more detail at the multi-model LEPS, the system seems able to identify the Reno river basin as an area likely to be affected by intense precipitation almost 4 days in advance. The multi-model LEPS provides better results with respect to COSMO-LEPS, being characterized by a larger spread at short range due to different model characteristics. At longer forecast ranges, the similar behaviour of the multi-model LEPS members indicates the relevant impact of the boundary conditions. The greater degree of diversity of the multi-model LEPS members seems to be the added value of the multi-model approach with respect to single-model COSMO-LEPS. Such conclusions require to be verified in additional case studies.

Acknowledgements. The Authors are grateful to the Italian regional agencies which provided observational data.

Edited by: B. Ahrens
Reviewed by: two anonymous referees

References

Buizza, R.: The value of probabilistic prediction, Atmospheric Sciences Letters, 9, 36–42, 2008.

Cloke, H. L. and Pappenberger, F.: Ensemble flood forecasting: a review, J. Hydrol., 375, 613–626, 2009.

Cuo, L., Pagano, T. C., and Wang, Q. J.: A review of quantitative precipitation forecasts and their use in short- to medium-range streamflow forecasting, J. Hydrometeorol., 12, 713–728, 2011.

Malguzzi, P., Grossi, G., Buzzi, A., Ranzi, R., and Buizza, R.: The 1966 "century" flood in Italy: A meteorological and hydrological revisitation, J. Geophys. Res., 111, D24106, doi:10.1029/2006JD007111, 2006.

Montani, A., Cesari, D., Marsigli, C., and Paccagnella, T.: Seven years of activity in the field of mesoscale ensemble forecasting by the COSMO-LEPS system: main achievements and open challenges, Tellus A, 63, 605–624, 2011.

Skamarock, W. C., Klemp, J. B., Dudhia, J., Gill, D. O., Barker, D. M., Wang, W., and Powers, J. G.: A description of the Advanced Research WRF Version 2. NCAR Technical Note, NCAR/TN-468+STR, 88, 100 pp. 2005.

Steppeler, J., Doms, G., Schattler, U., Bitzer, H. W., Gassmann, A., Damrath, U., and Gregoric, G.: Meso-gamma scale forecasts using the nonhydrostatic model LM, Meteorol. Atmos. Phys., 82, 75–96, 2003.

Todini, E. and Ciarapica, L.: The TOPKAPI model, in: Mathematical models of large watershed hydrology, edited by: Singh, V. P., Frevert and Littleton, D. K., Colorado, USA, Water Resources Publications, 914 pp., 2002.

24

Site-dependent decrease of odour-related peak-to-mean factors with distance

M. Piringer[1], **W. Knauder**[1], **E. Petz**[1], and **G. Schauberger**[2]

[1]Central Institute for Meteorology and Geodynamics, Vienna, Austria
[2]Unit for Physiology and Biophysics, University of Veterinary Medicine, Vienna, Austria

Correspondence to: M. Piringer (martin.piringer@zamg.ac.at)

Abstract. The peak-to-mean concept developed earlier by the authors to calculate odour-related separation distances is applied here to meteorological input for dispersion models provided by ultrasonic anemometers. In addition to conventional meteorological input parameters like wind direction, wind speed and stability classes, three-dimensional sonics provide also turbulence information via the Obukhov stability parameter and the variance of the wind speed, which can be used directly to determine peak-to-mean ratios depending on the distance from the source. The influence and importance of these site-specific peak-to-mean ratios on the resulting direction-dependent separation distances is investigated and discussed.

1 Introduction

Odour perception is a biological reaction of humans. The relevant time scale is the duration of a single human breath, i.e. 4 s on average (Kleemann et al., 2009). (Regulatory) dispersion models, which usually calculate half-hourly or hourly averages of concentrations, have to be adapted somehow to parameterize the relevant short-term odour concentrations. The peak-to-mean concept is a mean to parameterize short-term concentrations in dispersion models, used in several European countries and in Australian and US states (for an overview on regulations in different countries, see Piringer and Schauberger, 2013). Schauberger et al. (2012) discuss different definitions of the peak value and models (e.g. Best et al., 2001) of its reduction with distance.

For Austria, the authors developed a peak-to-mean approach depending on meteorological conditions used for the regulatory Austrian Gauss model; this algorithm is used in the Austrian Odour Dispersion Model (AODM) and described by Schauberger et al. (2000) and Piringer et al. (2007). In short, AODM consists of three modules: the first calculates the odour emission of a livestock building, taking also into account the diurnal variation caused by animal activity; the second estimates half-hourly or hourly ambient concentrations using a regulatory dispersion model, and the third transforms the mean odour concentrations of the

dispersion model to instantaneous values depending on the stability of the near-surface layer. The regulatory model is a Gaussian plume model applied for single stack emissions and distances from 100 m to 15 km. The model uses a traditional discrete stability classification scheme with dispersion parameters developed by Reuter (1970). It has been evaluated against a data set from the German environmental programme BWPLUS within the project "Odour emission and spread" (Baechlin et al., 2002); results are discussed in Piringer and Baumann-Stanzer (2009). In Piringer et al. (2007), the possibility to take the necessary meteorological information from three-dimensional (3-D) ultrasonic anemometers to arrive at site-specific attenuation curves for the peak-to-mean factor was already discussed. This approach is demonstrated herein for two Austrian sites with different meteorological conditions. The resulting direction-dependent separation distances are compared to those obtained from the original attenuation curves.

2 Method and sites

In the next section, the results are presented for two Austrian sites, namely the city of Linz and the village of Kittsee at the border to Slovakia and Hungary. Both sites are characterized by flat grass-covered terrain without nearby obstacles.

Table 1. Ratios of the standard deviations of the three wind components σ_u, σ_v and σ_w to the horizontal wind velocity u depending on the stability of the atmosphere. Values proposed by Robins (1979) are compared to site-specific ratios derived from 3-D ultrasonic anemometer measurements (OSP).

Stability class		σ_u/u		σ_v/u		σ_w/u	
		Robins	OSP	Robins	OSP	Robins	OSP
(a) Linz							
2	very unstable	0.2	0.53	0.2	0.53	0.3	0.27
3	unstable	0.2	0.44	0.2	0.42	0.2	0.20
4	neutral	0.2	0.37	0.2	0.31	0.1	0.17
5	slight stable	0.2	0.43	0.2	0.38	0.1	0.18
6	stable	0.2	0.46	0.2	0.43	0.1	0.18
7	very stable	0.2	0.46	0.2	0.43	0.1	0.16
(b) Kittsee							
2	very unstable	0.2	0.40	0.2	0.39	0.3	0.19
3	unstable	0.2	0.27	0.2	0.26	0.2	0.12
4	neutral	0.2	0.19	0.2	0.17	0.1	0.08
5	slight stable	0.2	0.19	0.2	0.15	0.1	0.08
6	stable	0.2	0.20	0.2	0.17	0.1	0.08
7	very stable	0.2	0.20	0.2	0.18	0.1	0.07

The ultrasonic anemometer was mounted on top of a 10 m high mast at both sites. At Linz, a site characteristic of the Austrian North-Alpine foreland, we can expect a west-east orientation of the wind regime. Westerly airflow in Linz is frequently connected with cyclonic conditions. The easterly wind directions generally lead to fair-weather conditions but can also be indicative of the flow ahead of cyclones arriving from the West. Kittsee can experience high wind speeds, mainly from northwesterly directions often associated with frontal systems and storms. The secondary most frequent wind direction is from north-east. This is explained by a topographical deflection of the regional flow in the area caused by the southernmost tip of the Carpathian mountains in the region of Bratislava, north of the site. These wind directions show on average lower wind speeds as they are mainly observed in anti-cyclonic conditions.

Besides wind data, dispersion models need parameters describing the vertical structure of the boundary layer (the atmospheric stability) as input. This information can be provided in its simplest form by discrete stability classes or by more sophisticated direct measures of atmospheric turbulence from 3-D sonics. An estimate of atmospheric stability is obtained using the standard deviations of the three wind components and the Obukhov stability parameter (OSP, in m^{-1}). Before 3-D ultrasonic anemometers were commonly available, assumptions of the relationship between the standard deviations of the three wind components and the mean wind speed were used; we followed those proposed by Robins (1979) and given in Table 1. The calculation of the peak-to-mean curves both for standard meteorological data

as well as for ultrasonic anemometer data is described in detail in Piringer et al. (2007, 2013).

3 Results and discussion

The ratios of the standard deviations of the three wind components to the horizontal wind velocity are displayed in Table 1. Besides the values from Robins (1979), which are independent of a specific site, also those derived from ultrasonic anemometer measurements (OSP) are given; the two sets of values differ considerably. At both sites, the ratios derived from ultrasonic anemometer measurements show a strong dependence of σ_u/u and σ_v/u on stability and generally larger values than suggested by Robins (1979). The dependence of σ_w/u on stability is generally in the range provided by Robins (1979), but lower values are mostly found, especially at Kittsee.

The peak-to-mean algorithm used in the AODM calculates large peak-to-mean ratios near the source. The reduction of the peak-to-mean ratio with distance due to turbulent mixing is described by exponential attenuation functions (Mylne and Mason, 1991; Mylne, 1992). The attenuation curves for the two sites are displayed in Fig. 1a–b (solid lines: Robins, 1979; dashed lines: OSP). Generally, all peak-to-mean curves approach the value of 1 after some distance from the source, usually within a few hundred meters or less. In unstable conditions, the use of OSP from sonics gives larger peak-to-mean factors, whereas in neutral and stable conditions, the differences between Robins (1979) and OSP are not as large and not systematic. The Robins' (1979) curves are of course the same in both figures.

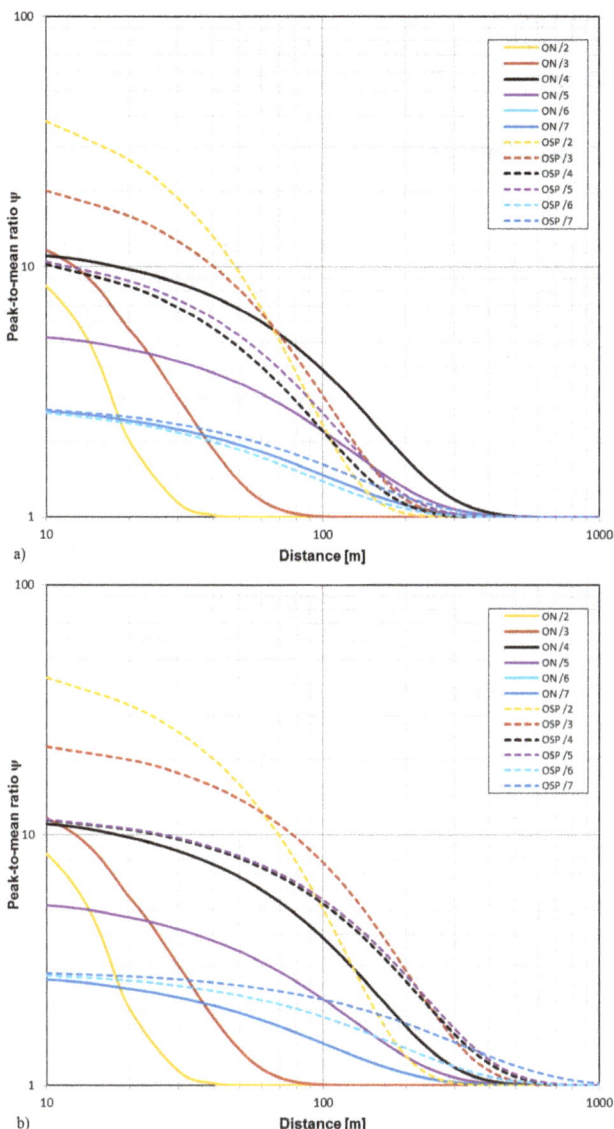

Figure 1. Peak-to-mean ratio attenuation curves dependent on atmospheric stability (classes 2 to 7) for (**a**) Linz and (**b**) Kittsee. The solid lines (ON) indicate the curves derived from Robins (1979) and the dashed lines (OSP) indicate those determined using the Obukhov stability parameter from the three-dimensional ultrasonic anemometer.

There are sometimes remarkable differences in the peak-to-mean curves between the two sites. The peak-to-mean ratios for unstable conditions (classes 2 and 3) using OSP are much greater near the source compared to the curves from Robins (1979), and their decrease with distance seems to be reduced, leading to peak-to-mean ratios at 100 m between about 2.5 and almost 10 (see Fig. 1). According to Robins (1979), a peak-to-mean ratio of only 1 is obtained at 100 m in unstable conditions. At Kittsee, the attenuation curve for class 3 indicates values above 1 several 100 m downwind. For neutral and stable conditions, the differences

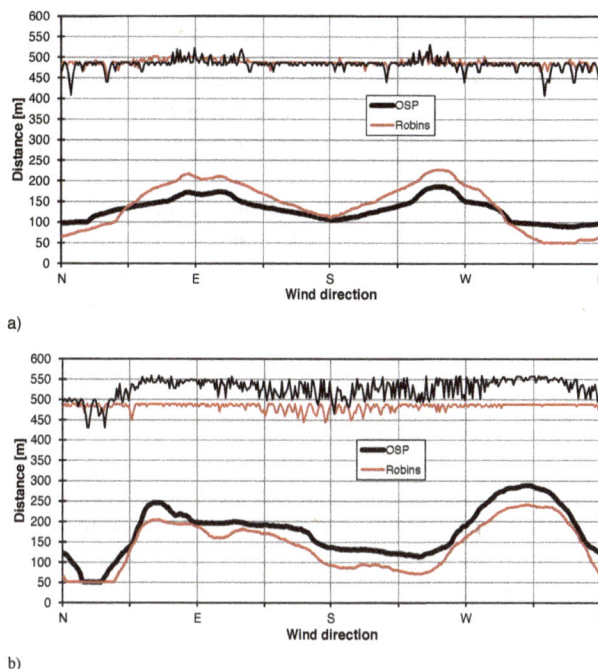

Figure 2. Separation distances for (**a**) Linz and (**b**) Kittsee. For details, see text.

Table 2. Source data for dispersion calculations with AODM.

Stack height	[m]	8.0
Stack diameter	[m]	2.7
Exit velocity	[m s^{-1}]	3.0
Volume flow	[m^3 h^{-1}]	60 000
Temperature	[°C]	20
Source strength	[OU$_E$ s^{-1}]	5200

between the curves are far less pronounced. The peak-to-mean ratios decrease less using OSP at Kittsee and more rapidly at Linz.

For odour protection, separation distances to protect the neighbours from odour nuisance can be defined by the so-called odour impact criteria, a combination of odour concentration (mostly a lower threshold of 1 OU (odour unit) per m^3) and a selected exceedence probability according to a land use category. The separation distances depending on wind direction are shown in Fig. 2. For each wind direction, a cumulative frequency distribution of separation distances is obtained (Piringer and Schauberger, 1999). The upper curves show the maximum separation distances (i.e. the separation distance for the 100 percentile per wind direction), the lower curves those for an exceedence probability of 3 % (or the 97 percentile) for 1 OU m^{-3}. For the two sites, the same source data are used (see Table 2).

At Linz, the maximum separation distances are mostly slightly below 500 m (see Fig. 2a), irrespective of the method used. This can be explained by the fact that the peak-to-mean ratio at this distance is one (see Fig. 1a). At Kittsee (see Fig. 2b), the maximum separation distances of more than 500 m are larger when determined by OSP, because at this distance peak-to-mean factors can be larger than 1 for OSP only. For practical applications, separation distances for the odour impact criteria, i.e. a combination of the odour threshold and a prescribed exceedence probability (here 3 %, equal to the 97 percentile), are more important than the maximum distances. The former are displayed in Fig. 2 as thick lines. As wind direction in meteorology is defined as a wind coming from a certain direction, the separation distance e.g. for East wind is stretching to the West of a source. The use of the 97 percentile leads to a considerable cut-off in each wind direction so that the maximum separation distances of about 500 m reduce to approximately 200 to 300 m, depending on the meteorological conditions on site. The largest separation distances are calculated for the main wind direction sectors. It is worth noting that this is also a result of the larger abundance of data in these sectors.

At Linz, the Robins' (1979) based site-independent attenuation curves display very often the largest separation distances, especially for the main wind directions. This is caused by the lower site-specific peak-to-mean factors at larger distances in neutral and stable conditions, compared to those derived from OSP (see Fig. 1a). However, the peak-to-mean factors in Kittsee are higher at larger distances than obtained from the standard procedure, causing the larger OSP-derived separation distances there.

For the meteorological interpretation of separation distances, peak-to-mean factors for large distances (at least 100 m) are most relevant. At shorter distances the implicit assumption in Gaussian plume models that the longitudinal diffusion is negligible compared to the lateral and vertical diffusion is no longer valid. At Linz, peak-to-mean factors beyond 100 m distance are about 3 at most and decrease rapidly to 1 with distance, irrespective of the method with which they are determined (see Fig. 1a). Hence, the influence of specific stability conditions on the separation distances should be small. The largest separation distances of up to 230 m are calculated for westerly wind (see Fig. 2a), which shows the highest frequency of large wind speeds. Only for the seldom occurring northerly wind, separation distances with Robins' (1979) approach are smaller than those obtained with the OSP method. This can be explained by the fact that this wind direction sector, in contrast to all others, is associated with low wind speeds and a high frequency of unstable dispersion categories. The latter are, however, relevant for larger separation distances in the OSP-derived approach with higher peak-to-mean ratios in unstable conditions (see Fig. 1a) compared to those from Robins (1979) with peak-to-mean ratios of about 1 beyond 100 m in unstable conditions.

At Kittsee (see Fig. 2b), the largest separation distances of 250 to 300 m are calculated for the wind directions with on average highest wind speeds (in this case wind from northwest). As the peak-to-mean factors determined from OSP for

this site (see Fig. 1b) are larger than the site-independent values from Robins (1979), the site-specific separation distances are larger by about 50 m for almost all wind direction sectors.

4 Concluding remarks

The comparison of separation distances derived with the site-independent Robins' (1979) approach and site-specific ultrasonic anemometer data has been undertaken for two sites in Austria in flat terrain without nearby obstacles. For neighbourhood protection it is important to use a method that will predict the likely largest separation distances. The results suggest that in low wind speed/elevated turbulence conditions, such as in Linz, the Robins' (1979) peak-to-mean factors tend to provide the largest separation distances (see Fig. 2a). In high wind speed/low turbulence conditions, such as in Kittsee, OSP determined from 3-D ultrasonic anemometers results in larger separation distances (see Fig. 2b). These results are preliminary, but more ultrasonic data will likely be available in the future to support or deny this hypothesis.

The use of site-specific attenuation curves will increase the reliability of dispersion calculations used for regulatory purposes (e.g. when licensing livestock farms) particularly under unstable conditions. This approach is recommended whenever applicable.

Acknowledgements. This research was funded by the Austrian ministry of Science and Research.

Edited by: C. Chemel
Reviewed by: I. Mavroidis and another anonymous referee

References

Baechlin, W., Ruehling, A., and Lohmeyer, A.: Bereitstellung von Validierungsdaten für Geruchsausbreitungsmodelle – Naturmessungen (A validation data set for odour dispersion models – field measurements), Forschungsbericht FZKA-BWPLUS, Förderkennzeichen BWE 20003, Ingenieurbüro Lohmeyer, Karlsruhe, Dresden, 183 pp., 2002.

Best, P., Lunney, K., and Killip, C.: Statistical elements of predicting the impact of a variety of odour sources, Water Sci. Technol., 44, 157–164, 2001.

Kleemann, A. M., Kopietz, R., Albrecht, J., Schöpf, V., Pollatos, O., Schreder, T., May, J., Linn, J., Brückmann, H., and Wiesmann, M.: Investigation of breathing parameters during odor perception and olfactory imager, Chem. Senses, 34, 1–9, 2009.

Mylne, K. R.: Concentration fluctuation measurements in a plume dispersing in a stable surface layer, Bound.-Layer Meteorol., 60, 15–48, 1992.

Mylne, K. R. and Mason, P. J.: Concentration fluctuation measurements in a dispersing plume at a range of up to 1000 m, Q. J. Roy. Meteorol. Soc., 117, 177–206, 1991.

Piringer, M. and Baumann-Stanzer, K.: Selected results of a model validation exercise, Adv. Sci. Res., 3, 13–16, 2009.

Piringer, M. and Schauberger, G.: Comparison of a Gaussian diffusion model with guidelines for calculating the separation distance between livestock farming and residential areas to avoid odour annoyance, Atmos. Environ., 33, 2219–2228, 1999.

Piringer, M. and Schauberger, G.: Dispersion modeling for odour exposure assessment, in: Odour Impact Assessment Handbook, edited by: Belgiorno, V., Naddeo, V. and Zarra, T., Wiley, Chichester, West Sussex, UK, 125–176, 2013.

Piringer, M., Petz, E., Groehn, I., and Schauberger, G.: A sensitivity study of separation distances calculated with the Austrian Odour Dispersion Model (AODM), Atmos. Environ., 41, 1725–1735, 2007.

Piringer, M., Petz, E., Groehn, I., and Schauberger, G.: Corrigendum to "A sensitivity study of separation distances calculated with the Austrian Odour Dispersion Model (AODM)" [Atmos. Environ. 41 (2007) 1725–1735], Atmos. Environ., 67, 461–462, 2013.

Reuter, H.: Die Ausbreitungsbedingungen von Luftverunreinigungen in Abhängigkeit von meteorologischen Parametern (Dispersion conditions of airborne pollutants in dependence on meteorological parameters), Archiv für Meteorologie, Geophys. Bioklimatol. A, 19, 173–186, 1970.

Robins, A. G.: Development and structure of neutrally simulated boundary layers, J. Indust. Aerodyn., 4, 71–100, 1979.

Schauberger, G., Piringer, M., and Petz, E.: Diurnal and annual variation of the sensation distance of odour emitted by livestock buildings calculated by the Austrian odour dispersion model (AODM), Atmos. Environ., 34, 4839–4851, 2000.

Schauberger, G., Piringer, M., Schmitzer, R., Kamp, M., Sowa, A., Koch, R., Eckhof, W., Grimm, E., Kypke, J., and Hartung, E.: Concept to assess the human perception of odour by estimating short-time peak concentrations from one-hour mean values, Reply to a comment by Janicke et al., Atmos. Environ., 54, 624–628, 2012.

25

Quality assessment of heterogeneous surface radiation network data

R. Becker and K. Behrens

Deutscher Wetterdienst, Meteorologisches Observatorium Lindenberg/Mark, Am Observatorium 12, 15848 Tauche, Germany

Correspondence to: R. Becker (ralf.becker@dwd.de)

Abstract. The DWD national radiation measurement network comprises 82 automatic sites, 29 manned sites with shaded and unshaded pyranometer and the BSRN station at Lindenberg. The quality assessment routinely applied takes into account the basic astronomical and empirical considerations as well as some interdependencies like total to diffuse flux relation and cross checking with sunshine duration.

A more advanced quality assessment approach attempts to routinely utilise timeseries of clear sky radiative transfer simulations for every site. For that purpose a link to cloud coverage obtained from Meteosat second generation geostationary satellite data, highly resolved in time and space, was established. There is a predefined calibration cycle of 30 month for automatic stations. Data analysis on this timescale allows for the detection of sensor degradation, wrong calibration or configuration and other possible local disturbances. Furthermore using satellite cloud mask enables the identification of larger clear sky regions characterized by similar atmospheric conditions. Thus, in a regionalization step correction or recalibration of moderate quality data to a higher level can be considered.

The paper provides an overview of DWD surface radiation network and the current activities to improve automatic quality assessment using remotely sensed data and clear sky modeling for the upgrading of radiation data.

1 Introduction

Irradiance is a key parameter in Earth's weather and climate system. Accurate observations of the components of the near surface radiation budget are therefore essential IPCC (2007). For the achievement of a global coverage the utilisation of satellite-based retrievals have become an important part. Such retrievals are developed and data are provided in the framework of EUMETSAT's Satellite Application Facilities like SAF on Climate Monitoring (Mueller et al., 2009) and LandSAF (Trigo et al., 2011). Even though not needed as a direct input but indispensable for testing, validation and adaptation of satellite-based algorithms are accurate ground-based observations. High quality ground-based measurements are performed in the context of the baseline surface radiation network (BSRN, Ohmura et al., 1998). The Deutscher Wetterdienst (DWD) is operating the national radiation network which is a heterogeneous collection of sites measuring downward shortwave radiation. The network is hosted by Meteorological Observatory Lindenberg.

This paper deals with the progress made in quality assessment of the radiation network data due to the following activities:

- by using satellite data input of cloud coverage

- by comparing measured quantities to simulated ones in clear-sky conditions

2 The ground-based network

The basic idea of the radiation network is to achieve a good coverage of whole Germany by means of radiation measurements, i.e. to avoid gaps. It comprises manned as well as fully automatic stations. Manned stations (in 2011: 29) are

Figure 1. Shaded pyranometer (left picture) and Scanning Pyrheliometer/Pyranometer (right) on the radiation measurement platform at Lindenberg observatory.

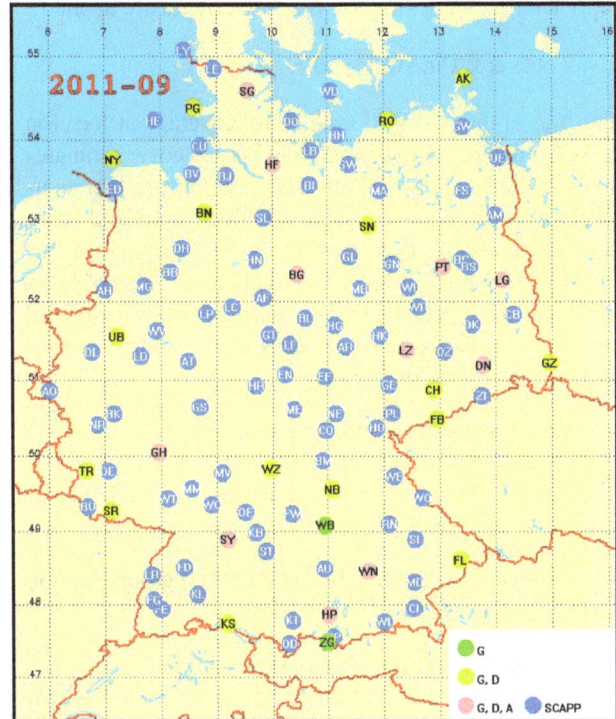

Figure 2. Ground based network for irradiance measurements, location and characterisation of sites. Green circles indicate shortwave total (G) and diffuse (D) component measured by pyranometers, pink indicates the same plus atmospheric longwave radiation (A). Blue: sites equipped with SCAPP.

equipped with two Kipp & Zonen CM11 or CM21 pyranometer (Kipp & Zonen, 2004) to measure diffuse (shaded) and total (unshaded) downward shortwave radiation (0.3–2.8 µm). Instrumentation is categorised as *secondary standard* according to ISO 9060 or *high quality* with respect to WMO classification. CM 21 response time is 5 s for 95 % of responses, less than 15 s for CM 11, respectively. 11 sites are equipped additionally with Kipp & Zonen CG4 pyrgeometer to cover the spectral range from 4.5 to 45 µm (Kipp & Zonen, 2001). SCAPP (scanning pyrheliometer and pyranometer) is a low-cost instrument to measure diffuse and direct irradiances developed by DWD and manufactured by Siggelkow Germany (Fig. 1). It was designed for automatic mode and operates without outer sun shading and outer moving parts. The spectral coverage is limited to 0.3 to 1.1 µm due to the use of a silicon detector. For daily sums of downward shortwave radiation an accuracy of 10 % is expected (20 % for hourly sums, Bergholter and Dehne, 1994). Figure 2 provides an overview of the DWD ground-based radiation measurement network.

The calibration of the field/network pyranometers is performed at the RRC/NRC at Meterological Observatory Lindenberg (MOL) indoor referring to ISO 9847:1992 "Solar energy – Calibration of field pyranometers by comparison to a reference pyranometer". SCAPPs are calibrated at MOL locally too, but outdoor only between the Northern Hemisphere spring and autumn equinoxes using direct and diffuse irradiance measured by the station pyrheliometer and shaded pyranometer. All calibrations are tracable to **W**orld **R**adiation **R**eference (WRR).

Indoor calibration capability – to get independent from the season – is under development. A more detailed analysis of SCAPP functionality is given in Behrens and Grewe (2005).

3   Scheme for automatic quality assessment

The core conventional data analysis currently operated refers to the proposal given by Long and Dutton (2002) with regard to BSRN (Ohmura et al., 1998). This will be complemented by a procedure checking overall data status (sensor degra-

dation, systematic disturbances like restrictions of full sky view) described in the following sections. It makes use of an independant approach for detection of clear-sky episodes at measurement site operating daily (Sect. 3.1). For the sites labelled as clear simulation of expected downward shortwave radiation will be performed using a simplified radiative transfer model setup (Sect. 3.2). First results using BIAS-corrected reference values are given in Sect. 4.

3.1   Utilisation of satellite data

Current geostationary satellite imagery is provided with temporal resolution of 15 min and a footprint size of 3 km at nadir view (Schmetz et al., 2002). EUMETSAT's Satellite Application Facility on Nowcasting and Very Short Range Forecasting is developing algorithms to routinely process data of MSG-SEVIRI. Retrieval of cloud mask and cloud type is based on pixel-by-pixel analysis on original satellite projection by means of multispectral thresholding method (Derrien and LeGleau, 2005). Static and dynamic thresholds are precomputed from geographical, climatological and forecasted data.

MSG cloud mask and cloud type in a pixel window centered at ground site are used to identify insolation episodes at noon. Then following simple extraction criteria are applied:

$$N_{ave} < 2 \text{ octa } \&\& \text{ DatAvail} \geq 66\%$$

where N_{ave} is the averaged total cloud coverage at location and DatAvail is the ratio of valid MSG cloud cover estimates in the predefined time window. This is to ensure that averaged cloud cover represents the whole time frame and is not biased due to greater lack of data. Figure 3 displays examplary output of total cloud cover calculation. Referring to satellite data to characterise the cloud coverage at site instead of taking the radiation measurements itself yields some advantages:

- conventional methods like evaluation of the ratio of diffuse and total irradiation might fail if at least one parameter is disturbed, potential detection of such misbehaviour could be prevented

- a homogeneous method is applied for all ground station independant of its instrumentation

- evaluation of satellite pixels enables a full sky view, i.e. all sky segments as seen by upward-looking instrument are checked for cloud contamination

Some potential drawbacks need to be mentioned here, too. The footprint (sampling distance) of SEVIRI radiometer is 3 km at nadir view and resolution gets coarser towards north and south. It might effect in an underestimation of small cumulus clouds (*Cumulus humilis*). To exclude such cases from clear-sky evaluation the variability of the ground-based signal is checked additionally. Secondly, very thin cirrus from time to time remains undetected, but no strong affection is expected here.

3.2   Simulation vs. measurement

Clouds strongly affect the radiation field and therefore influence the measurements of shortwave total and diffuse irradiance. Simulations of arbitrary meteorological scenes requires detailed knowledge of microphysical cloud properties that can hardly be provided in appropriate quality for the whole region of interest. Best matches between calculated and measured quantities can be expected for clear-sky conditions when direct component of solar radiation dominates the total (global) radiation. The focus was set on the insolation time frame ranging from 10:00 to 13:00 UTC. Calculation of incoming shortwave radiation is done using radiation transfer code Streamer (Key, 2001). The setup for the model only roughly distinguishes between winter and summer mid-latitude atmosphere. It runs with climatological ozone and background tropospheric aerosol.

For the comparison study data of pyranometer sites of the period May 2010 to August 2011 were selected, resulting in a total number of 560 cases. Time series of calculated total radiation in general shows a good correspondence to measured

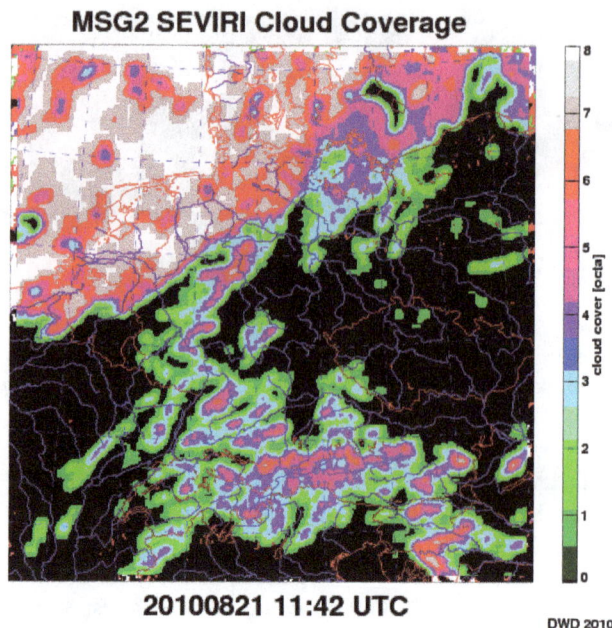

Figure 3. Total cloud coverage from MSG based on pixel-by-pixel analysis using NWCSAF algorithm, mapped onto COSMO-DE model domain and calculated for a 9×9 pixel matrix.

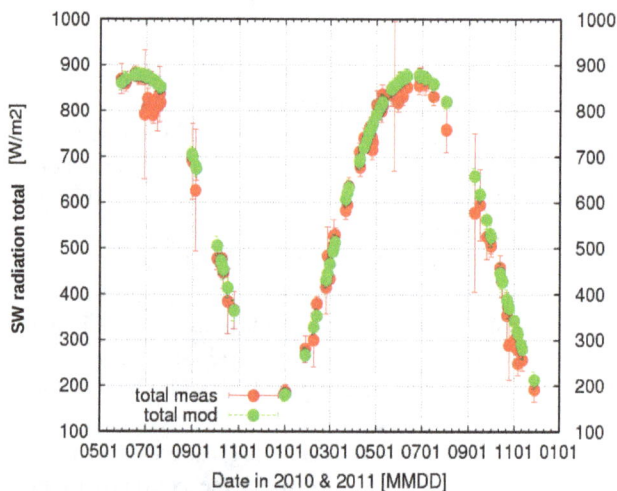

Figure 4. Timeserie of calculated and measured shortwave downward radiation for the pyranometer site Arkona, located at the coast of the Baltic Sea. Bad weather periods cause data gaps, like in August and November/December 2010 and August 2011 as well.

quantities (Fig. 4). Periods of enduring bad weather conditions cause a further reduction of evaluable data. The distribution of differences simulated vs. measured is displayed in the left of Fig. 5. The model tends to give higher estimates even though some negative deviations are present, too. Same is indicated by the relative bias of 1.02.

Figure 6 provides a more station specific view on the relation of calculations and measurements for pyranometer sites.

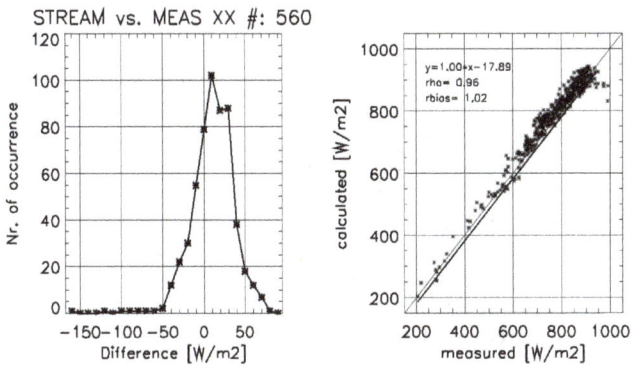

Figure 5. Systematic deviations of calculated and measured irradiances averaged from 10:00–13:00 UTC for pyranometer sites. A distinct peak can be found at positive differences (measured subtracted from modeled).

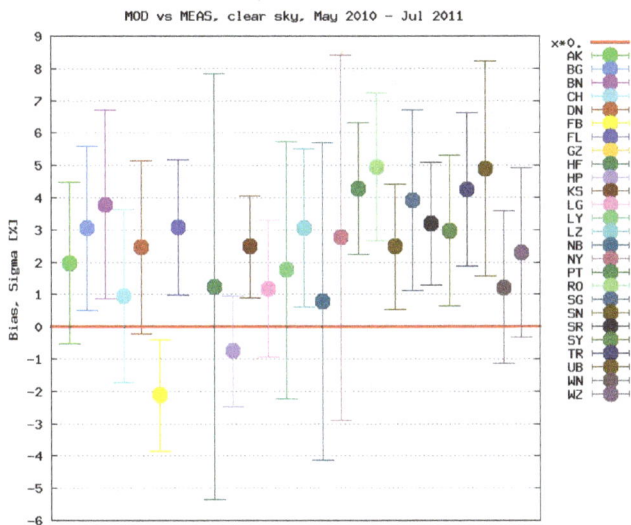

Figure 6. Systematic discrepancy and standard deviation of model calculations of downward total shortwave radiation from measured quantities, given for all pyranometer sites, expressed in percent.

Almost all stations show an overestimation by the model, except two mountaineous sites located on top of low mountain summits (Fichtelberg 1214 m a.s.l., Hohenpeissenberg 988 m). In other words, lowland stations are receiving less radiation than calculated, however mountaineous get more. It is explained by the fact that calculations were performed assuming a fixed height above sea level (120 m a.s.l.). This finding means that the model setup will need further adjustment for actual station elevation.

A summary of the results is provided in Table 1. The model generally slightly overestimates, but filtering for instrument category draws a different picture for pyranometer and SCAPP: the systematic deviation is almost doubled for the latter. This is possibly due to less long-term stability of the sensors, but needs to be further investigated with regard to calibration procedure too. However, the higher variability

Table 1. Statistics of comparison model-measurement, instrument classes treated separately. The model overestimates for both categories, but difference is larger for SCAPP.

	Pyranometer	SCAPP
Bias	+2.40 %	+4.29 %
Sigma	3.36 %	4.61 %

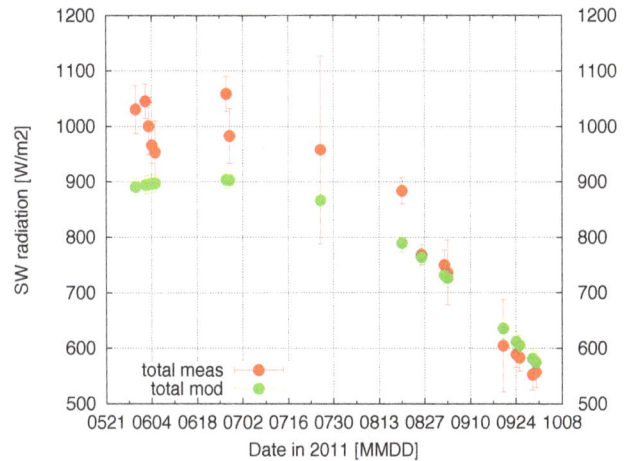

Figure 7. Shortwave radiation (total) measured at site Magdeburg/Saxony-Anhalt with SCAPP (red) and calculated (green). Note that identification of strong overestimation in July possibly due to wrong station setup caused a release of instrument substitution end of August.

meets our expectations and is inline with instrument specifications (Bergholter and Dehne, 1994).

4 First applications

Monitoring the differences in the timeseries of calculated and measured shortwave radiances reveales possible errors in station configuration, wrong calibration or sensor degradation. For SCAPP a calibration cycle of 30 month was predefined according to manufactorers advices. Several years of operational service have shown that in practice the cycle must be set shorter. The comparison model-measurement helps to identify the right time for instrument substitution (see Fig. 7).

Instrument status is a valuable information for operators and basic level quality assessment, too. Figure 8 displays an example for a graphical realisation of instrument status for site Magdeburg (Fig. 8).

5 Conclusions and outlook

The study presented here investigates relations between measured downward shortwave radiation and calculated values in clear sky conditions to improve automatic data quality checks.

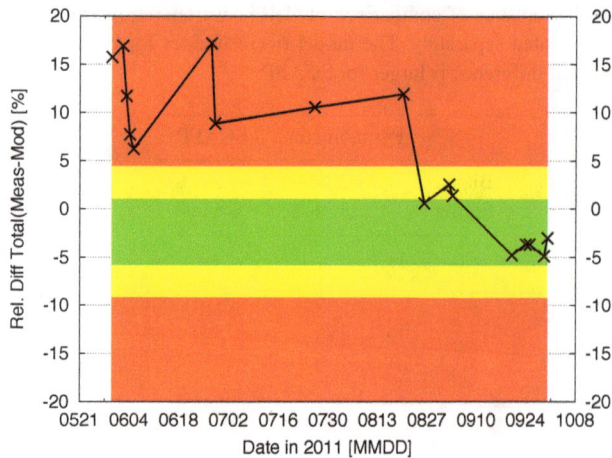

Figure 8. Relative Difference of measured and calculated short-wave radiation at Magdeburg/Saxony-Anhalt. Green area indicates $\pm 1\sigma$, yellow $\pm 2\sigma$.

Based on a dataset of real clear-sky conditions identified using Meteosat/SEVIRI cloud coverage distributed over more than one year dedicated relations to simulations according to instrument group (pyranometer, SCAPP) could be found. These slight but systematic differences can be considered and used to assess the quality of realtime data. Currently a simple setup for the model is used. It can be expected that feeding of more realistic atmospheric profiles achieved by providing radiosoundings, filter radiometry (if available) or forecasted/analysed profiles of numerical weather prediction model will lead to more refined and precised results. The future consideration of the site elevation will surely cause a positive bias at the elevated stations (Fichtelberg, Hohenpeissenberg), too. As a consequence the spread would decrease.

The approach presented here will be checked for applicability for overcast sky conditions.

Acknowledgements. Thanks to Michael Sommer (DWD-MOL) for fruitful discussions concerning the use of Streamer.

Edited by: I. Auer
Reviewed by: two anonymous referees

References

Behrens, K. and Grewe, R.: A comparison of SCAPP radiation data with global, diffuse and direct solar radiation, in: Papers Presented at the WMO Technical Conference on Meteorological and Environmental Instruments and Methods of Observation (TECO-2005), Vol. 1265, Bucharest, Romania, 2005.

Bergholter, U. and Dehne, K.: SCAPP – a compact scanning pyrheliometer/pyranometer system for direct, diffuse and global solar radiation, Tech. Rep. 57, WMO-IOM, 1994.

Derrien, M. and LeGleau, H.: MSG/SEVIRI cloud mask and type from SAFNWC, Int. J. Remote Sens., 26, 4707–4732, 2005.

IPCC: Climate Change 2007: The Physical Science Basis. Contribution of Working Group I to the Fourth Assessment Report of the Intergovernmental Panel on Climate Change, edited by: Solomon, S., Qin, D., Manning, M., Chen, Z., Marquis, M., Averyt, K. B., Tignor, M., and Miller, H. L., Tech. rep., Cambridge Univ. Press, UK, 2007.

Key, J.: Streamer User's Guide, Tech. rep., Cooperative Institute for Meteorological Satellite Studies, 96 pp., 2001.

Kipp & Zonen: Instruction Manual CG4 version 0304, Download center, discontinued products, www.kippzonen.com, 2001.

Kipp & Zonen: Instruction Manual CM21 version 1004, Download center, discontinued products, www.kippzonen.com, 2004.

Long, C. and Dutton, E.: BSRN Global Network recommended QC tests, V2.0, Tech. rep., available as PDF at: http://www.bsrn.awi.de, 2002.

Mueller, R., Matsoukas, C., Gratzki, A., Behr, H., and Hollmann, R.: The CM-SAF operational scheme for the satellite based retrieval of solar surface irradiance – a LUT based eigenvector approach, Remote Sens. Environ., 113, 1012–1024, 2009.

Ohmura, A., Dutton, E., Forgan, B., Froehlich, C., Gilgen, H., Hegner, H., Heimo, A., Koenig-Langlo, G., McArthur, B., Mueller, G., Philipona, R., Pinker, R., Whitlock, C., Dehne, K., and Wild, M.: Baseline surface radiation network (BSRN/WCRP): new precision radiometry for climate research, B. Am. Meteorol. Soc., 79, 2115–2136, 1998.

Schmetz, J., Pili, P., Tjemkes, S., Just, D., Kerkmann, J., Rota, S., and Ratier, A.: An introduction to Meteosat Second Generation (MSG), B. Am. Meteorol. Soc., 83, 977–992, 2002.

Trigo, I. F., DaCamara, C., Viterbo, P., Roujean, J., Olesen, F., Barroso, C., Camacho-de Coca, F., Carrer, D., Freitas, S., García-Haro, J., Geiger, B., Gellens-Meulenberghs, F., Ghilain, N., Melia, J., Pessanha, L., Siljamo, N., and Arboleda, A.: The Satellite Application Facility on Land Surface Analysis, Int. J. Remote Sens., 32, 2725–2744, 2011.

26

Development of adaptive IWRM options for climate change mitigation and adaptation

W.-A. Flügel

Institute for Geography, University of Jena, Germany

Abstract. Adaptive Integrated Water Resources Management (IWRM) options related to the impacts of climate change in the twinning basins of the Upper Danube River Basin (UDRB) and the Upper Brahmaputra River Basin (UBRB) are developed based on the results obtained in the different work packages of the BRAHMATWINN project. They have been described and discussed in Chapter 2 till Chapter 9 and the paper is referring to and is integrating these findings with respect to their application and interpretation for the development of adaptive IWRM options addressing impacts of climate change in river basins. The data and information related to the results discussed in Chapter 2 till 8 have been input to the RBIS as a central component of the IWRMS (Chapter 9). Meanwhile the UDRB has been analysed with respect to IWRM and climate change impacts by various projects, i.e. the GLOWA-Danube BMBF funded project (GLOWA Danube, 2009; Mauser and Ludwig, 2002) the UBRB has not been studied so far in a similar way as it was done in the BRAHMATWINN project. Therefore the IWRM option development is focussing on the UBRB but the methodology presented can be applied for the UDRB and other river basins as well. Data presented and analysed in this chapter have been elaborated by the BRAHMATWINN project partners and are published in the project deliverable reports available from the project homepage http://www.brahmatwinn.uni-jena.de/index.php?id=5311&L=2.

1 Introduction and objectives

The development of IWRM adaptation options to account for impacts from climate change must apply a holistic system's approach (Flügel, 2009) comprising a thorough hydrological system analysis (Flügel, 2000). Both require a methodology that integrates tools provided by Geoinformatics (Flügel, 2010) and a central data and information platform. The latter was established for the BARAHMATWINN project by implementing the River Basin Information System (RBIS) developed by the FSU-Jena (Flügel, 2007; Kralisch et al., 2009). The IWRMS (Chapter 9) provides the means to analyse the deliverables of the BRAHMATWINN project, like the measured data time series, the results of the climate modelling exercises (Dobler and Ahrens, 2008, 2010), modelled data time series as output from the DANUBIA modelling system (Mauser and Bach, 2009), socio-economic vulnerability studies (Kienberger et al., 2009), and the analysis done by means of the NetSyMod – mulino decision support system (mDss) (Giupponi et al., 2008).

The *overall objective* of the IWRM adaptation options development was to demonstrate the knowledge based potential of such an integrated system analysis using the results obtained from interdisciplinary research cooperation between natural, socio-economic and engineering sciences. This will be demonstrated by examples derived from the

- climate modelling studies presented in Chapter 2,

- natural and socio-economic system assessment done in Chapter 3 and 4,

- IWRM assessment given in Chapter 5,

- DANUBIA hydrological modelling results given in Chapter 7,

- socio-economic vulnerability analysis discussed in Chapter 6, and the

- development of stakeholder approved "what-if?" scenarios in Chapter 8.

The ILWRMS presented in Chapter 9 has been used to carry out the analysis based on the data and information provided by the numerous BRAHMATWINN deliverables available from the RBIS database.

2 Role within the integrated project

The development of alternative IWRM options to adapt to impacts of climate change is the integral component of the BRAHMATWINN project. It relies on the IWRMS development that provides the software techniques and methodologies for such an integrated analysis and the results obtained from the natural and socio-economic system assessment complemented by the scenario based modelling studies. They are presented and discussed in Chapter 2 till Chapter 9.

3 Scientific methods applied

Neither does the design of adaptive IWRM options appear from the holistic systems approach applied in the BRAHMATWINN project per se nor is there a "perfect" strategy to be proposed as the optimum solution for IWRM options adapting this process to climate change impacts. Instead professional expert knowledge is required to properly define IWRM challenges related to climate change and identify the appropriate project results to be applied and analyse them in an integrated way. The IWRMS in this regard provides the technical means from Geoinformatics but needs the expert knowledge and professional expertise that is required for the integrated analysis.

In result such an exercise will deliver a set of alternative IWRM options that have to be discussed with stakeholders and decision makers to find a solution to the defined IWRM challenge. The selected option should receive the broadest level of acceptance by the stakeholder communities that have to implement and support the respective IWRM strategy.

4 Results achieved and IWRM options proposed

4.1 IWRM options due to climate change

The climate modelling results and their discussion and analysis presented in Chapter 2 of this publication for the UDRB and UBRB shows that both twinning basins have a temperate climate. In the UBRB it is dominated by the monsoon system which supplies the region with up to 80% of the annual total rainfall meanwhile the UDRB has a rainfall pattern distributed over all months of the year. Both basins have in common to receive winter precipitation as snow that is melting in the following spring time and summer season.

4.1.1 Model projections

The modelled projected temperature confirm average temperature increase up to 5 °C in 2100 in the UBRB with the higher values in the region of the Tibetan Plateau and up to 4 °C in the UDRB. Thus, processes that directly dependent on temperature, like potential evapotranspiration melting of snow and glacier ice will show similar trends with consequent impacts on the hydrology of the river basins. The annual average precipitation in the UBRB is without a significant trend because an increasing trend in the summer is compensated by opposite trends in others seasons. The results of the modelled rainfall projections for the IPCC scenarios A1B and B2 are aggregated in Table 4 of Chapter 2 and reveal:

1. Precipitation trends in both basins are negative with a higher trend in the UDRB than in the UBRB.

2. Different climate change indicators, like the length of the longest dry periods, indicate more frequent and prolonged droughts.

3. The projected increasing amount of (1-day and 5-day) spring precipitation in the UDRB in combination with increased spring snow melt due to higher temperatures in the Alps is likely to increase the magnitude and frequency of floods.

4. In Assam the positive trend in the number of consecutive dry days in the monsoon season indicates longer monsoon breaks in the forthcoming decades.

5. An increase in the number of consecutive dry days and in the maximum 5-day precipitation amount in the region of the Tibetan Plateau for the monsoon season. The complement temperature impacts of trends indicate the UBRB as a highly sensitive region to future climate changes.

4.1.2 Analysis of measured climate

An intensive climate data collection was done during the course of the BRAHMATWINN project and hundreds of stations have been input into the BrahmaRBIS and DanubeRBIS respectively. For the UBRB the quality assessment revealed that most of these stations unfortunately have significant data gaps and therefore cannot be used for a comparative assessment. After a thorough screening the remaining stations presented the regional climate trends for the UBRB in Fig. 1. They can be described as follows:

1. Throughout the UBRB, i.e. from the semi-arid western till the monsoon driven eastern part of the basin there is a positive trend projected for air temperature.

2. The precipitation trend is negative in the arid western and Tibetan part of the UBRB and turns into a positive trend in the eastern part of the UBRB, i.e. in Assam and the windward located slopes of the Himalaya mountain ridge.

As showing in Figs. 2 and 3 for the station Dibrugarh (India/Assam) the increase of temperature and precipitation has also been measured in Assam. Both trends are significant on the annual and monthly scale, as the monthly averages after 1990 show considerably higher values as those from the previous 15 year time span.

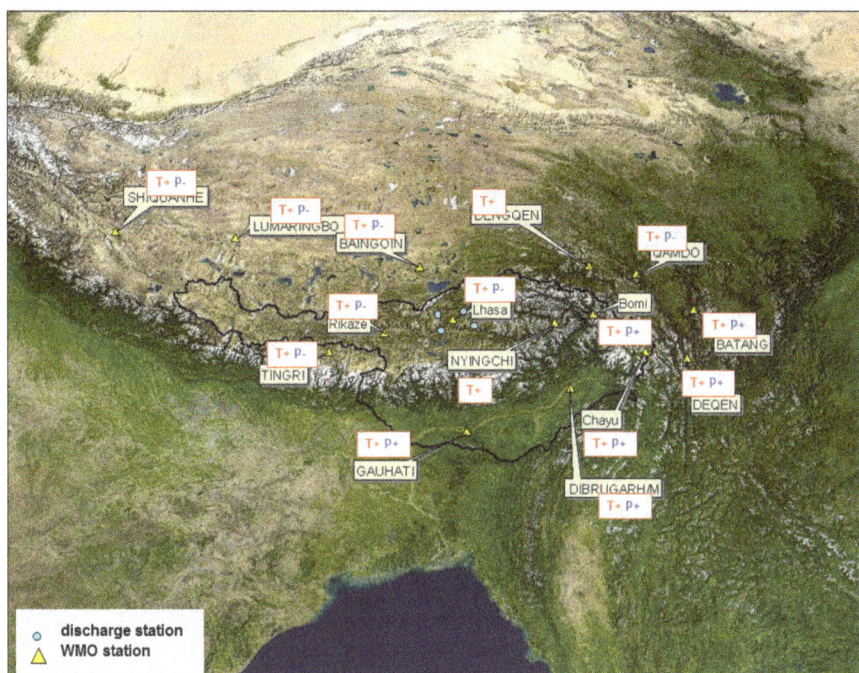

Figure 1. Trends of temperature and precipitation measured at stations in the UBRB.

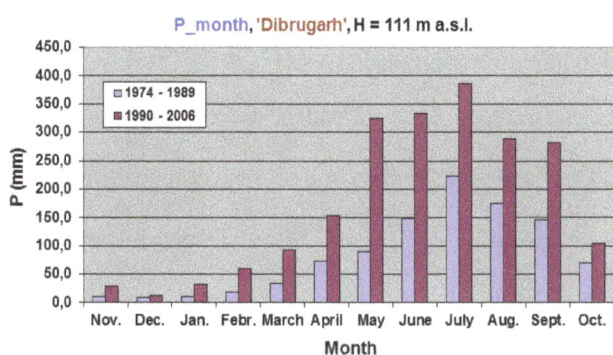

Figure 2. Positive trend of annual precipitation at Dibrugarh, Assam, NE-India.

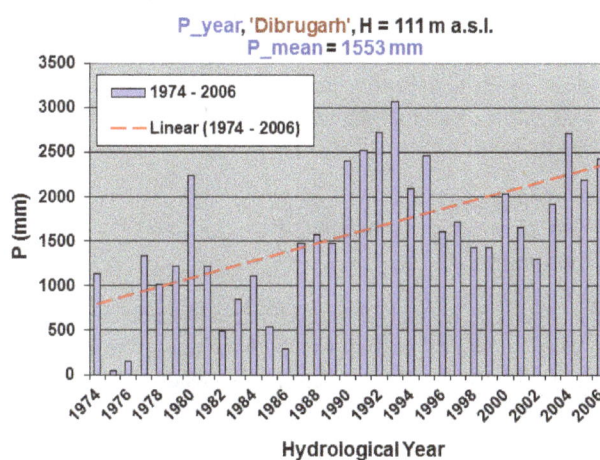

Figure 3. Positive trend of monthly mean precipitation at Dibrugarh, Assam, NE-India.

4.1.3 Consequences for adaptive IWRM options

Although the climate projections are still coarse in resolution and are not directly of use for IWRM planning they can be analysed together with the measured climate trends with respect to IWRM as follows:

1. There is a clear indication that climate warming is most likely to continue further and evapotranspiration will increase resulting *firstly* in increasing melt of glaciers, permanent snow fields and permafrost, and *secondly* in a reduction of runoff generation in the hot summer months.

2. Precipitation has a declining trend in the north-western part of the UBRB and complemented by the increasing temperatures the presently semi-arid climate of this region will most likely get drier and more arid. As a result the annual runoff yield will decrease as well so that less water can be expected for distribution to irrigation schemes and for hydropower generation.

3. In the monsoon driven North Eastern Region (NER) of India, i.e. in Assam the climate modelling results indicate longer dry spells during the monsoon period.

Table 1. Modelled scenario trends till 2080 for precipitation in the UDRB and UBRB (adapted from Dl_2, 2010).

Acronym	Description	UDRB		UBRB	
		A1B	B2	A1B	B2
PFRE	Number of precipitation days	−27	−19	−19	−11
PRECIP	Mean annual precipitation	−22	−16	−11	−6
PX5D	Max. 5-day precipitation period	0	−5	6	6
PCDD	Longest period of consecutive dry days	34	22	24	16

Table 2. Glacier reduction in the UBRB and UDRB between 1970 and 2000 (adapted from Dl_3, 2010).

Area of glacier cover (AGC)	River Basin (RB)			
	Lhasa River	Wang Chu	UBRB	Salzach
1970 (km^2)	535	60	17580	95
2000 (km^2)	429	50	14400	79
ΔAGC (km^2)	−106	−10	−3180	−16
ΔAGC (%)	−19.8	−16.7	−18.1	−16.8

4. Higher precipitation is projected for the NER during the monsoon period and this is supported by station time series from this region. They can exaggerate the already threatening flood inundations and consequent bank erosion. Both will further put valuable farm land at risk and call for the implementation of effective river training measures.

4.2   IWRM options due to melting glaciers and permafrost

Melting and retreating glaciers are obviously associated with the positive temperature trend in both twinning basins and this has been validated from change analysis done by means of satellite images between 1980 and today (Chapter 3).

The results of such an change detection analysis are listed in Table 2 and reveal that glacier melt and their retreat is almost of the same magnitude in the alpine mountains of the UDRB and the UBRB respectively ranging between 16.7% and 19.8%. The following interpretation can be derived from the results given in Table 2:

1. The glacier area in the Lhasa river catchment is about 8 times larger than in the Wang Chu catchment. The glacier area change, however, is similar in both catchments with a decline of about −7% per decade. The significant debris cover of the glacier tongues in the Wang Chu is buffering this process that reduces glacier mass loss and retreat.

2. The glacier area loss in the Salzach catchment during ~1970 to ~2000 was similar, and altogether glacier loss

in the four catchments between the 1970s and 2000 ranged between 16.7% and 19.6%.

Consequences for adaptive IWRM options

The consequences for IWRM can be described as follows:

1. Discharge from glaciers and permanent snow fields will increase due to progressive melting of these permanent storages till a new balance between precipitation input and consequent melting is reached.

2. Melt water from glacier ice and snow is stored in glacier lakes in front of ice core moraines and significantly enhances the threat of glacier lake outburst floods (GLOFs) which frequently occurred in the past in both twinning basins (Subba, 2001).

3. After a new ice and snow balance has been established their contribution to river runoff most probably will be considerably smaller than in the 20th century.

4. Especially during the summer time consequent water shortages for irrigation agriculture are likely and farmers either need to adapt in their cropping pattern or additional infrastructures must be built to provide the storage capacity to sustain irrigation during that period.

5. Melting permafrost is reducing the slope stability as the ice that is cementing the weathered debris is disappearing. Consequent landslides, mudslides, and rock falls can dam rivers and add further sediment to the rivers threatening irrigation infrastructures and reservoirs.

Table 3. Vulnerability of wetlands to socio-economic development (SED) and climate change (CC) in river basins in Assam, Bhutan and India (adapted from Dl_3, 2010).

River Basin	Vulnerability to	alluvial	floodplain	swamp	lake	alpine meadow	beel
Lhasa River, Tibet	SED			low		low	
	CC			high		high	
Wang Chu, Bhutan	SED	low	high	low	medium		high
	CC			high			
Assam, North-East India	SED						
	CC						

4.3 IWRM options due to the transfer of wetlands

Types and functions of wetlands have been described and classified according to their distribution and hydrological dynamics for the Lhasa River, the Wang Chu and for Assam in India in Chapter 3. Projected climate change (Chapter 2) indicates that runoff variability will increase in terms of extremens, i.e. floods and dry weather runoff. Increased sediment load is likely due to landslides slipping off from slopes that became instable due to melting permafrost (Chapter 3). These are negative impacts on wetlands which need a sufficient period of flooding of water with little sediment load for fish breeding.

In addition wetlands are under continuous pressure by getting transferred into settlements or agriculture fields to satisfy the needs to accommodate and feed the ever growing population. Although the importance of wetlands for the buffering of floods and the support of dry weather base flow as well as for the biodiversity are often appreciated by planners and water managers these destructive processes are continuing in unchanged dynamics.

Consequences for adaptive IWRM options

Draining of wetlands to generate new areas for settlements or agriculture has impacts for IWRM that cannot completely be quantified at present. In cases that small wetlands are destroyed the impact on biodiversity outweighs the impact on the hydrological dynamics. This might be the case in the Wang Chu in Bhutan and to a certain degree also for the flood plains in the western part of the Yarlung Tsangpo.

In the flood plains of Assam, however, adaptive IWRM initiatives are required to ensure that their hydrological as well as their environmental functions and services can be sustained. Otherwise it is likely that essential ecosystem functions (ESF), i.e. flood retention or biodiversity regeneration will be impacted. Furthermore vital livelihood capacities for the local population that depend on respective ecosystem services (ESS), i.e. food supply from fishery and the purification of flood water for rural water supply will become at risk.

The risk matrix for the wetlands classified in the BRAHMATWINN project is listed in Table 3 derived from detailed weighting analysis done in Chapter 3 applying the Millennium Ecosystem Assessment (2005). Table 3 defines the ranked magnitudes of likely impacts that can be expected from climate change and consequent socio-economic pressure on ESS and ESF. Especially the *Beels* which are permanent flooded wetlands representing the majority of wetlands in Assam are under strong pressure and in almost all aspects will be impacted in terms of ESS, ESF, biodiversity and flood peak buffering if impacts from climate change continue and are progressively exaggerated by human activities.

4.4 IWRM options due to water balance modelling and bank erosion

The results of the water balance modelling exercises done by means of the DANUBIA model have been reported in Chapter 7. The maybe most alarming result is the projected decline of the runoff in the Brahmaputra River at Guwahati (Fig. 10 of Chapter 7), and this trend is obvious in all four IPCC scenarios applied. They are corresponding to the projected temperature increase and the respective decline of precipitation and for the snow precipitation in particular (Fig. 12 of Chapter 7). Projected lower rainfall input (Table 1) and increasing temperature (Fig. 4 of Chapter 2) result in higher evapotranspiration that in turn is reducing runoff generation and discharge height.

This process is not only impacting the already semi-arid western part of the Yarlung Tsangpo in Tibet but also the monsoon dominated runoff generation in the NER of India, i.e. in Assam. According to the model results shown in Fig. 13 of Chapter 7 the A1B and B1 scenarios agree in this trend and both areas are expected to show more severe dry periods with related impacts to agriculture, fisheries and socio-economy in general. If compared with the historical past (1971–2000) the magnitude of the projected discharge decline for the A1B scenario ranges between 15% for the period 2011–2040 and 28% for the period 2051–2080 meanwhile the B1 scenario has projected declines of 15% and 23% respectively (Table 4, line 2).

Table 4. Integrated indicators to evaluate quantifying SRES based model projections for climate change impact on sustainable IWRM (adapted from Dl_7, 2010).

Integrated Indicator		Scenarios and Projections		
Domain (subdomain)	Indicator	Base Line (1971–2000)	A1B	B1
env (climate)	air temperature (T) precipitation (P) evapotranspiration (ET)	Different regional trends depend on seasonal dynamics	$T + 2$–$6\,°C$ $P +$ in Bhutan $P +$ in Tibet $P -$ in Assam	$T + 2$–$3\,°C$ $P +$ in Bhutan $P +$ in Tibet $P -$ in Assam
env (hydrology)	surface runoff (Sr) interflow (Int) groundwater flow (Gf) snow and glacier melt (SGM)	Changes of flow volume and seasonal flow distribution	15% till 28% less mean annual discharge and changing runoff regimes	15% till 23% less mean annual discharge and changing runoff regimes
env (glaciology)	Δ Area of Glacier Cover (AGC)	ΔAGC (1970–2000) −17% till −20%	glaciers and permafrost will melt away	less pace of glacier and permafrost melt
env (hydrobiology)	wetlands with eco-system services (ESS) and ecosystem functions (ESF)	regional diversity with functioning ESS and ESF	strong pressure from GDP and population growth on ESS and ESF	less pressure from GDP and population growth on ESS and ESF
soc-econ (vulnerability)	Gross domestic product (GDP) population pressure (PP)	high vulnerability in flood prone areas with high PP	high GDP development decreases vulnerability	lower plus of GDP reduces decrease of vulnerability
soc (governance)	governance and policy	IWRM is not in place but first attempts are made on a transnational level	less good with respect to sustainability but governance levels as good as in B1	best setup for management of the whole river basin and IWRM

4.4.1 Changing runoff components

If one compares the modelled mean monthly discharge distribution shown in Fig. 15 of Chapter 7 of the historical past from 1971 to 2000 with the modelled runoff projections of the A1B scenario the following IWRM relevant information can be extracted:

1. Meanwhile the monthly runoff is almost normal distributed in the historical past it becomes skewed in the projected period 2011–2040 but still reaches similar peak discharge in July.

2. The rising limb of the monthly hydrograph for the A1B scenario in the projected period 2011–2040 has significant lower discharges if compared to the historical past, which indicates less precipitation input and a smaller contribution of snow and glacier melt from reduced snow and glacier storages in the alpine mountains.

3. In the second projected period of the modelled A1B scenario the reduced rainfall and the increased temperatures are becoming fully effective and reduce the hydrograph peak but keep the shape skewed towards the end of the monsoon period.

The modelled change of snow and glacier melt runoff components have been discussed in Chapter 7 for the Lhasa River basin and are shown in Fig. 16 of Chapter 7. Significant

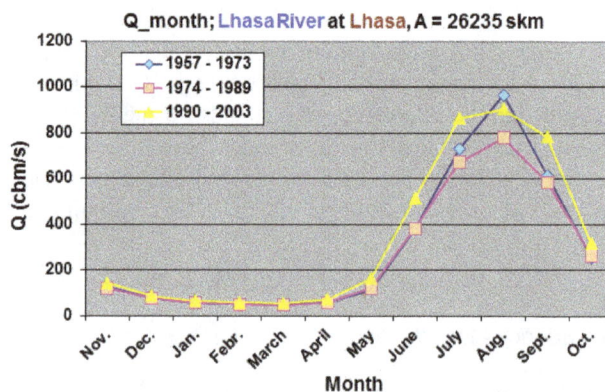

Figure 4. Mean monthly discharge of the Lhasa River at the gauging station Lhasa.

changes in the contribution of glacier melt runoff can be expected according to the modelled projections. The modelling results are supported by measured discharge values given as monthly averages from 1957 onwards for the gauging station at Lhasa in Fig. 4 and reveal:

1. For the two time periods 1957–1973 and 1974–1989 the distribution of mean monthly discharge is almost the same and only the peak discharge is different most likely due to variations in rainfall.

Figure 5. Bank erosion from slipping unstable river banks (left) and unfertile flood sand deposits destroying fertile farm land (Photo: Nayan Sharma).

2. The third time period 1990–2003 is giving a different distribution, as the hydrograph has higher values in the rising and falling limb and is of a broader shape.

This obvious change in runoff dynamics and seasonal distribution can be accounted for a higher contribution of snow and glacier melt in the past 14 years which due to climate warming is starting earlier, contributes more runoff then in previous decades and supports runoff even after the summer rainfall season.

4.4.2 Bank erosion

Bank erosion is an obvious threat in the floodplains of the Brahmaputra River and its tributaries. Especially after the last severe earth quake in 1950 this process has reached disastrous extremes eroding about $4000 \, km^2$ of rural farmland and leaving almost a million of people home- and landless. The erosion process in many cases is related to the changing hydrostatic pressure between high and low water levels during flood and low flow respectively. This change of pressure balance is destabilizing the steep sandy banks which then slip down into the river (Fig. 5, left). If floods inundate fertile farm land they deposit thick layers of unfertile sands thus making this former farm land unfertile for years (Fig. 5, right).

As shown in Fig. 6 the Brahmaputra River had broadened its river bed considerably during the last three decades and at present is having an uncontrolled dynamics with the tendency of extending towards the southern banks bordering fertile farm land regions.

4.4.3 Consequences for adaptive IWRM options

The hydrological modelling of the UBRB water balance is integrating the results obtained from Chapter 2 and 3 and confirms to a large extent the findings from the glacier and permafrost studies. They clearly show the need for additional validation and require the access to hydro-meteorological station time series, which at present are still ranked as classified and are not available to the research community. Especially the regional distribution of precipitation is crucial for the water balance and runoff modelling studies.

Declining discharge of the Brahmaputra River is most alarming and calls for further validation concentrating on the NER of India, which at present is receiving the bulk of the monsoon rainfall and generates a large portion of the Brahmaputra discharge. In this region adaptive IWRM options must be focussing on the following components which have to be combined in an integrated way:

1. A comprehensive hydrological system analysis is required to obtain enhanced understanding of the Brahmaputra River system in the NER of India complemented by more detailed hydrological modelling of the tributary runoff dynamics contributing to the discharge and sediment load along the main river stretch.

2. The hydrological system analysis and modelling exercises must be complemented by a respective computational hydrodynamic flow analysis linked with the hydrological modelling into an integrated river basin model of the Brahmaputra River in the NER of India.

3. The buffering of flood peaks and the support of base flow is essential and must include the exploration of wetland retention potential, i.e. the flooding management of *Beels* integrating aspects of biodiversity and socio-economic development.

4. Complement flood retention infrastructures should be considered and respective conceptions must integrate in a balanced way aspects of hydro-power generation, irrigation water supply, environmental flow requirements, and biodiversity. Scale and distribution of infrastructure concepts must account for the prevailing geo-tectonic activity in the NER of India.

5. Bank erosion must be addresses and to a large extend can be prevented by means of properly designed and

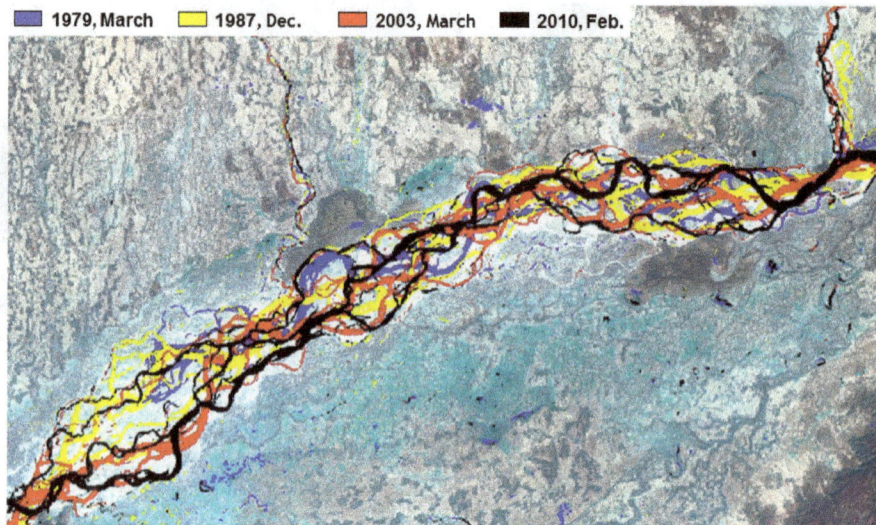

Figure 6. Braided flow paths of the Brahmaputra River system between 1979 and 2010 (analysis and graphic provided by Bettina Böhm).

effective river training measures. The installation of these measures has to apply the findings from the thorough hydrological system analysis and complement hydrodynamic flow studies.

6. River training measures should be based on expertise obtained from Indian rivers, i.e. the Ganga River. They must be designed in such a way that they are easy to build, make use of locally available resources, i.e. bamboo, and create jobs for the rural poor when being implemented and maintained.

4.5 Vulnerability and governance analysis

Vulnerability against floods was analysed in Chapter 4 and a comprehensive governance analysis was discussed in Chapter 8.

4.5.1 Vulnerability analysis

The vulnerability against floods was established for the UDRB and UBRB using the test regions of the Brahmaputra River floodplain in Assam and the Salzach River in Austria for the UBRB and the UDRB respectively. The results are discussed in Chapter 4 and here Fig. 6 reveals:

1. High vulnerabilities are strongly related to population centres like in the west of Assam at the city of Guwahati in Assam and of the city Salzburg in Austria.

2. Higher vulnerabilities can be found along the main stem of the Brahmaputra River and its tributaries because of bank erosion and flooding of the *Beels* which are used for complement food supply and grazing.

Vulnerability projection modelling for the SRES (A1, B1) was done for both case regions in the UDRB and UBRB and

discussed in Chapter 8. They show that GDP and population growth impacts both household and community factors that control socio-economic vulnerability to climate hazards. Factors that are relevant in this regard are the proportions of the population working in agriculture, of households with a television, houses with burnt brick walls and of households using firewood for cooking.

4.5.2 Governance analysis

Each of the storylines for the SRES scenarios A1B and B1 makes certain assumptions about the balance of the socio-economic drivers in place in 2050 but fails to make comments regarding the governance regimes expected to support the scenarios. Governance scenarios might be difficult to project but there is no doubt that the extent to which IWRM options adapting to impacts of climate change will be effective or not depends on the governance and policy positions in place (Ministerial Declaration, 2000). It is therefore necessary to assess the characteristics and suitability of the governance system that would be needed to support adaptive IWRM options as proposed response options.

4.5.3 Consequences for adaptive IWRM options

Adaptive IWRM options must account for socio-economic developments and constraints as it is the human dimension in which they will be implemented. Moreover their successful implementation and effectiveness strongly depends on the appreciation and willingness of people to accept them as a valuable support for socio-economic development and sustainable use of water resources that both account for environmental preservation. Accounting for these findings should be done as follows:

1. IWRM as a process must support the growth of GDP which contributes to reduce the flood vulnerability, i.e. by a better maintenance and management of protection measures and management operations.

2. Substantial growth in GDP is supporting the slowing down of the present growth in population, thus must be a complement objective of IWRM options in reducing levels of vulnerability.

3. It is necessary to assess the suitability of adaptive IWRM options with respect to their functional potential within the governance system to support its implementation.The characteristics of the governance system that would be needed to support the storylines must be determined and addressed by the IWRM strategy design.

5 Contributions to sustainable IWRM

The analysis given in this chapter is interpreting the findings from the research studies presented in Chapter 2 till Chapter 9 by an integrated approach towards sustainable IWRM. They have been quantified by numerous indicators out of which by means of an expert assessment (Chapter 6) those have been selected that quantify climate change impacts for the "what-if?" scenarios developed in Chapter 8 for the A1B and B1 scenarios. They have been named "integrated indicators" as they comprise meanings for interdisciplinary interpretation of the natural environment and its socio-economic development. Furthermore they are considered relevant for vulnerability and governance analysis and the development of adaptive IWRM options respectively. They have been grouped in three domains of sustainability: environmental (env), socio-economic (soc-econ) and social (soc) indicators and are listed together in Table 4.

Required data for the indicator calculation have been collected and analysed in the work packages (WP) WP2 till WP5 and WP7 of the BRAHMATWINN project and respective information is available for the A1B, B1 and B2 scenarios from the IWRMS BrahmaRBIS and DanubeRBIS (http://www.brahmatwinn.uni-jena.de/) both implemented for the twinning basins in WP9. Data and information also originates from the modelling exercises in WP2 and WP7 as well as from the stakeholder workshops and expert assessments done in WP4 till WP6 and WP8.

6 Conclusions and recommendations

Climate change is a complex problem and the assessment of respective impacts requires a holistic system's approach comprising the interdisciplinary cooperation and integration of natural, socio-economic and engineering sciences. Summarizing the discussions presented herein the following conclusions are presented:

1. When applying the holistic system's approach it is necessary to address the problem of scales. Data and information provided relate to different scales and it is necessary to appreciate the potential and limitations of downscaling procedures.

2. Modelling is an essential methodology and used in almost all disciplines that have been involved in the BRAHMATWINN project. However, the types of data which are produced by the natural and socio-economic models are not always compatible and need expert interpretation.

3. Applied Geoinformatics via the RBIS has provided data and information to all disciplines and research teams involved, and is implementing the knowledge obtained in Decision Information Support Tools (DIST) like the Integrated Water Resources Management System (IWRMS).

4. "What-if?" scenario based model projections are essential tools for IWRM analysis and respective options development. The storylines of the "what-if?" scenarios must include the human dimension based on stakeholder integration.

5. Socio-economic and governance analysis are as important as the natural and engineering studies as they provide the information about the human dimension to implement IWRM options successfully within a given governance system.

From these experiences and the constraints met during the BRAHMATWINN project the following recommendations can be formulated:

1. There is still a lack of hydro-meteorological information in the UBRB. This situation could be significantly improved if measured hydro-meteorological data time series are not any longer classified but made available to research projects like BRAHMATWINN.

2. Downscaling of climate data that are relevant for hydrological modelling studies is an essential part of a climate impact assessment study and needs further research. Software packages that have been developed for this purpose should be well documented with respect to their input data demands and constraints. They should be available to the scientific community for testing and evaluation. Impact analysis of climate change on river basin water resources must be based on a holistic approach comprising land and water resources management to develop adaptive ILWRM strategies applying innovative software toolsets like the Integrated Land Management System (ILMS) described by Flügel (2010).

Acknowledgements. The author greatly appreciates and acknowledges the support from all BRAHMATWINN partners and stakeholders who contributed to the subjects presented herein during the project workshops. By means of the BRAHMATWINN deliverable reports Dl_2 till Dl_10 (http://www.brahmatwinn.uni-jena.de/5311.0.html) they also provided the data, information and research results for the integrated IWRM analysis. Acknowledgement is also given to the EC which funded the IWRMS development in the BRAHMATWINN project in the 6th Framework Programme under the contract number 036952.

The interdisciplinary BRAHMATWINN EC-project carried out between 2006–2009 by European and Asian research teams in the UDRB and in the UBRB enhanced capacities and supported the implementation of sustainable Integrated Land and Water Resources Management (ILWRM).

References

Dobler, A. and Ahrens, B.: Precipitation by a regional climate model and bias correction in Europe and South Asia, Meteor. Z., 17, 499–509, 2008.

Dobler, A. and Ahrens, B.: Analysis of the Indian summer monsoon system in the regional climate model COSMO-CLM, J. Geophys. Res., 115, D16101, doi:10.1029/2009JD013497, 2010.

Flügel, W.-A.: Systembezogene Entwicklung regionaler hydrologischer Modellsysteme, Wasser und Boden, 52, 3, 14–17, 2000.

Flügel, W.-A.: The Adaptive Integrated Data Information System (AIDIS) for global water research, Water Resources Management (WARM) Journal, 21, 199–210, 2007.

Flügel, W.-A.: Applied Geoinformatics for sustainable IWRM and climate change impact analysis, Technology, Resource Management and Development, 6, 57–85, 2009.

Flügel, W.-A.: Climate impact analysis for IWRM in Man-made landscapes: Applications for Geoinformatics in Africa and Europe, Initiativen zum Umweltschutz, 79, 101–134, 2010.

Giupponi, C., Sgobbi, A., Mysiak, J., Camera, R., and Fassio, A.: NetSyMoD – An Integrated Approach for Water Resources Management, in: Integrated Water Management, edited by: Meire, P., Coenen, M., Lombardo, C., Robba, M., and Sacile, R., Springer, Netherlands, 69–93, 2008.

GLOWA-Danube: www.glowa-danube.de (last access: 27 March 2011), 2009.

IPCC, Intergovernmental Panel on Climate Change: Emissions Scenarios, A Special Report of IPCC Working Group III, 27 pp., 2000.

Kienberger, S., Lang, S., and Zeil, P.: Spatial vulnerability units – expert-based spatial modelling of socio-economic vulnerability in the Salzach catchment, Austria, Nat. Hazards Earth Syst. Sci., 9, 767–778, doi:10.5194/nhess-9-767-2009, 2009.

Kralisch, S., Zander, F., and Krause, P.: Coupling the RBIS Environmental Information System and the JAMS Modelling Framework, in: Proc. 18th World IMACS/and MODSIM09 International Congress on Modelling and Simulation, edited by: Anderssen, R., Braddock, R., and Newham, L., Cairns, Australia, 902–908, 2009.

Mauser, W. and Bach, H.: PROMET – Large scale distributed hydrological modelling to study the impact of climate change on the water flows of mountain watersheds, J. Hydrol., 376, 362–377, 2009.

Mauser, W. and Ludwig, R.: GLOWA-DANUBE – a research concept to develop integrative techniques, scenarios and strategies regarding global changes of the water cycle, in: Climatic Change: Implications for the Hydrological Cycle and for Water Management, Advances in Global Change Research, Vol. 10, edited by: Beniston, M., Kluwer Academic Publishers, Dordrecht and Boston, 171–188, 2002.

Millennium Ecosystem Assessment: Ecosystems and Human Wellbeing: Wetlands and Water Sythesis, World Resources Institute, Washington DC, 2005.

Ministerial Declaration of the Hague on Water Security in the 21st Century, The Hague, The Netherlands, available at: http://www.waternunc.com/gb/secwwf12.htm, 22 March 2000.

Nevo, D. and Chan, Y. E.: A Delphi study of knowledge management systems: Scope and requirements, Inform. Manage., 44, 583–597, 2007.

Subba, B.: Himalayan Waters, The Panos Institute, South Asia, 286 pp., 2001.

27

Meteorological observations of the coastal boundary layer structure at the Bulgarian Black Sea coast

D. Barantiev[1], M. Novitsky[2], and E. Batchvarova[1,3]

[1]National Institute of Meteorology and Hydrology, Bulgarian Academy of Sciences, Sofia, Bulgaria
[2]Research and Production Association "Typhoon" – Obninsk, Russian Federal Service on Hydrometeorology and Environmental Monitoring, Moscow, Russia
[3]Risoe National Laboratory for Sustainable Energy, RISOE DTU, Roskilde, Denmark

Abstract. Continuous wind profile and turbulence measurements were initiated in July 2008 at the coastal meteorological observatory of Ahtopol on the Black Sea (south-east Bulgaria) under a Bulgarian-Russian collaborative program. These observations are the start of high resolution atmospheric boundary layer vertical structure climatology at the Bulgarian Black Sea coast using remote sensing technology and turbulence measurements. The potential of the measurement program with respect to this goal is illustrated with examples of sea breeze formation and characteristics during the summer of 2008. The analysis revealed three distinct types of weather conditions: no breeze, breeze with sharp frontal passage and gradually developing breeze. During the sea breeze days, the average wind speed near the ground (from sonic anemometer at 4.5 m and first layer of sodar at 30–40 m) did not exceed 3–4 m s^{-1}. The onset of breeze circulation was detected based on surface layer measurements of air temperature (platinum sensor and acoustic), wind speed and direction, and turbulence parameters. The sodar measurements revealed the vertical structure of the wind field.

1 Introduction

As well known, the breeze circulation and the formation of thermal internal boundary layer (TIBL) over land in coastal areas lead to specific meteorological conditions and specific air pollution problems (Batchvarova et al., 1999; Novitzky et al., 1992; Zhong and Takle, 1992). For this reason, conducting detailed meteorological observations in coastal areas is very important issue both from scientific and practical point of view.

During the last two decades of XX century, a number of experimental campaigns were carried out using Doppler wind radars, lidars, sodars, instrumented tall masts, mesoscale networks of ground and aerological stations (Alpert and Rabinovich-Hadar, 2003; Batchvarova et al., 1999; Batchvarova and Gryning, 1998; Wilczak et al., 1991; Zhong and Takle, 1992). The aim of such studies is to provide data for evaluation of mesoscale models performance in coastal areas and to develop further the parameterisations used in them. The variety of physical, geographical and cli-mate conditions related to sea breeze circulations, as well as weather patterns in coastal regions is huge (Simpson, 1994), so mesoscale models need to be constantly and vastly evaluated. As noted by Wilczak et al. (1996), the sea breeze is well known and well studied boundary-layer phenomenon, but there still remain issues for investigation about its structure and dynamics, especially in regions of complex or sloping topography.

The present day remote sensing technologies developed robust instruments allowing continuous monitoring and new quality of data for model evaluations. The reliability of these measurements is confirmed by a number of studies devoted to comparisons of remote sensing, tall masts and radiosoundings data (Floors et al., 2011; Novitzky et al., 2010; O'Connor et al., 2010).

In this paper, sodar and ultra sonic anemometer data at a costal site are explored to elucidate the structure of the coastal boundary layer at breeze circulation. Presently, these measurements of turbulence and the vertical wind field structure are unique in Bulgaria. The study area and instruments are presented in Sect. 2; the measurements and analysis – in Sect. 3; and the conclusions – in Sect. 4 of the paper.

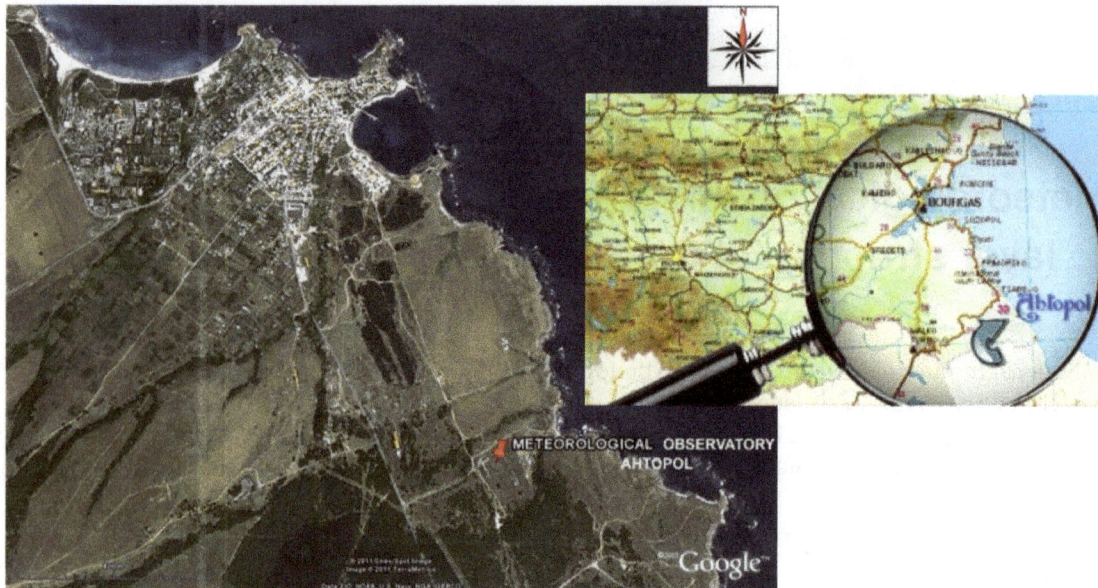

Figure 1. Location of the Meteorological Observatory of Ahtopol (60 km south-east of Burgas) and Google map of the region.

2   Study area, instruments and data

The measurements at the meteorological observatory of Ahtopol on the southeast Black Sea coast of Bulgaria (Fig. 1) started on 18 July 2008 under a joint research project between the National Institute of Meteorology and Hydrology – Bulgarian Academy of Sciences (NIMH-BAS) and the Research and Production Association (RPA) "Typhoon" in Obninsk – Russian Federal Service on Hydrometeorology and Environmental Monitoring (Roshydromet). The site is located in flat grassland, 30 m above sea level, about 500 m inland from a steep about 10 m high coast. The coast line is stretching out from north-north-west to south-south-east direction, therefore the winds from the sector 0–150 degrees are representing marine conditions. In the 1970s and 1980s the site was used for the launch of stratospheric rockets. These activities ended in the 1990s and the site was transformed to a meteorological observatory in the system of NIMH.

As shown in Fig. 2 (left) in the present setup, the ultra sonic anemometer and the sensor for solar radiation are mounted on a meteorological mast at height of 4.5 m; and the air temperature and the humidity sensors are installed at 2 m height within a thermometer screen. These sensors and a global solar radiation sensor form an automatic meteorological station named MK-15. During few days in July 2008, the regular manual measurements at the observatory and MK-15 were compared and showed differences within 0.2 K for air temperature, 3 % for humidity and 0.2 hPa for pressure. The ultrasonic anemometer is developed by "Typhoon" in Russia; it is produced in limited numbers and is used in the network

of Roshydromet. Based on parallel measurements at Obninsk observatory, this instrument (with vertical axis and 3 paths of 125 mm length at angle 45 degrees from it) is found to give very close results to those of available commercial instruments, Gill Instruments Ltd Wind master, for example (Mazurin and Kolijnikova, 2008; Mazurin et al., 2010). The first prototype of this instrument was used during the International turbulence comparison experiment (ITCE-81), Tsvang et al. (1985). The frequency of measurements of MK-15 is 0.5 Hz and records are made every 10 s.

The sodar is located on the roof of the administrative building at about 4.5 m high, Fig. 2 (right panel). It is a Scintec Flat Array middle range instrument (MFAS) with frequency range 1650–2750 Hz; 9 emission/reception angles ($0°$, $\pm22°$, $\pm29°$); maximum 100 vertical layers; range between 500–1000 m; accuracy of horizontal wind speed 0.1–0.3 m s^{-1}; range of horizontal wind speed ±50 m s^{-1}; accuracy of vertical wind speed 0.03–0.1 m s^{-1}; range of vertical wind speed ±10 m s^{-1}; accuracy of wind direction 2–3 degrees.

In summer 2008, the sodar is set to measure in regime "optimized pulses for resolution" at 47 levels from 30 to 500 m with resolution of 10 m. Ten fixed frequencies are emitted sequentially in a standard mode. The averaging time is 20 min and the records are made every 10 min, thus presenting running 20-min averages.

Concerning data availability, problems with MK-15 occur in winter 2008/2009. Thus data are available for the periods July to October 2008 and from June 2009 onwards.

Apart from electricity shut down problems, the sodar data are available during the days for the entire period since July 2008. During the summers of 2008 and 2009 the night

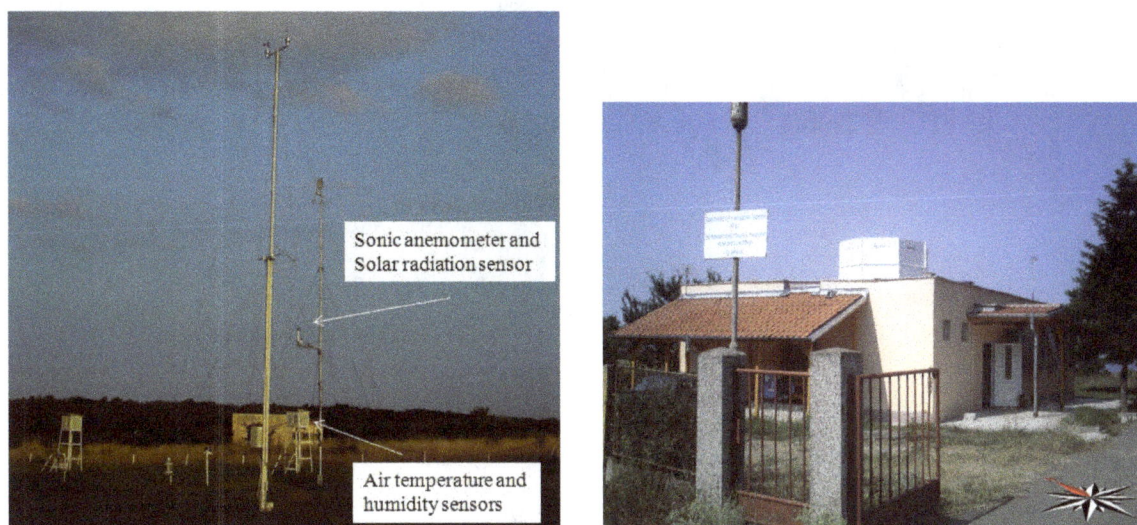

Figure 2. Location of sensors at the meteorological observatory (MO) Ahtopol (left). The SODAR is mounted on the roof of the administration building (right).

operation was suspended. Concerning the period of the present analysis (18 July to 5 September 2008) it can be specified that the sodar was performing measurements from 07:30 until 18:00 in July and August 2008; and from 07:30 until 23:50 in September 2008 in all days except 2 and 3 September. Full diurnal operation and more stable electricity network were achieved since October 2009. The availability of sodar data at different heights was analysed for the period July 2008–June 2009. At 50 m height, data are available between 83 % of the time in December 2008 and 97 % of the time in May 2009. At 200 m height, the availability is between 82 % (March 2009) and 96 % (May 2009). In 2010 the performance was more stable.

All measurements included in the analysis are referring to official time, GMT + 3 h in summer. The Meteorological Observatory Ahtopol is located at 42.08 N, 27.95 E, and 29 m above sea level. The local sunrise in August is 18 min earlier compared to Sofia.

3 Measurements and analysis

Synoptic analysis shows that the combination of weak high pressure field with warm air mass and clear weather leads to breeze circulation development. The onset of the sea breeze is identified in the records of the mean values and turbulent parameters of all measured meteorological variables. The breeze front passage is expressed in wind direction change, increased wind speed after relatively calm period, increased turbulence parameters (friction velocity, standard deviation of the vertical velocity, standard deviation of the acoustic temperature, sensible heat flux), levelling the air temperature, decrease and/or levelling of relative humidity, decrease of standard deviation of wind direction. In this study, we

analyse the summer of 2008, after the start of measurements on 19 July. Data for 2 days with sharp breeze onset, 1 day with gradually developing breeze and 1 day of no breeze are used for illustration.

On 2 August, a small diurnal variation of air temperature (only 3 K) was observed. In addition, almost stationary wind speed (2–3 m s^{-1}) and direction (35–40 degrees) were recorded (Fig. 3). No breeze circulation developed on this day. The flow was north easterly (from the sea) during the nights and during the day. A high pressure centre was located north of the Black Sea (Met Office Archive, 2008), Fig. 9a.

On 6 August, the sea breeze was developing gradually because of interaction of large and local scale forcing. The records show sharp increase in temperature between 07:30 and 08:00 a.m. with simultaneous drop in relative humidity, drop in wind speed and change in wind direction by about 90 degrees from 225 to 315 degrees (Fig. 3). This wind direction is from the sea, but the northerly component is due to advection. On that day, a cold front was approaching the Black sea from the Northwest (Met Office Archive, 2008 – Fig. 9c). In the afternoon at about 1 p.m. another abrupt change of wind direction occurs and easterly wind (sea breeze) is established. A reason to consider the day as a type of breeze day is the pronounced land breeze during both nights. A south-westerly wind is observed in the morning probably related to the clear breeze circulation of the previous day and is re-established after 8 p.m. when again south-south-westerly wind was measured. Such gradual change of wind direction clockwise from north through east and south to west is seen often for the Black Sea coast of Bulgaria from synoptic observations and can be associated with the interaction between local and large scale processes. The new observation set up will allow more detailed analysis of such situations.

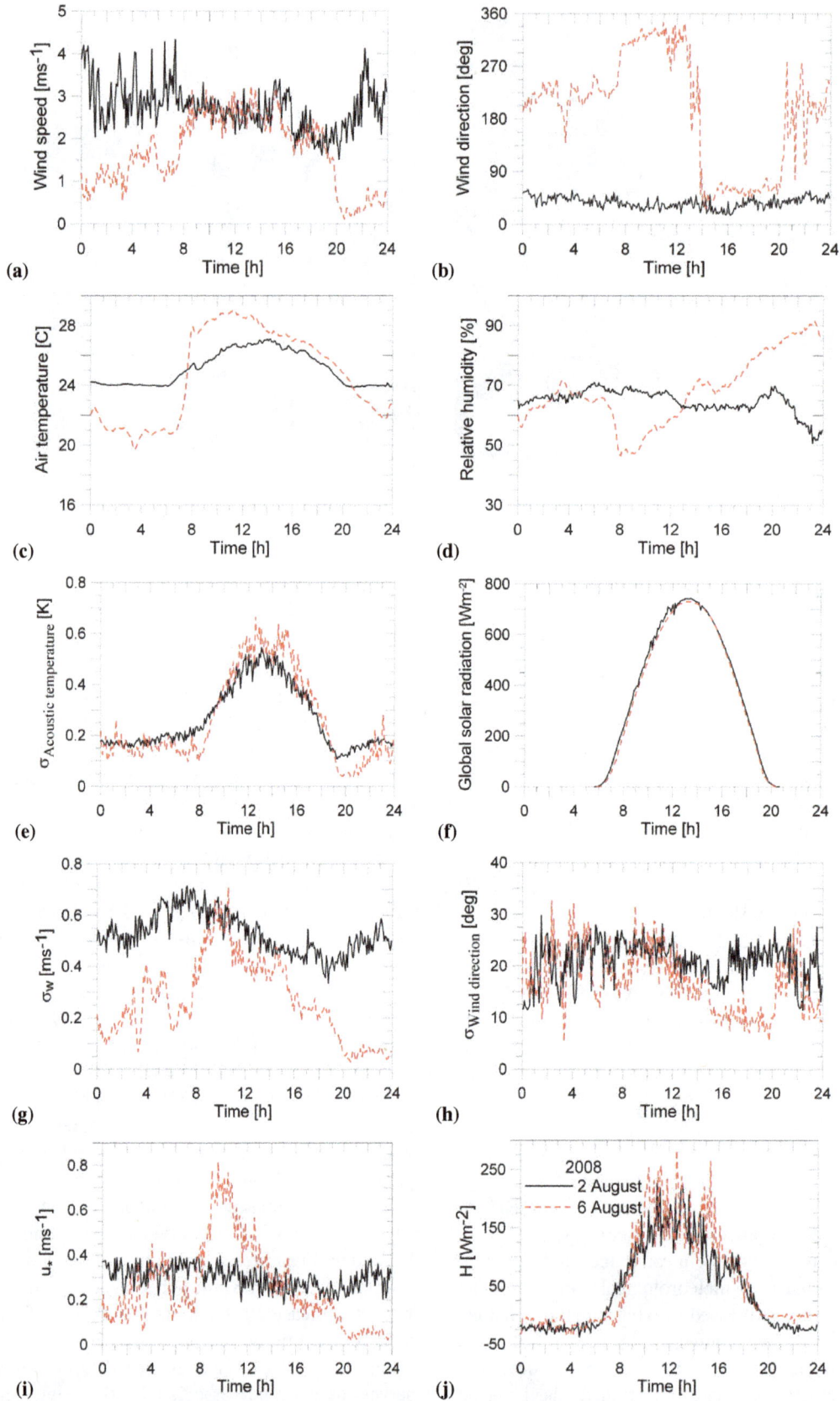

Figure 3. Horizontal wind speed (**a**); wind direction (**b**); air temperature (**c**); relative humidity (**d**); standard deviation of acoustic temperature (**e**); global solar radiation (**f**); standard deviation for the vertical wind speed (**g**); standard deviation for the wind direction (**h**); friction velocity (**i**); and sensible heat flux (**j**) on 2 August 2008 (black solid lines) and on 6 August (red dashed lines) at MO Ahtopol. Five-minute averaged values.

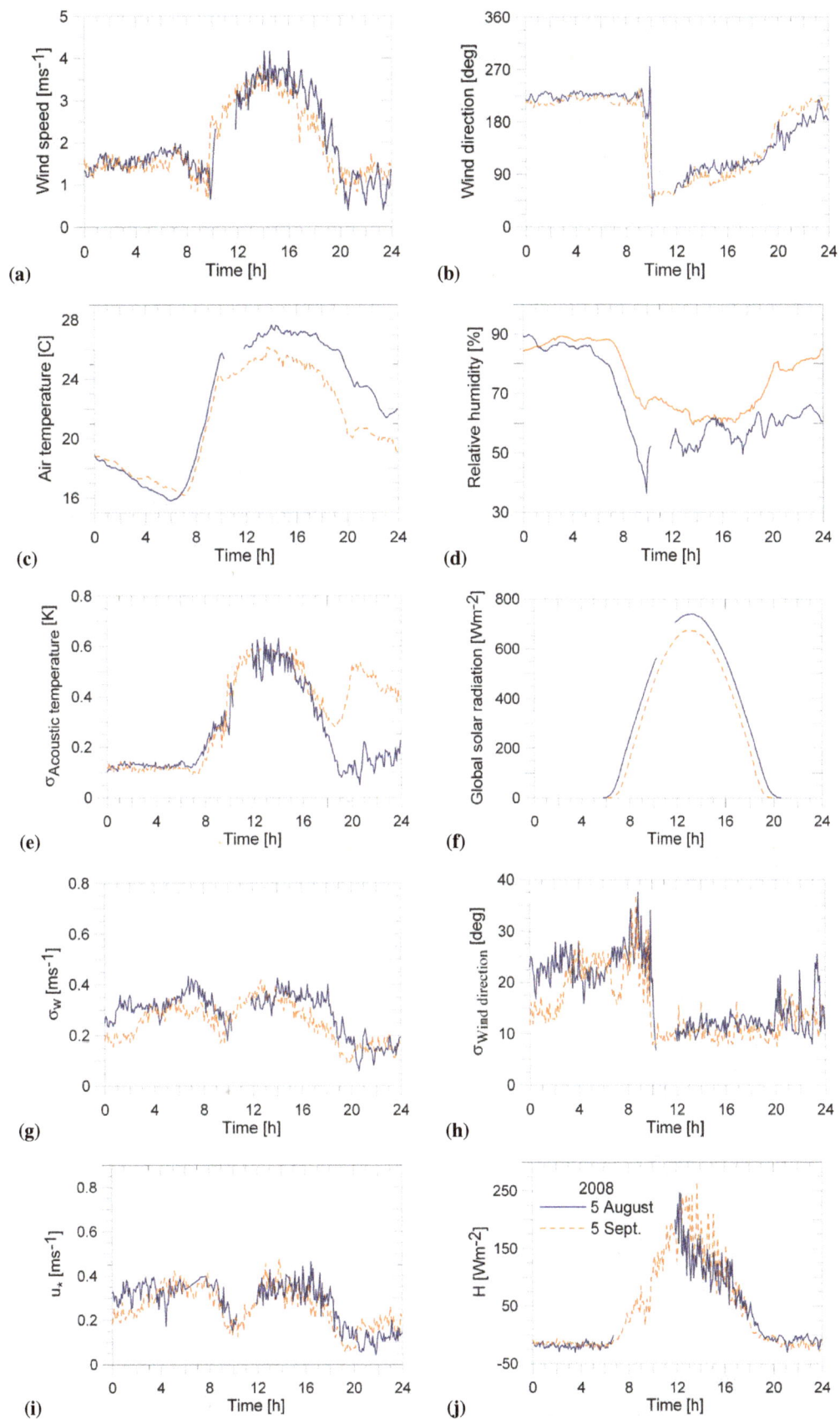

Figure 4. Notation is as in Fig. 3, but for 5 August 2008 (dark blue solid lines) and 5 September 2008 (orange dashed lines).

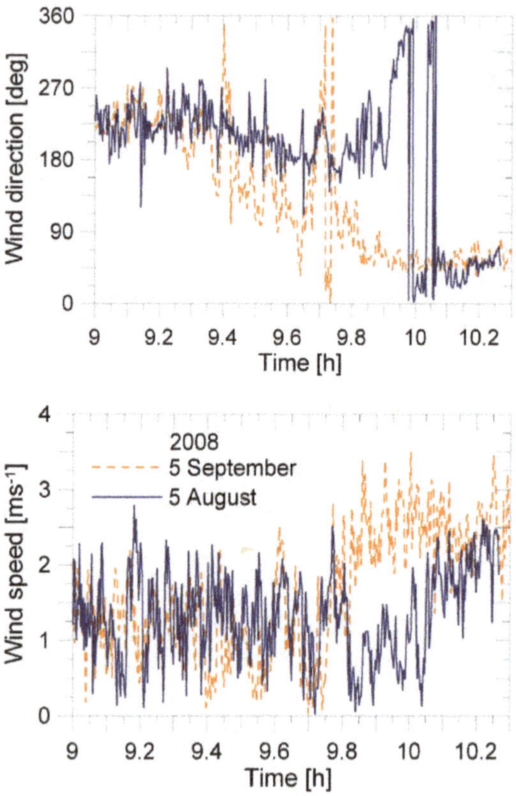

Figure 5. The sharp onset of the sea breeze on 5 August 2008 (dark blue solid lines) and 5 September 2008 (orange dashed lines) in wind direction (upper panel) and wind speed (lower panel) based on the 10-s data.

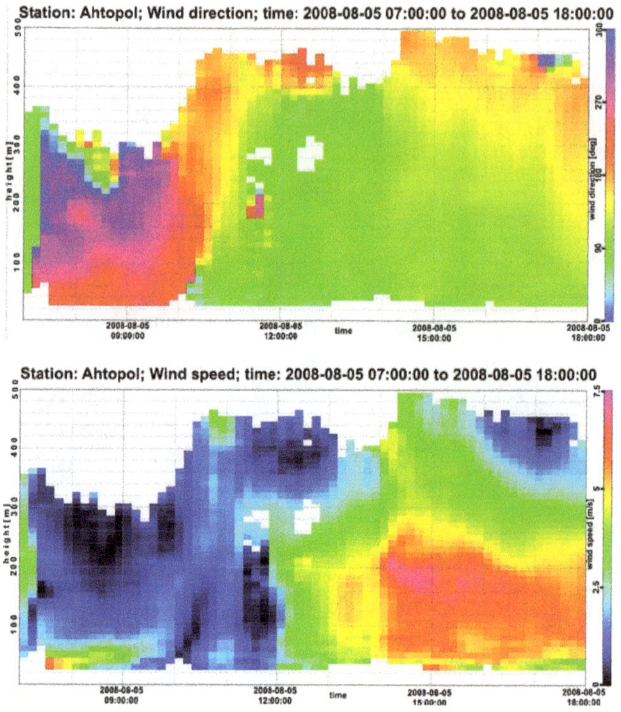

On 5 August and 5 September, the breeze front passes rapidly resulting in abrupt change in wind direction, wind speed and relative humidity, Fig. 4. The temperature slightly decreases and then levels. The diurnal amplitude for the air temperature is about 12 K on 5 August and 10 K on 5 September. The solar radiation is strong reaching above 750 W m^{-2} and the sensible heat flux reaches about 250 W m^{-2}, Fig. 4. Especially clear sea breeze front passage is observed on 5 August. At around 10 a.m. a rapid increase in wind speed from 1.5 m s^{-1} to 3.0 m s^{-1} is observed and a sharp turn of wind direction by 180 degrees in less than 10 minutes, Fig. 5. On 5 September the change of wind direction happens within 35–40 min in two steps. On these days, a change in the values of the turbulence parameters during the sea breeze can be followed. The friction velocity and the standard deviation of the vertical velocity increase after a drop before the front passage. The standard the deviation of the acoustic temperature and the sensible heat flux fluctuate largely during the sea breeze. The standard deviation of wind direction decreases. The synoptic situation on both days is characterized by high pressure conditions and small pressure gradients (Met Office Archive, 2008 – Fig. 9b and d).

Figure 6. Wind direction (upper panel) and wind speed (lower panel) from sodar at MO Ahtopol on 5 August 2008.

All the parameters shown in Figs. 3 and 4 are calculated as 5 min averaged values. Partly cloudy conditions in the morning of 2 August are traceable from the small variations of the global radiation before noon.

The soil temperature and the sea water temperature have not been measured at the MO Ahtopol. The closest sea water temperature measurement is performed by NIMH at MO Burgas, 60 km to the North (Fig. 1), 40 m offshore using water mercury thermometer at about 30 cm depth at 09:00, 15:00 and 21:00 local summer time. The sea water temperature record for a period in the beginning of August 2008 was of the range of 25–26 °C and showed typical diurnal variation 0.6–1.8 K.

Based on surface layer meteorological parameters the onset and characteristics of the breeze circulation can be studied near the ground, but for the vertical extend of the phenomenon, remote sensing technology is required. Sodar backscatter is illustrated here for both days with sharp sea breeze front passage (5 August and 5 September 2008, Figs. 6 and 7) and one day with gradual sea breeze onset (6 August 2008, Fig. 8).

When sharp sea breeze front is observed near the ground, the abrupt change in wind direction happens simultaneously at all levels within a 400–500 m deep layer (Figs. 6 and 7, upper panels). The wind speed is about 2 m s^{-1} and constant with height within the 400 m layer for about 2 h after the front passage, but later on stratifies and reaches a maximum of

Station: Ahtopol; Wind direction; time: 2008-09-05 07:30:00 to 2008-09-05 21:40:00

Station: Ahtopol; Wind speed; time: 2008-09-05 07:30:00 to 2008-09-05 21:40:00

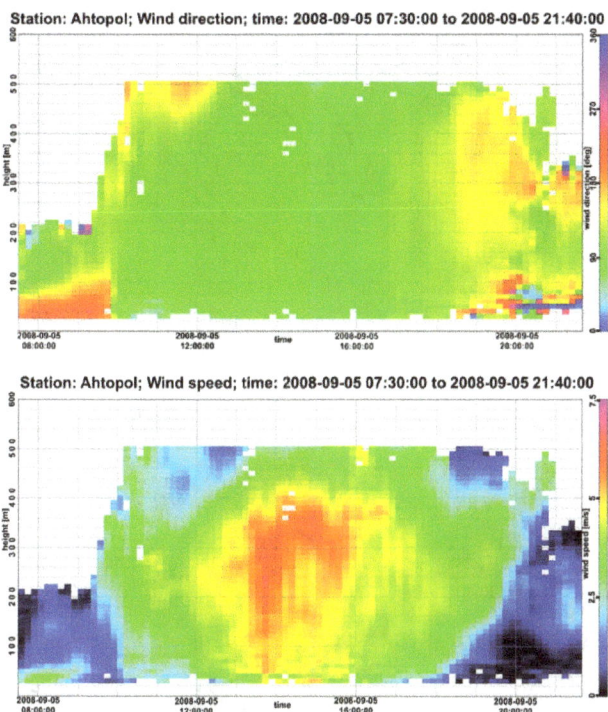

Figure 7. Notation is as in Fig. 6, but for 5 September 2008.

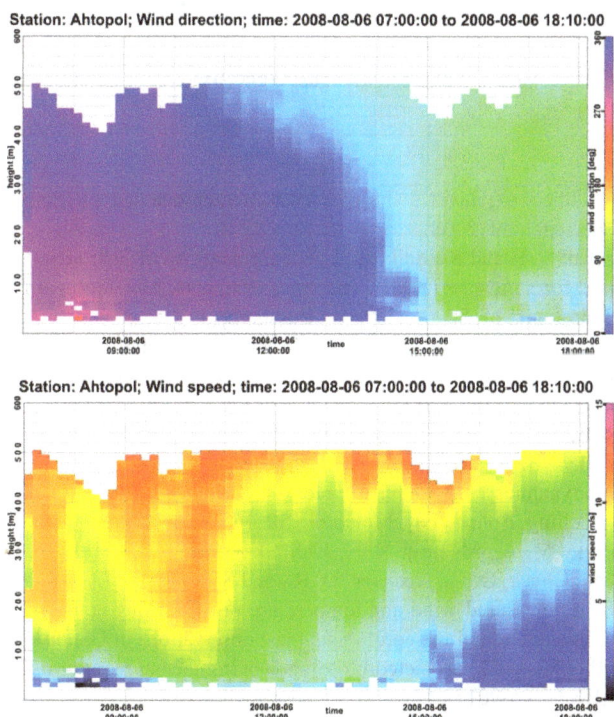

Station: Ahtopol; Wind direction; time: 2008-08-06 07:00:00 to 2008-08-06 18:10:00

Station: Ahtopol; Wind speed; time: 2008-08-06 07:00:00 to 2008-08-06 18:10:00

Figure 8. Notation is as in Fig. 6, but for 6 August 2008.

5–$7\,\mathrm{m\,s^{-1}}$ in the layer 50–250 m on 5 August (Fig. 6, lower panel). On 5 September these features are less pronounced having increase of wind speed with height already at the on-set of the sea breeze, reaching 5–$7\,\mathrm{m\,s^{-1}}$ in a deeper layer 40–420 m and only between 1 and 4 p.m. (Fig. 7, lower panel).

On 6 August, the westerly flow gradually turns to northerly with height and time from the onset around 8 a.m. until 1 p.m. when a change to easterly flow occurs reaching the entire depth covered by the sodar (500 m) at 3 p.m. (Fig. 8, upper panel). The wind speed on that day is reaching $10\,\mathrm{m\,s^{-1}}$ at the higher levels (Fig. 8, lower panel).

This analysis shows that on 6 August, a larger scale northerly flow was modified both by the land breeze during the night (early morning westerly flow) and the sea breeze during the day (early afternoon easterly flow). The presence of this large scale flow is also suggested by the synoptic maps (Met Office Archive, 2008 – Fig. 9c). On 5 August and 5 September, no large scale flow was present.

4 Conclusions

The data analysis has shown that the combination of remote sensing and eddy correlation measurements is a powerful tool for coastal atmospheric boundary layer studies. With long records of data using the presented instrumental set-up, it will be possible to develop a classification of the sea breeze types at the southern Bulgarian Black Sea coast taking into account the information on turbulence and the vertical structure of the coastal boundary layer. Further, the analysis will continue with comparisons of the measurements with mesoscale models predictions, assessments of sea water temperature from satellite images, introduction of soil measurements. Depending on external funding an extension of the measurements programme is planned to cover several sites in the region and to investigate the penetration of the sea breeze. In addition soil and sea water temperature measurements will be introduced. An important message of this paper is also, that now high quality data are available at a Bulgarian Black Sea coastal site for model validation based on vertical wind profiles.

Acknowledgements. This work is result of intergovernmental agreement for scientific cooperation between Bulgaria and Russia, and in particular it is a part of a research project initiated between the National Institute of Meteorology and Hydrology – Bulgarian Academy of Sciences (NIMH-BAS) and the Research and Production Association (RPA) "Typhoon" – Russian Federal Service on Hydrometeorology and Environmental Monitoring (Roshydromet). The contribution of E. Batchvarova is also supported by the Danish Research Agency Strategic Research Council (Sagsnr. 2104-08-0025) "Tall wind project" and the EU FP7-People-IEF VSABLA (PIEF-GA-2009-237471). The work is also related to COST Actions ES0702 (EG-CLIMET) and ES1002 (WIRE).

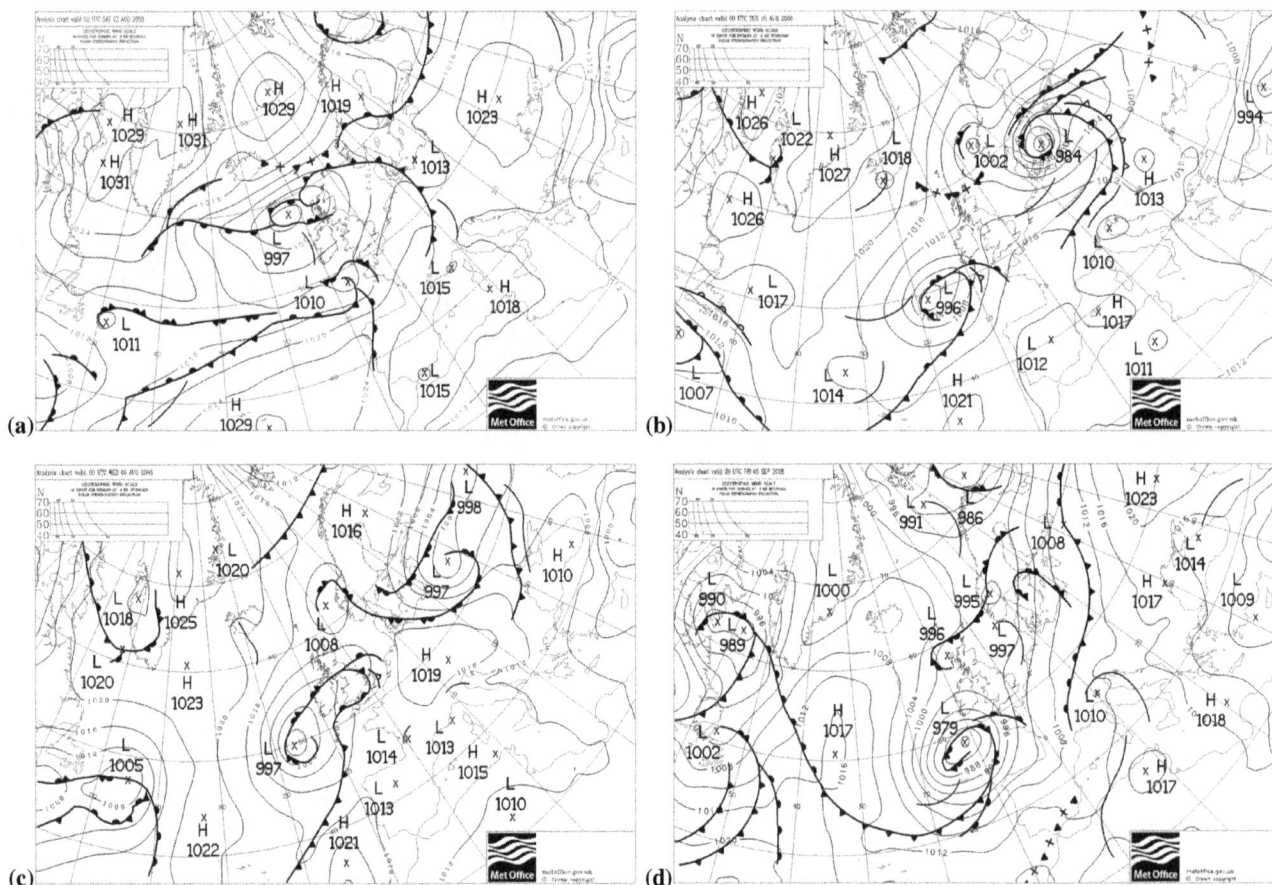

Figure 9. Analysis charts from UK Met Office Archive (2008) at 00:00 UTC on 2 August 2008 (**a**); 5 August 2008 (**b**); 6 August 2008 (**c**) and 5 September 2008 (**d**).

Edited by: F. Beyrich
Reviewed by: two anonymous referees

References

Alpert, P. and Rabinovich-Hadar, M.: Pre- and post-sea-breeze frontal line – a meso-γ-scale analysis over South Israel, J. Atmos. Sci., 60, 2994–3008, 2003.

Batchvarova, E. and Gryning, S.-E.: Wind climatology, atmospheric turbulence and internal boundary layer development in Athens during the MEDCAPHOT – TRACE experiment, Atmos. Environ., 32, 2055–2069, 1998.

Batchvarova, E., Cai, X., Gryning, S.-E., and Steyn, D.: Modelling internal boundary layer development in a region with complex coastline, Bound.-Lay. Meteorol., 90, 1–20, 1999.

Floors, R., Batchvarova, E., Gryning, S.-E., Hahmann, A. N., Peña, A., and Mikkelsen, T.: Atmospheric boundary layer wind profile at a flat coastal site – wind speed lidar measurements and mesoscale modeling results, Adv. Sci. Res., 6, 155–159, doi:10.5194/asr-6-155-2011, 2011.

Met Office Archive: http://www.wetterzentrale.de/topkarten/ fsfaxsem.html, 2008.

Mazurin, N. F. and Kulijnikova, L. K.: Comparison of instruments used for atmospheric turbulence measurements, Meteorologia i Gydrologia, 11, 90–96, 2008 (in Russian).

Mazurin, N. F., Matzkevich, M. K., Milchenko V. T., and Novitzky, M. A.: Atmospheric surface layer turbulence measurements in urban conditions, Meteorologia i Gydrologia, 6, 38–49, 2010 (in Russian).

Novitsky, M. A., Reible, D. D., and Corripio, B. M.: Modeling the dynamics of the land-sea breeze circulation for air quality modelling, Bound.-Lay. Meteorol., 59, 163–175, 1992.

Novitzky M. A., Mazurin N. F., Kulijnikova, L. K., Tereb L. A., Kalinicheva, O. U., Nechaev, D. R., and Safronov, V. L.: Comparing measured wind and temperature profiles using commercial sodar and tall meteorological mast observations in Obninsk. – Problems of hydrometeorology and environmental monitoring, ISBN 978-5-901579-21-3, 4, 122–136, 2010 (in Russian).

O'Connor, E. J., Illingworth A. J., Brooks, I. M., Westbrook, C. D., Hogan, R. J., Davies, F., and Brooks, B. J.: A method for estimating the turbulent kinetic energy dissipation rate from a vertically pointing Doppler lidar, and independent evaluation from balloon-borne in situ measurements, J. Atmos. Ocean. Technol., 27, 1652–1664, 2010.

Simpson J. E: Sea breeze and local wind, Cambridge University Press, 234 pp., 1994.

Tsvang, L. R., Zubkovskii, S. L., Kader, B. A., Kallistratova, M. A., Foken T., Gerstmann, V., Przadka, A., Pretel, Ya., Zeleny, Ya., and Keder, J.: Internationa turbulence comparison experiment (ITCE-81), Bound.-Lay. Meteorol., 31, 325–348, 1985.

Wilczak, J. M., Dabberdt, W. F., and Kropfli, R. A.: Observations and numerical model simulations of the atmospheric boundary layer in the Santa Barbara coastal region, J. Appl. Meteorol., 30, 652–673, 1991.

Wilczak, J. M., Gossard, E. E., Neff, W. D., and Eberhard, W. L.: Ground-based remote sensing of the atmospheric boundary layer: 25 years of progress, Bound.-Lay. Meteorol., 78, 321–349, 1996.

Zhong, S. and Takle, E. S.: An observational study of sea- and land-breeze circulation in an area of complex coastal heating, J. Appl. Meteorol., 31, 1426–1438, 1992.

Permissions

List of Contributors

M. Journée
Royal Meteorological Institute of Belgium, Brussels, Belgium

C. Bertrand
Royal Meteorological Institute of Belgium, Brussels, Belgium

H. Nakayama
Japan Atomic Energy Agency, Ibaraki, Japan

T. Takemi
Disaster Prevention Research Institute, Kyoto University, Kyoto, Japan

H. Nagai
Japan Atomic Energy Agency, Ibaraki, Japan

V. Giannini
Ca' Foscari University of Venice, Italy
Fondazione Eni Enrico Mattei, Venice, Italy

C. Giupponi
Ca' Foscari University of Venice, Italy
Fondazione Eni Enrico Mattei, Venice, Italy

E. I. F. de Bruijn
Royal Netherlands Meteorological Institute, De Bilt, The Netherlands

W. C. de Rooy
Royal Netherlands Meteorological Institute, De Bilt, The Netherlands

J. Spinoni
JRC-IES, Ispra, Italy

T. Antofie
JRC-IES, Ispra, Italy

P. Barbosa
JRC-IES, Ispra, Italy

Z. Bihari
Hungarian Meteorological Service, Budapest, Hungary

M. Lakatos
Hungarian Meteorological Service, Budapest, Hungary

S. Szalai
Szent Istvan University, Gödöllo, Hungary

T. Szentimrey
Hungarian Meteorological Service, Budapest, Hungary

J. Vogt
JRC-IES, Ispra, Italy

F. Kalinka
Goethe-University Frankfurt am Main, Institute for Atmospheric and Environmental Sciences, Germany
LOEWE BiK-F research centre Frankfurt am Main, Germany

B. Ahrens
Goethe-University Frankfurt am Main, Institute for Atmospheric and Environmental Sciences, Germany

H. Breuer
Eötvös Loránd University, Department of Meteorology, 1117, Pázmány P. s. 1/a, Budapest, Hungary

F. Ács
Eötvös Loránd University, Department of Meteorology, 1117, Pázmány P. s. 1/a, Budapest, Hungary

Á. Horváth
Hungarian Meteorological Service, 8600 Vitorlás utca 17, Siófok, Hungary

P. Németh
Hungarian Meteorological Service, Marcell György Observatory, P.O. Box 39, 1675 Budapest, Hungary

K. Rajkai
Institute for Soil Sciences and Agricultural Chemistry, Centre for Agricultural Research, Hungarian Academy of Sciences, 1022 Herman Ottó 15, Budapest, Hungary

C. Demain
Royal Meteorological Institute of Belgium, Brussels, Belgium

M. Journèe
Royal Meteorological Institute of Belgium, Brussels, Belgium

C. Bertrand
Royal Meteorological Institute of Belgium, Brussels, Belgium

P. Blanc
MINES ParisTech, PSL Research University, Centre Observation, Impacts, Energy, BP 204, 06905 Sophia Antipolis CEDEX, France

C. Coulaud
ADEME, Valbonne, France

L. Wald
MINES ParisTech, PSL Research University, Centre Observation, Impacts, Energy, BP 204, 06905 Sophia Antipolis CEDEX, France

V. Giannini
CáFoscari University of Venice, Italy
Fondazione Eni Enrico Mattei, Venice, Italy

L. Ceccato
Ca' Foscari University of Venice, Italy

C. Hutton
GeoData Institute, Southampton, UK

A. A. Allan
Centre for Water Law, Policy and Science, University of Dundee, UK

S. Kienberger
Centre for Geoinformatics, University of Salzburg, Salzburg, Austria

W.-A. Flügel
Department of Geoinformatics, Hydrology and Modelling, Friedrich-Schiller University Jena, Germany

C. Giupponi
Ca' Foscari University of Venice, Italy
Fondazione Eni Enrico Mattei, Venice, Italy

W. Weng
Department of Earth and Space Science and Engineering, York University, 4700 Keele Street, Toronto,Ontario, M3J 1P3, Canada

P. A. Taylor
Department of Earth and Space Science and Engineering, York University, 4700 Keele Street, Toronto,Ontario, M3J 1P3, Canada

W. Wandji Nyamsi
MINES ParisTech, PSL Research University, O. I. E. – Centre Observation, Impacts, Energy, Sophia Antipolis CEDEX, France

B. Espinar
MINES ParisTech, PSL Research University, O. I. E. – Centre Observation, Impacts, Energy, Sophia Antipolis CEDEX, France

P. Blanc
MINES ParisTech, PSL Research University, O. I. E. – Centre Observation, Impacts, Energy, Sophia Antipolis CEDEX, France

L. Wald
MINES ParisTech, PSL Research University, O. I. E. – Centre Observation, Impacts, Energy, Sophia Antipolis CEDEX, France

W.-A. Flügel
Institute for Geography, University of Jena, Germany

C. Busch
Codematix GmbH, Jena, Germany

J. Estèvez
University of Córdoba, Projects Engineering, Córdoba, Spain

P. Gavilán
IFAPA Center "Alameda del Obispo", Junta de Andalućia, Córdoba, Spain

A. P. Garćıa-Marı́n
University of Córdoba, Projects Engineering, Córdoba, Spain

R. Magno
LaMMA Consortium, Sesto Fiorentino, Florence, Italy
Institute of Biometeorology, National Research Council, Florence, Italy

L. Angeli
LaMMA Consortium, Sesto Fiorentino, Florence, Italy

M. Chiesi
Institute of Biometeorology, National Research Council, Florence, Italy

M. Pasqui
LaMMA Consortium, Sesto Fiorentino, Florence, Italy

A. Bartosch
Department of Geoinformatics, Friedrich Schiller University Jena, Germany

S. Asharaf
Institute for Atmospheric and Environmental Sciences, Goethe University, Frankfurt, Germany

A. Dobler
Institute for Atmospheric and Environmental Sciences, Goethe University, Frankfurt, Germany

B. Ahrens
Institute for Atmospheric and Environmental Sciences, Goethe University, Frankfurt, Germany

J. Mazón
Applied Physics Department, BarcelonaTech (UPC), Barcelona, Spain

D. Pino
Applied Physics Department, BarcelonaTech (UPC), Barcelona, Spain
Institute for Space Studies of Catalonia (IEEC-UPC), Barcelona, Spain

S. Gaztelumendi
Basque Meteorology Agency (Euskalmet), Vitoria-Gasteiz, Spain
TECNALIA, Meteo Unit, Vitoria-Gasteiz, Spain

J. Egāna
Basque Meteorology Agency (Euskalmet), Vitoria-Gasteiz, Spain
TECNALIA, Meteo Unit, Vitoria-Gasteiz, Spain

K. Otxoa-de-Alda
Basque Meteorology Agency (Euskalmet), Vitoria-Gasteiz, Spain
TECNALIA, Meteo Unit, Vitoria-Gasteiz, Spain

R. Hernande
Basque Meteorology Agency (Euskalmet), Vitoria-Gasteiz, Spain
TECNALIA, Meteo Unit, Vitoria-Gasteiz, Spain

J. Aranda
Basque Government, Interior Dept., Directorate of Emergencies and Meteorology, Vitoria-Gasteiz, Spain

P. Anitua
Basque Government, Interior Dept., Directorate of Emergencies and Meteorology, Vitoria-Gasteiz, Spain

M. Casaioli
Italian National Institute for Environmental Protection and Research (ISPRA), Rome, Italy

F. Catini
Italian Interuniversity Consortium High Performance Systems (CINECA), Rome, Italy

R. Inghilesi
Italian National Institute for Environmental Protection and Research (ISPRA), Rome, Italy

P. Lanucara
Italian Interuniversity Consortium High Performance Systems (CINECA), Rome, Italy

P. Malguzzi
Institute of Atmospheric Sciences and Climate-Italian National Research Council (ISAC-CNR), Bologna, Italy

S. Mariani
Italian National Institute for Environmental Protection and Research (ISPRA), Rome, Italy

A. Orasi
Italian National Institute for Environmental Protection and Research (ISPRA), Rome, Italy

M. Sastre
Dept. de Geof´ısica y Meteorolog´ıa, Universidad Complutense de Madrid, Spain

C. Yagüe
Dept. de Geofísica y Meteorología, Universidad Complutense de Madrid, Spain

C. Román-Casćon
Dept. de Geofíísica y Meteorologíia, Universidad Complutense de Madrid, Spain

G. Maqueda
Dept. de Astrofísica y Ciencias de la Atmósfera, Universidad Complutense de Madrid, Spain

F. Salamanca
Lawrence Berkeley National Laboratory (LBNL), Berkeley (CA), USA

S. Viana
Agencia Estatal de Meteorología (AEMET), Delegación Territorial de Cataluña, Barcelona, Spain

S. Pietzsch
Deutscher Wetterdienst, Offenbach, Germany

P. Bissolli
Deutscher Wetterdienst, Offenbach, Germany

S. Davolio
Institute of Atmospheric Sciences and Climate, National Research Council, Bologna, Italy

T. Diomede
HydroMeteoClimate Regional Service of ARPA Emilia Romagna, Bologna, Italy

C. Marsigli
HydroMeteoClimate Regional Service of ARPA Emilia Romagna, Bologna, Italy

M. M. Miglietta
Institute of Atmospheric Sciences and Climate, National Research Council, Lecce, Italy
Institute of Ecosystem Study, National Research Council, Verbania Pallanza, Italy

A. Montani
HydroMeteoClimate Regional Service of ARPA Emilia
Romagna, Bologna, Italy

A. Morgillo
HydroMeteoClimate Regional Service of ARPA Emilia
Romagna, Bologna, Italy

M. Piringer
Central Institute for Meteorology and Geodynamics,
Vienna, Austria

W. Knauder
Central Institute for Meteorology and Geodynamics,
Vienna, Austria

E. Petz
Central Institute for Meteorology and Geodynamics,
Vienna, Austria

G. Schauberger
Unit for Physiology and Biophysics, University of
Veterinary Medicine, Vienna, Austria

R. Becker
Deutscher Wetterdienst, Meteorologisches Observatorium
Lindenberg/Mark, Am Observatorium 12,15848 Tauche,
Germany

K. Behrens
Deutscher Wetterdienst, Meteorologisches Observatorium
Lindenberg/Mark, Am Observatorium 12,15848 Tauche,
Germany

D. Barantiev
National Institute of Meteorology and Hydrology,
Bulgarian Academy of Sciences, Sofia, Bulgaria

M. Novitsky
Research and Production Association "Typhoon" -
Obninsk, Russian Federal Service on Hydrometeorology
and Environmental Monitoring, Moscow, Russia

E. Batchvarova
National Institute of Meteorology and Hydrology,
Bulgarian Academy of Sciences, Sofia, Bulgaria
Risoe National Laboratory for Sustainable Energy, RISOE
DTU, Roskilde, Denmark